Joachim Weiss
Handbook of Ion Chromatography

Further Titles of Interest:

G. Gauglitz, T. Vo-Dinh (Eds.)

Handbook of Spectroscopy

2 Volumes
2003, ISBN 3-527-29782-0

H. Günzler, A. Williams (Eds.)

Handbook of Analytical Techniques

2 Volumes
2001, ISBN 3-527-30165-8

J. S. Fritz, D. T. Gjerde

Ion Chromatography

2000, ISBN 3-527-29914-9

Joachim Weiss

Handbook of Ion Chromatography

Third, completely revised and updated edition

Volume 2

Translated by Tatjana Weiss

WILEY-VCH

WILEY-VCH Verlag GmbH & Co. KGaA

Dr. Joachim Weiss
Dionex GmbH
Am Wörtzgarten 10
65510 Idstein
Germany

■ All books published by Wiley-VCH are carefully produced. Nevertheless, author and publisher do not warrant the information contained in these books, including this book, to be free of errors. Readers are advised to keep in mind that statements, data, illustrations, procedural details or other items may inadvertently be inaccurate.

Library of Congress Card No. applied for.

British Library Cataloguing-in-Publication Data:
A catalogue record for this book is available from the British Library.

Bibliographic information published by Die Deutsche Bibliothek
Die Deutsche Bibliothek lists this publication in the Deutsche Nationalbibliografie; detailed bibliographic data is available in the Internet at <http://dnb.ddb.de>.

© 2004 WILEY-VCH Verlag GmbH & Co. KGaA, Weinheim

Printed on acid-free paper.

All rights reserved (including those of translation into other languages). No part of this book may be reproduced in any form – by photoprinting, microfilm, or any other means – nor transmitted or translated into a machine language without written permission from the publishers. Registered names, trademarks, etc. used in this book, even when not specifically marked as such, are not to be considered unprotected by law.

Composition   pagina media gmbh, Hemsbach
Printing   betz-druck gmbh, Darmstadt
Bookbinding   J. Schäffer GmbH i.G., Industrie- und Verlagsbuchbinderei, Grünstadt

Printed in the Federal Republic of Germany.
ISBN   3-527-28701-9

Foreword

Over the past few years, ion chromatography has developed into a significant chromatographic technique for ion analysis within the field of separation science. In addition to publications on the basic principles, it is above all the applied research and the broad applicability of ion chromatography that has made this technique practically indispensable for the analytical chemist. In view of the ever-growing number of publications in this field, the numerous international conferences, as well as the diversity of applications, qualitative, quantitative, and quality-assured data acquisition become extremely important. It is thus of great importance to have a fundamental book on ion chromatography, such as this one by Dr. Joachim Weiss.

Today, in addition to the classical inorganic ions, the field of ion chromatography and the economical significance of this method cover such organic ionic compounds as organic acids, carbohydrates, and glycoproteins, to name but a few.

The previous books by Dr. Weiss in the field of ion chromatography are already regarded as classics. This new edition again accommodates all possible needs in terms of the science and industrial applications involved in this technique.

The volume describes in great detail the theoretical principles, considerations regarding the separation mechanisms, new stationary phases for anion- and cation-exchange chromatography, ion-exclusion chromatography, and ion-pair chromatography along with new detection methods. Even the statistical data acquisition together with the related fundamental principles, which is indispensable nowadays for a modern certified analytical laboratory, is outlined, as are special applications important, for example, to the semiconductor industry.

I trust that the present book, succeeding the previous successful works by Dr. Joachim Weiss, will once again gain recognition as an important reference work in science, the laboratory, and industry.

I wish the author and the publishing house much success with this auspicious work.

Innsbruck, January 2004

Dr. Günther Bonn
Professor for Analytical Chemistry
University of Innsbruck (Austria)

Table of Contents

Foreword V

Preface XIII

Volume 1

1	**Introduction** 1	
1.1	Historical Perspective 1	
1.2	Types of Ion Chromatography 3	
1.3	The Ion Chromatographic System 5	
1.4	Advantages of Ion Chromatography 7	
1.5	Selection of Separation and Detection Systems 9	
2	**Theory of Chromatography** 13	
2.1	Chromatographic Terms 13	
2.1.1	Asymmetry Factor A_s 14	
2.2	Parameters for Assessing the Quality of a Separation 15	
2.2.1	Resolution 15	
2.2.2	Selectivity 16	
2.2.3	Capacity Factor 17	
2.3	Column Efficiency 17	
2.4	The Concept of Theoretical Plates (Van-Deemter Theory) 19	
2.5	Van-Deemter Curves in Ion Chromatography 24	
3	**Anion Exchange Chromatography (HPIC)** 27	
3.1	General Remarks 27	
3.2	The Ion-Exchange Process 27	
3.3	Thermodynamic Aspects 29	
3.4	Stationary Phases 34	
3.4.1	Polymer-Based Anion Exchangers 35	
3.4.1.1	Styrene/Divinylbenzene Copolymers 35	
3.4.1.2	Polymethacrylate and Polyvinyl Resins 47

3.4.2	Latex-Agglomerated Anion Exchangers	54
3.4.2.1	Review of Different Latex-Agglomerated Anion Exchangers	57
3.4.3	Silica-Based Anion Exchangers	86
3.4.4	Other Materials for Anion Separations	86
3.4.4.1	Crown Ether and Cryptand Phases	86
3.4.4.2	Alumina Phases	95
3.5	Eluants for Anion Exchange Chromatography	98
3.6	Suppressor Systems in Anion Exchange Chromatography	103
3.6.1	Suppressor Columns	104
3.6.2	Hollow Fiber Suppressors	108
3.6.3	Micromembrane Suppressors	111
3.6.4	Self-Regenerating Suppressors	115
3.6.5	Suppressors with Monolithic Suppression Beds	120
3.7	Anion Exchange Chromatography of Inorganic Anions	121
3.7.1	Overview	121
3.7.2	General Parameters Affecting Retention	123
3.7.3	Experimental Parameters Affecting Retention When Applying Suppressor Systems	123
3.7.3.1	Choice of Eluant	124
3.7.3.2	Eluant Concentration and pH Value	137
3.7.3.3	Influence of Organic Solvents	144
3.7.4	Experimental Parameters Affecting Retention When Applying Direct Conductivity Detection	145
3.7.4.1	Choice of Eluant	145
3.7.4.2	System Peaks	153
3.7.4.3	Eluant Concentration and pH Value	156
3.7.5	Polarizable Anions	159
3.8	Anion Exchange Chromatography of Organic Anions	168
3.8.1	Organic Acids	168
3.8.2	Polyvalent Anions	184
3.9	Gradient Elution Techniques in Anion Exchange Chromatography of Inorganic and Organic Anions	191
3.9.1	Theoretical Aspects	192
3.9.2	Choice of Eluants	194
3.9.3	Possibilities for Optimizing Concentration Gradients	201
3.9.4	Isoconductive Techniques	204
3.10	Carbohydrates	205
3.10.1	Sugar Alcohols	209
3.10.2	Monosaccharides	212
3.10.3	Oligosaccharides	226
3.10.4	Polysaccharides	231
3.10.5	Carbohydrates Derived from Glycoproteins	236

3.10.5.1	Compositional and Structural Analysis	240
3.10.5.2	Chosen Examples	254
3.11	Amino Acids	257
3.12	Proteins	267
3.13	Nucleic Acids	275

4 Cation Exchange Chromatography (HPIC) *279*

4.1	Stationary Phases	279
4.1.1	Polymer-Based Cation Exchangers	280
4.1.1.1	Styrene/Divinylbenzene Copolymers	280
4.1.1.2	Ethylvinylbenzene/Divinylbenzene Copolymers	282
4.1.1.3	Polymethacrylate and Polyvinyl Resins	296
4.1.2	Latexed Cation Exchangers	298
4.1.3	Silica-Based Cation Exchangers	303
4.2	Eluants in Cation Exchange Chromatography	308
4.3	Suppressor Systems in Cation Exchange Chromatography	310
4.3.1	Suppressor Columns	310
4.3.2	Hollow Fiber Suppressors	311
4.3.3	Micromembrane Suppressors	312
4.3.4	Self-Regenerating Suppressors	313
4.3.5	Suppressors with Monolithic Suppression Beds	318
4.4	Cation Exchange Chromatography of Alkali Metals, Alkaline-Earth Metals, and Amines	318
4.5	Transition Metal Analysis	325
4.5.1	Basic Theory	326
4.5.2	Transition Metal Analysis with Non-Suppressed Conductivity Detection	330
4.5.3	Transition Metal Analysis with Spectrophotometric Detection	331
4.6	Analysis of Polyamines	344
4.7	Gradient Techniques in Cation Exchange Chromatography of Inorganic and Organic Cations	349

5 Ion-Exclusion chromatography (HPICE) *359*

5.1	The Ion-Exclusion Process	359
5.2	Stationary Phases	361
5.3	Eluants in Ion-Exclusion Chromatography	365
5.4	Suppressor Systems in Ion-Exclusion Chromatography	366
5.5	Analysis of Inorganic Acids	368
5.6	Analysis of Organic Acids	372
5.7	HPICE/HPIC-Coupling	375
5.8	Analysis of Alcohols and Aldehydes	379

5.9	Amino Acid Analysis	382
5.9.1	Separation of Amino Acids	384
5.9.2	Post-Column Derivatizations of Amino Acids	389
5.9.3	Sample Preparation	391

6	**Ion-Pair Chromatography (MPIC)**	**393**
6.1	Survey of Existing Retention Models	394
6.2	Suppressor Systems in Ion-Pair Chromatography	399
6.3	Experimental Parameters that Affect Retention	400
6.3.1	Type and Concentration of Lipophilic Counter Ions in the Mobile Phase	400
6.3.2	Type and Concentration of the Organic Modifier	403
6.3.3	Inorganic Additives	406
6.3.4	pH Effects and Temperature Influence	408
6.4	Analysis of Surface-Inactive Ions	409
6.5	Analysis of Surface-Active Ions	424
6.6	Applications of the Ion-Suppression Technique	440
6.7	Applications of Multi-Dimensional Ion Chromatography on Multimode Phases	444

7	**Detection Methods in Ion Chromatography**	**461**
7.1	Electrochemical Detection Methods	461
7.1.1	Conductivity Detection	461
7.1.1.1	Theoretical Principles	462
7.1.1.2	Application Modes of Conductivity Detection	469
7.1.2	Amperometric Detection	474
7.1.2.1	Fundamental Principles of Voltammetry	475
7.1.2.2	Amperometry	478
7.2	Spectrometric Detection Methods	498
7.2.1	UV/Vis Detection	498
7.2.1.1	Direct UV/Vis Detection	498
7.2.1.2	UV/Vis Detection in Combination with Derivatization Techniques	499
7.2.1.3	Indirect UV Detection	510
7.2.2	Fluorescence Detection	514
7.3	Other Detection Methods	521
7.4	Hyphenated Techniques	522
7.4.1	IC-ICP Coupling	523
7.4.2	IC-MS Coupling	530

Index   I 1

Volume 2

8	**Quantitative Analysis** 549	
8.1	General 549	
8.2	Analytical Chemical Information Parameters 550	
8.3	Determination of Peak Areas 551	
8.3.1	Manual Determination of Peak Areas and Peak Heights 552	
8.3.2	Electronic Peak Area Determination 554	
8.4	Statistical Quantities 557	
8.4.1	Mean Value 558	
8.4.2	Standard Deviation 558	
8.4.3	Scatter and Confidence Interval 559	
8.5	Calibration of an Analytical Method (Basic Calibration) 560	
8.5.1	Acquisition of the Calibration Function 561	
8.5.1.1	Method Characteristic Parameters of a Linear Calibration Function 561	
8.5.1.2	Method Parameters of a Calibration Function of 2nd Degree 564	
8.5.2	Testing of the Basic Calibration 565	
8.5.3	Testing the Precision 566	
8.5.3.1	Homogeneity of Variances 566	
8.5.3.2	Outlier Tests 567	
8.5.4	Calibration Methods 569	
8.5.4.1	Area Normalization 569	
8.5.4.2	Internal Standard 570	
8.5.4.3	External Standard 571	
8.5.4.4	Standard Addition 572	
8.6	Detection Criteria, Limit of Detection, Limit of Determination 574	
8.6.1	Determination of Detection Criteria, Limit of Detection 574	
8.7	The System of Quality Control Cards 578	
8.7.1	Types of Quality Control Cards and their Applications 579	
9	**Applications** 587	
9.1	Ion Chromatography in Environmental Analysis 588	
9.2	Ion Chromatography in Power Plant Chemistry 626	
9.2.1	Analysis of Conditioned Waters 634	
9.2.2	Cooling Water Analysis 640	
9.2.3	Flue Gas Scrubber Solutions 644	
9.2.4	Analysis of Chemicals 648	
9.3	Ion Chromatography in the Semiconductor Industry 651	
9.3.1	High-Purity Water Analysis 652	
9.3.2	Surface Contaminations 654	
9.3.3	Solvents 658	

9.3.4	Acids, Bases, and Etching Agents 664
9.3.5	Other Applications 673
9.4	Ion Chromatography in the Electroplating Industry 678
9.4.1	Analysis of Inorganic Anions 679
9.4.2	Analysis of Metal Complexes 686
9.4.3	Analysis of Organic Acids 688
9.4.4.	Analysis of Inorganic Cations 689
9.4.5	Analysis of Organic Additives 690
9.5	Ion Chromatography in the Detergent and Household Product Industry 697
9.5.1	Detergents 697
9.5.2	Household Products 704
9.6	Ion Chromatography in the Food and Beverage Industry 711
9.6.1	Beverages 713
9.6.2	Dairy Products 733
9.6.3	Meat Processing 739
9.6.4	Baby Food 742
9.6.5	Groceries and Luxuries 746
9.6.6	Sweeteners 752
9.7	Ion Chromatography in the Pharmaceutical Industry 756
9.7.1	Fermentation 772
9.8	Ion Chromatography in Clinical Chemistry 781
9.9	Oligosaccharide Analysis of Membrane Coupled Glycoproteins 797
9.10	Other Applications 803
9.11	Sample Preparation and Matrix Problems 823

Index 871

Preface to the Third Edition

Precisely ten years have passed since the publication of the second edition of this book in 1994. Over this decade the method of ion chromatography has rapidly developed and been further established. One can name the many new separator columns with partly extraordinary selectivities and separation power in this connection. Nowadays, new grafted polymers, compatible with large volume injections, enable ion analysis down to the sub-µg/L range without time-consuming pre-concentration. Of particular importance is the development of continuous and contamination-free eluant generation by means of electrolysis, which considerably facilitates the use of gradient elution techniques in ion chromatography. Furthermore, only de-ionized water is used as a carrier with ion chromatography systems configured for the use of such technology. Along with this development, hydroxide eluants, which are particularly suitable for concentration gradients in anion exchange chromatography, are increasingly replacing the classical carbonate/bicarbonate buffers predominantly used so far. In contrast to carbonate/bicarbonate buffers, which will still be used for relatively simple applications, higher sensitivities are now achieved with hydroxide eluants. This trend is supported by an exciting development of hydroxide-selective stationary phases. In line with classical liquid chromatography, hyphenation with atomic spectrometry and molecular spectrometry, such as ICP and ESI-MS, is becoming increasingly important in ion chromatography, too. The section *Hyphenated Techniques* underlines this importance. Because carbohydrates, proteins, and oligonucleotides are also analyzed by ion-exchange chromatography, but only carbohydrates were included in the second edition, the sections *Proteins* and *Oligonucleotides* have been added in this new edition. In combination with integrated amperometry as a direct detection method, ion-exchange chromatography also revolutionized amino acid analysis. Thus, in many application areas ion chromatography has become almost indispensable for the analysis of inorganic and organic anions and cations.

Since the publication of the second edition, all these developments have made it necessary to rewrite major parts, such that this third edition can be confidently regarded as a new text. Almost every chapter has been renewed or significantly revised. For better clarity, the previous chapter *Ion-Exchange Chromatography* is now split into the chapters *Anion Exchange Chromatography* and *Cation Exchange Chromatography*. The remaining structure of this book proved to be of value, and has thus remained unchanged. The sections *Carbohydrates Derived from Glycoproteins* (Section 3.10.5), *Proteins* (Section 3.12), *Nucleic Acids* (Section 3.13), and *Oligosaccharide Analysis of Membrane-Coupled Glycoproteins* (Section 9.9) have been completely rewritten by an expert, Dr. Dietrich Hauffe, to whom I would like to express my sincere gratitude. The chapter Quantitative Analysis was also rewritten and expanded with information on validation parameters and quality control cards. In addition, the chapters on detection and applications were significantly expanded with new material and with numerous practical examples in the form of chromatograms, while applications of ion chromatography in the petrochemical industry and in the pulp and paper industry were also added.

The objective for this third edition is the same as for the previous two editions: The author addresses analytical chemists, who wish to familiarize themselves with this method, as well as practitioners who employ these techniques on a day-to-day basis and are looking for a reference book that can help to facilitate method development and provide an overview on existing applications.

At this point, I would like to express my sincere gratitude to many of my colleagues in all parts of the world, who contributed their experience and knowledge to the preparation of this third edition. I am particularly grateful to Dr. Detlef Jensen (Germany) for his willingness to always discuss the various aspects of ion chromatography and for his valuable suggestions, and to Jennifer Kindred (Sunnyvale, USA) for her incredible effort, patience, and diligence in editing the translated manuscript. I would be grateful for any criticisms or suggestions that could serve to improve future editions of this book.

Finally, many thanks to my wife and children who for quite some time, with an amazing amount of understanding and tolerance, did not see much of their husband or dad, who spent many evenings, weekends, and public holidays at the computer.

Idstein, February 2004 *Joachim Weiss*

8
Quantitative Analysis

8.1
General

The following discussion deals with the various steps necessary to develop a procedure for the quantitative analysis [1] of a compound. It is assumed that the chromatographic conditions for the separation of this compound have been set and optimized.

A quantitative analysis may be subdivided into the following individual steps:
- Sampling
- Sample preparation
- Separation
- Detection
- Signal processing

Chemical analysis starts with the sampling procedure, which is aimed at obtaining a sample that represents the bulk composition. To ensure sample quality, only samples with a known and documented origin should be analyzed. For every single sample it must be possible to reconstruct
- when, where and how sampling was carried out
- who took the sample
- how the sample was treated at the sampling location
- how and in what kind of container the sample was transported
- where and for how long the sample was stored
- how the sample was preserved
- whether the sample was homogenized prior to any treatment
- how the sample was split for analysis
- how positive identification was ensured

Depending on the purpose of the analysis, the origin of the sample, and the analytical method being used, regulations and directives are already in effect and standard procedures have been written up that describe and regulate sampling as well as the storage and preservation of samples [2-12]. The common practice (sending samples labeled with a number to a laboratory together with a catalogue of parameters to be analyzed) can lead, as experience shows, to the application of unsuitable analytical methods. The consequence of such imprecise practices is nonsensical analytical results; the analytical chemist does not have enough

Handbook of Ion Chromatography, Third, Completely Revised and Enlarged Edition. Joachim Weiss
Copyright © 2004 WILEY-VCH Verlag GmbH & Co. KGaA, Weinheim
ISBN: 3-527-28701-9

information about the analytical problem and is, therefore, unable to check the plausibility based on sample characteristics and background information. The majority of errors occur are encountered in obtaining a representative sample, storing the sample, and preparing the sample. The sampling of inhomogeneous solids such as soil, detergents, etc. poses a special problem. Physiological fluids such as urine are usually collected over a longer period of time to obtain a representative sample. Errors in sample storage are often encountered with volatile, reactive, or oxygen-sensitive samples. Likewise, adsorption at and desorption from container walls must be considered as possible sources of inaccuracy.

Sample preparation embraces all processes that convert the analyte into a form suitable for analysis. This includes comminution, homogenization, digestion, dissolution, and filtration of the sample. Ideally, the sample should be dissolved in the eluant being used, because then, for example, the negative peaks observed in the system void when applying conductivity detection do not occur. Before injecting the sample solution, it should be membrane-filtered (0.22 or 0.45 µm) to prevent particulate matter from entering the column. An overview of sample preparation techniques relevant for ion chromatography can be found in Section 9.11.

Sample injection is typically performed via a sample loop. This allows the reproducible injection of the sample volume. Also, the injection process may easily be automated.

The subsequent chromatographic separation also involves a series of error sources. Major reasons are the co-elution of ions and their possible decomposition in the mobile phase or at the stationary phase.

The chromatographic separation is followed by a measurement of a physical property (electrical conductivity, UV absorption, etc.), which must correlate in a defined way with the chemical composition and structure of the sample. The measured signals are then processed and listed in an analysis report.

8.2
Analytical Chemical Information Parameters

In the evaluation of the measured signals, one has to determine quantities that provide information about the chemical composition of the sample. These quantities are called information parameters because they carry analytical chemical information.

In chromatography, a series of signals from the detector output is registered as a chromatogram. The qualitative information is derived from the retention time, t_{ms}, which is determined by the chromatographic process. Retention time is also dependent on the thermodynamic properties of both the stationary phase and the solutes. The quantitative information stems from the area under the signal. It is determined by the measurement process and depends on the detector properties. Thus, it is assumed that the sample component elutes quantitatively from the analytical column.

According to Kucera [13] and Grushka [14], the area of peak i corresponds to the zero moment, m_{0i}, of the distribution function describing the peak.

$$m_{0i} = \int_{-\infty}^{+\infty} y_i(t)\,dt \tag{218}$$

The first moment normalized to m_0 corresponds to the retention time:

$$m_{1i} = \int_{-\infty}^{+\infty} t \cdot y_i(t)\,dt / m_{0i} \tag{219}$$

m_{0i} Peak area A_i
m_{1i} Retention time of the component i
y_i Intensity of the component signal i as a function of time
t Time

The second moment, m_{2i}, of a peak i with the distribution function $y_i(t)$ is related to the retention time, m_{1i}, and as the central moment $\overline{m_{2i}}$ represents the peak time variance σ_{ti}^2:

$$\overline{m_{2i}} = \int_{-\infty}^{+\infty} (t - t_{ms})^2 \cdot y(t)\,dt / m_{0i} \tag{220}$$

The interpretation of the retention time as the first moment, m_1, is more comprehensive than the use of the peak time (t_{max}); m_1 indicates the position of the peak center on the time scale. In the case of unsymmetric peaks, that position may deviate significantly from the position of the peak maximum. The thermodynamically correct retention time is actually derived from the position of the Gaussian component in a deconvoluted total peak profile. According to the more recent opinion of Jönsson [15], the correct retention time equals the median value of the peak.

There are several approaches (Dorsey et al. [16]; Yau [17]; Grushka [18]) to determining moments using the exponential peak function (retrievable from chromatography data systems) that yield slightly different results.

8.3
Determination of Peak Areas

The first step in the quantitative evaluation of a chromatogram is peak area determination or the measurement of quantities that are proportional to the peak area (peak height, peak area by multiplication of peak height and half width, triangulation). The various methods for determining peak areas or quantities proportional to the peak area differ with respect to:
- Accuracy
- Precision
- Applicability for area determination
- Time expenditure/costs.

8.3.1
Manual Determination of Peak Areas and Peak Heights

Peak Height

As seen in Fig. 8-1, the peak height is measured as the distance between baseline and peak maximum. Baseline drift is compensated for by interpolation of the baseline between peak start and peak end. The determination of the peak height yields a quantity that is proportional to the peak area only in the case of constant peak shapes and, therefore, has the character of an approximation method. In overlapping peaks, the measurement of the peak height is superior to the peak area measurement, because the same accuracy is obtained at lower resolution. The precision of determination via peak height measurement is – depending on the quality of the calibration – between 1% and 2% rel.

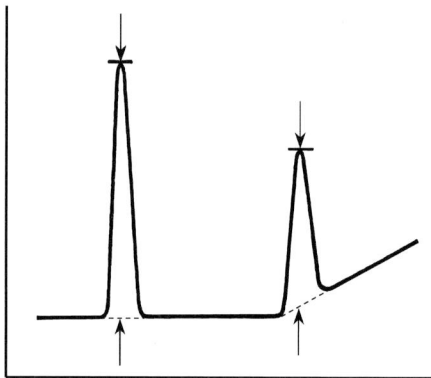

Fig. 8-1 Peak height measurement.

Peak Area by Multiplication of Peak Height and Half Width

Often, a chromatographic signal may be approximated by a triangle. The calculation of the area is performed as shown in Fig. 8-2:

$$A = H \cdot w_{1/2} \tag{221}$$

H Peak height
$w_{1/2}$ Half width of the peak

Because the peak width on the baseline may be affected by absorption and tailing effects, the peak's half width is used in the area calculation. This procedure for peak area calculation should only be applied to symmetric signals. The area thus calculated is proportional to the concentration but slightly smaller than the true value. For an accurate determination of the half widths, a high chart speed must be chosen.

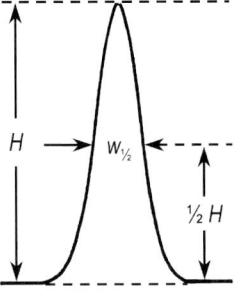

Fig. 8-2. Determination of the peak area by multiplication of peak height and half width.

Triangulation

This procedure is also based on the approximation of a chromatographic signal by a triangle. For area calculation, tangents in the two inflection points of the peak are positioned. As seen in Fig. 8-3, the peak height is measured as the distance between the baseline and the point of intersection of both tangents. The peak width is given by the baseline section as determined by the inflectional tangents. The peak area is then calculated as follows:

$$A = \frac{1}{2} B \cdot H \tag{222}$$

B   Peak width on the baseline as determined by the inflectional tangents
H   Peak height

The procedure of triangulation is subject to the same limitations as the approaches described above. It has an additional drawback, in that inflectional tangents have to be drawn. Small errors in the positioning of these inflectional tangents significantly affect the peak height measurement; therefore, this procedure is not recommended.

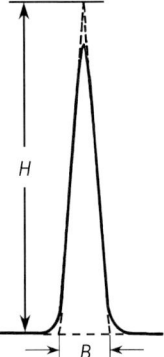

Fig. 8-3. Triangulation.

8.3.2
Electronic Peak Area Determination

For the electronic evaluation of chromatographic data, one uses either dedicated systems, integrators, or personal computers equipped with appropriate software.

Digital Integrators

A digital electronic integrator is an instrument that enables automatic measurement of peak areas and retention times and compensation for baseline drifts. With the aid of a voltage-to-frequency-converter, an output signal is generated with a given frequency that is proportional to the input signal. During integration, all pulses from the V/F-converter are counted. The total number of pulses within a peak represents the peak area. By means of an integrator, peaks from several microvolts up to about one volt may be detected; thus, the linear range covers more than five orders of magnitude. In addition, the recorder signal can be attenuated independent of the integrator signal. Digital integrators offer the advantage of high precision and the conversion of the chromatographic signal into a numeric result. However, if the control parameters for the integrator are erroneously set, there is a possibility that trends in the form of baseline drifts and offsets will not be recognized and, consequently, the baseline will not be adjusted. This could lead to inaccuracies in the area calculation of non-resolved or tangential peaks.

In addition to enabling normal integration and the printing of chromatograms, modern compact integrators provide the ability to customize calculations and analysis reports. This includes linear or non-linear multi-level calibration using peak areas or peak heights, the calibration of peak areas via an internal standard, peak area normalization, and the calculation of simple statistical quantities such as the relative standard deviation. Nearly all commercial instruments can be programmed in BASIC and are equipped with a power-failure back-up system to prevent the loss of data in case of short-term black-outs. An optional 256 K RAM can be used for storing chromatograms for blockwise re-integration, and for storing raw data from a second channel that may be connected. Many digital integrators can be equipped with a timed-function module that, with the aid of a BASIC program, can be used to control external instrument modules such as an autosampler. Data may be transferred to other computer systems via an RS-232C-interface.

A survey of methods for determining peak areas and quantities proportional to peak area is given in Table 8-1.

Table 8-1. Survey of various methods for calculating peak areas.

Method	Required time	Max. precision in routine operation σ [%]	Accuracy	Applicability to unsymmetric peaks
Peak height measurement	small	0.5 (depending on peak symmetry)	good (with symmetric peak shape)	not suited
Peak height × half width	small	2	good (with constant and almost symmetric peak shape)	rather unsuited
Triangulation	small	3	good (with constant and almost symmetric peak shape)	rather unsuited
Electronic integration	zero	0.1	very good (depending on the algorithm for curve smoothing, peak detection and baseline drift)	suited

Personal Computer

While digital integrators serve to measure peak areas and retention times, they are of limited use for instrument control purposes and advanced data processing. Therefore, computer-assisted chromatography systems have almost completely replaced digital integrators. The personal computer used in these systems has to fulfill the following tasks:
- Control and monitoring of the chromatographic system
- Calculation of peak areas and retention times
- Calculation of the sample composition based on the obtained data and the various calibration methods

The required software is usually based on the Windows platform and features a user friendly graphical control interface. At present, the most modern chromatography management systems are client/server systems [18] that offer a high degree of flexibility. Client/server systems can be configured either for only one workstation with up to six independent instruments or for several users bi-directionally via an Ethernet-LAN network or RS-232. In such systems, the client is the user interface through which the chemist views chromatograms, enters commands, starts or finishes an analytical sequence, etc. The server controls the instruments and aquires data. Client and server communicate with each other via so-called Remote Procedure Calls (RPC's). This communication can be local (client and server are on the same PC) or via RPC's from one PC to another. Client and server can be connected to each other via LAN (Local Area Network), WAN (Wide Area Network), ISDN or Modem. In this way, even large distances between the instrument and the chromatography management system are tolerable.

The chromatography BIOS(CBIOS) concept allows system control and data acquisition under one platform. On the CBIOS driver level, general chromatography commands, for example a NaOH gradient, are converted to system-specific commands. The concept works bi-directionally, that is, the instruments are under direct control and give positive feedback, so the user is continuously informed about the status of a single module or the complete system.

Object-oriented data systems focus on the user with his samples. As with Windows Explorer or MS-Office applications such as MS-Excel, operations are facilitated by meaningful tool bars, context-sensitive menus, and drag-and-drop combinations such as "Strg+C" for "Copy" and "Strg+V" for "Paste".

An indispensable requirement of a modern data management software is a detailed audit trail that complies with GLP/GMP. Every event related to instrument control and data acquisition is automatically captured in event logs. Every change in a baseline, a quantification method, or a calibration is tracked in the modification history. And all security-related events such as logins, password changes, privilege changes, and electronic signature events are recorded in a detailed security system log. The display can be filtered down to records matching a specific user, time period, or event type. The filtered and sorted list can then be printed or embedded in the analytical report. Built-in system security features permit assigned password access to individual screens. This ensures that the users see only the information and controls that are needed. Thorough system security, including user access to data and permitted software functions, can be established and finely tuned by the system administrator. Usually, a large number of chromatography privileges can be allocated as appropriate to any number of Privilege Groups. One can also control access to specific instruments and data folders by defining any number of Access Groups. Other GLP functions such as the Electronic Signature feature turn the vision of a paperless laboratory into reality. Modern data management software has all the tools needed to implement an efficient electronic record-keeping system that complies with 21 CFR Part 11 requirements. All of the processes for document creation, review, and approval fit right into the natural laboratory workflow, so that paperwork can be eliminated without having to completely restructure operations.

Moreover, the validation of a chromatography system is of great importance today. The U.S. FDA (**F**ood and **D**rug **A**dministration) defines validation as "submission of documented proofs, which ensure the generation of a product by a specific process with a high degree of certainty to comply with certain specifications and quality features that were determined beforehand." The integral parts of the validation process include Vendor Qualification (VQ), Installation Qualification (IQ), Operational Qualification (OQ), Performance Qualification (PQ), and a validated computer system. Installation Qualification, Operational Qualification, and Performance Qualification can be carried out by the operator at any time, because modern data management software fully automates these procedures for instruments and software and provides detailed reports. Method validation is easy, too, due to versatile System Suitability Testing and statistical analysis features. In addition to standard USP and EP tests like efficiency, resolution, tailing, and reproducibility, any number of tests can be defined to automatically check any reportable variables across any set of samples in the respective sequences.

Client/server systems usually incorporate a relational database and support an MS Access, Oracle, or MS-SQL server, which gives the analyst more power to analyze and interpret the data. As injections are performed, they are automatically linked into the database. Users can locate data by querying chromatography results, not just sample input information. Unlike conventional database queries, which limit the choice to basic criteria like sample name and analysis date, data management software lets the analyst search *ad hoc* using any combination of criteria, including calculated analytical results. Data can be arranged according to any hierarchy structure (e.g., according to a project, a product, a user, an instrument, a laboratory, etc.).

Last but not least, modern data management software is equipped with a powerful report generator. An integrated Excel-compatible spreadsheet, for example, makes it easy to analyze data and to present results in the way the consumer wants to see them. Report workbooks can include result tables, chromatograms, calibration plots, spectra, audit trails, and even custom equations and charts. Every cell, table, and chart updates instantly if any of the source data changes. Thus, the operator never has to worry about consistency.

8.4
Statistical Quantities

When a sample is repeatedly analyzed in the laboratory using the same analytical method, results collected will deviate from each other to some extent. The deviations, representing a scatter of individual values around a mean value, are denoted as statistical or random errors, a measure of which is the *precision*. Deviations from the true content of a sample are caused by systematic errors. An analytical method only provides true values if it is free of systematic errors. Random errors make an analytical result less precise while systematic errors give incorrect values. Hence, the precision and accuracy of an analytical method have to be discussed separately. Statements regarding the accuracy are only feasible if the true value is known.

For an assessment of empirical distributions, the data resulting from an experiment are characterized by numerical indicators. These include mean values and mean variations. A frequency distribution should be defined with these two quantities. Often, only mean values are stated in the result, which is not very expressive. For this reason, results should be supplemented by a statement of the statistical error.

8.4.1
Mean value

In ion chromatography data processing the arithmetic mean is used exclusively. The arithmetic mean value $\bar{x}$ of a random test with n individual measurements $x_1, x_2, ..., x_n$ is calculated according to Eq. (223):

$$\bar{x} = \frac{x_1 + x_2 + x_3 + \cdots + x_n}{n}$$
$$= \frac{1}{n} \cdot \sum_{i=1}^{n} x_i \tag{223}$$

Only data resulting from comparable measurements should be used to calculate the mean value. In general, the mean value should be based on at least three individual measurements. Significantly smaller or larger measured values must not be omitted, unless they are proved to be outliers. This strict rule can be softened if the measurements are carried out in one and the same laboratory. In this case, one deviating measured value can be replaced by three others. Mean value calculation is not allowed if the timely sorted measured values exhibit upward or downward trends.

8.4.2
Standard Deviation

The reproducibility is typically expressed as the standard deviation of the mean value. At an infinite number of measurements, the distribution of individual results is represented by a Gaussian curve. Therefore, the distance of the inflection points from the true value μ is used as a measure for the scatter. This is denoted as theoretical standard deviation σ. However, in practice, neither an infinite number of measuring values exists nor is the true value μ of a sample known. The measured distribution of a limited number of individual values no longer corresponds to a normal distribution but rather to a t distribution. At n repetitions of a measurement, one obtains s as an estimate for the standard deviation [19]:

$$s = \sqrt{\frac{1}{n-1} \cdot \sum_{i=1}^{n} (x_i - \bar{x})^2} \tag{224}$$

As the number of measurements increases, the estimated values $\bar{x}$ and s approach the true values μ and σ, respectively.

The standard deviation is a measure for random error of an analysis, which depends on the method and the sample composition. The relative random error ε serves as a measure for the precision of an analysis: the smaller the error, the higher the precision.

$$\varepsilon = \sigma/\mu \tag{225}$$

Using the relative random error, an analyst may quickly prove whether the obtained measuring values, independently of their magnitude, are within the precision range of the method.

8.4.3
Scatter and Confidence Interval

The statistical reliability of a result is expressed by the *scatter range T* and the *confidence interval*. The confidence interval means that $P\%$ of all individual measurements, whose mean value $\bar{x}$ is stated as a result, can be expected to be within the range $\bar{x} + T$ and $\bar{x} - T$. Using details about T, one judges the quality of the raw data material without having to know the uninteresting individual data. The scatter range T is calculated by:

$$T = s \cdot t \tag{226}$$

t Student factor

The Student t-factor depends on the statistical reliability P and the number of degrees of freedom $f = n - 1$. Because a certain statistical reliability P must be chosen for the calculation of T, which has to be taken into account when interpreting the result, the result is given in % P:

Result = $\bar{x} \pm T$ (unit); ($\pm s$; $P\%$; n)

In contrast to the scatter range T of individual values, the scatter range of the arithmetic mean value is denoted as confidence interval $T/\sqrt{n}$.

If a series of m arithmetic mean values μ_i is determined each from n individual measurements, the following relation results for the estimated value $s_{\mu i}$ of the standard deviation of the arithmetic mean values:

$$s_{\mu i} = s/\sqrt{m} \tag{227}$$

This means that the estimated value of the standard deviation of the arithmetic mean is lowered with m relative to the standard deviation of the individual measurements.

8.5
Calibration of an Analytical Method (Basic Calibration)

In an analytical method such as ion chromatography that requires a calibration, the application of a physical measuring principle does not directly lead to the analytical result. The obtained measured value only represents a physical result that has to be converted into the analytical result via method parameters previously obtained in a calibration experiment [20]. The use of an *analysis function*

$$\hat{x} = f(\hat{y}) \tag{228}$$

$\hat{x}$ Component content
$\hat{y}$ Measured value

is based on the application of a *calibration function* obtained in a calibration experiment

$$y = f(x) \tag{229}$$

x Component content of the standard solution
y Its measuring value

and its precision parameters for the determination of an unknown component amount in a sample to be analyzed. The calibration function — when solved to x — becomes the analysis function, which yields the analytical result $\hat{x}$ after inserting the measured value $\hat{y}$ of the sample to be analyzed. In the quantitative evaluation of chromatograms, the peak areas A_i (or quantities being proportional to the peak areas) are correlated with the quantities Q_j of the separate sample components:

$$A_i = \overline{f(Q_i)} = \bar{S}_i \cdot Q_i \tag{230}$$

$S_i = (dS/dQ)_{iQ}$ Sensitivity of the detection and registration system for the sample component i with a defined quantity Q
$\bar{S}_i$ Mean of the sensitivity, defined by the slope of the regression line $A = A_0 + \bar{S}_i \cdot Q$

The value of $\bar{S}_i$ has to be determined via calibration. The average sensitivity $\bar{S}_i$ is usually not constant. Because it depends on the quantity Q_j, a single calibration using an optional calibration quantity Q_{j0} is insufficient. Either the calibration quantity Q_{j0} is approximated to the unknown sample quantity Q_{jx} or a calibration function with n calibration quantities $Q_{j1}, ..., Q_{jn}$, is used, which can be applied for samples with different quantities Q_j within the calibrated range.

In this case, the calibration function describes the connection between the peak area and the amount of sample component, and is generally not precisely linear [21]. Often, the calibration function can be approximated by the linear function described in Eq. (230), although deviations from the linear function are observed with increasing span of the calibrated range.

Every calibration starts with the choice of a temporary working range which is application oriented. The working range should cover a large application range with the middle being as close as possible to the most frequent sample concentration. Moreover, the measured values at the lower working range limit have to be significantly different from the method blank value. A lower working range limit only makes sense, when it is at least equal to if not larger than the method detection limit. In order to apply a linear regression, the analysis precision has to be constant over the whole working range [22] (homogeneity of variance), and sample amount and measured value have to be in a linear relation. When a variance inhomogeneity is disregarded, the non-precision of the subsequently obtained analytical results increases. In the case of variance inhomogeneity or non-linearity, the chosen working range has to be reduced until those conditions are fulfilled. Alternatively, a more sophisticated regression model such as a regression function of a higher grade must be selected.

8.5.1
Acquisition of the Calibration Function

Based on the measured values of the standards, the provisional calibration functions of the first and second grade are calculated, yielding method parameters necessary for further statistical tests.

8.5.1.1 Method Characteristic Parameters of a Linear Calibration Function

As already mentioned in Section 8.4, the calibration function describes the relation between the peak area and the amount of sample component. As systematic deviations from a linear function may occur, a measure for the linearity of a calibration function is needed. For this purpose, one can use the maximum relative deviation of the linearized calibration function from the true calibration function in a given range of signals. The linear range is defined by the maximum signal value, up to which the relative deviation does not exceed a predefined value. The straight line equation $y = a + b \cdot x$ is obtained by linear regression of the individual data. The regression coefficients are calculated by the Gaussian least-squares method. With n pair of data (x_i, y_i), it is demanded that

$$\sum_{i=1}^{n}(y_i - Y_i)^2 = \sum_{i=1}^{n}[y_i - (a + b \cdot x_i)^2] \tag{231}$$

is to be minimized. From this follows:

$$b = \frac{\sum\limits_{i=1}^{n}(x_i - \bar{x})(y_i - \bar{y})}{\sum\limits_{i=1}^{n}(x_i - \bar{x})^2}$$

$$= \frac{\sum\limits_{i=1}^{n} x_i y_i - \frac{1}{n}\left[\sum\limits_{i=1}^{n} x_i \cdot \sum\limits_{i=1}^{n} y_i\right]}{\sum\limits_{i=1}^{n} x_i^2 - \frac{1}{n}\left[\sum\limits_{i=1}^{n} x_i\right]^2} \tag{232}$$

$$a = \bar{y} - b \cdot \bar{x} \tag{233}$$

The quantities $\bar{x}$ and $\bar{y}$ represent the mean values of the values x_i and y_i. The slope b is a measure for the sensitivity of a method. The larger the slope, the more sensitive the method is and, thus, the more easily small concentration differences can be measured. This means that a small concentration change in the samples will lead to an utmost change of the measured value. This demand is especially important for trace analysis.

An intercept different from zero means that a signal is measured even when the concentration is zero. The blank value represents a constant, systematic error. Although this blank value can be subtracted from all measured values, it is much safer to work with the calculated calibration function including the intercept.

The measure for the relation between x_i and y_i is given by the correlation coefficient $r_{x,y}$:

$$r_{x,y} = \frac{\sum\limits_{i=1}^{n}(x_i - \bar{x})(y_i - \bar{y})}{\sqrt{\sum\limits_{i=1}^{n}(x_i - \bar{x})^2 \cdot \sum\limits_{i=1}^{n}(y_i - \bar{y})^2}}$$

$$= \frac{\sum\limits_{i=1}^{n} x_i y_i - \frac{1}{n}\left[\sum\limits_{i=1}^{n} x_i \cdot \sum\limits_{i=1}^{n} y_i\right]}{\sqrt{\sum\limits_{i=1}^{n} x_i^2 - \frac{1}{n}\left(\sum\limits_{i=1}^{n} x_i\right)^2} \cdot \sqrt{\sum\limits_{i=1}^{n} y_i^2 - \frac{1}{n}\left(\sum\limits_{i=1}^{n} y_i\right)^2}} \tag{234}$$

This coefficient is independent of the unit of the characteristics and may have any value between -1 and $+1$. If $r_{x,y} = 1$, the relation is directly linear. The correlation coefficient is an important criteria for linearity. However, its information value should be regarded in a more distinguished way, because a high absolute value represents a necessary, but not sufficient, measure for linearity. The statement "method A has a larger correlation coefficient than method B, therefore, method A is better" is *per se* not correct. Even if r is equally high, we do not know much about the quality of two different methods. It has to be

8.5 Calibration of an Analytical Method (Basic Calibration)

emphasized that r is a measure for the quality of a correlation of two values by a mathematical function. A large value of r only means that large values of x lead to large values of y and small values of x to small values of y. The correlation between x and y is described very well with this mathematical function. This is not the case when a steady increase of x leads to a varying increase of y, resulting in a random scatter of the points around a straight line or around a curve. Consequently, a large correlation coefficient can describe the circumstances with a straight line as well as with a curve.

The method characteristic parameters [20, 23-26] can be determined with the aid of regression coefficients. This includes the *residual standard deviation* S_y

$$S_y = \sqrt{\frac{\sum_{i=1}^{n}(y_i - \bar{y}_i)^2}{N-2}} \tag{235}$$

N   Number of calibration points

The residual standard deviation is the scatter of measured values around the regression line and, consequently, a precision measure for the calibration. Normalization of the residual standard deviation to the middle of the working range, accomplished by dividing by the slope, yields the *method standard deviation* S_{x0}, which is a measure of performance:

$$S_{x0} = S_y/b \tag{236}$$

If the measured values are related to equal concentrations, the *relative method standard deviation* V_{x0} allows a comparison between different methods.

$$V_{x0} = \frac{S_{x0} \cdot 100}{\bar{x}} \; [\%] \tag{237}$$

The relative method standard deviation is a suitable evaluation criteria for the robustness of an analytical method that requires calibration. If it stays constant, the method is robust for the evaluated period of time.

In order to calculate the confidence interval for the regression coefficient b, a confidence number γ (95%, 99%) is chosen. Then the solution c of the equation

$$F(c) = \frac{1}{2} \cdot (1 + \gamma) \tag{238}$$

is determined from the table with Student t-factors for $n - 2$ degrees of freedom. Calculate

$$1 = c \cdot \frac{\sqrt{a}}{s} \cdot \sqrt{(n-1)(n-2)} \tag{239}$$

The confidence interval is then

$$\text{Conf}\{b - 1 \leq b \leq b + 1\}$$

and is depicted in Fig. 8-4. The relative error L of the slope is given by:

$$L = 1/b \cdot 100 \ [\%] \tag{240}$$

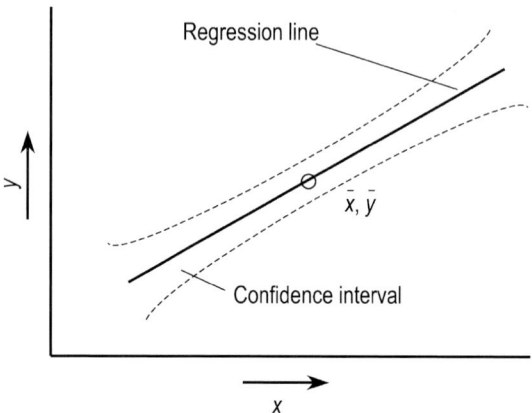

Fig. 8-4. Representation of the confidence interval.

8.5.1.2 Method Parameters of a Calibration Function of 2nd Degree

Regression analysis yields the 2nd degree calibration function

$$y = a + bx + cx^2 \tag{241}$$

with its method parameters. The function coefficients a, b, and c are calculated as follows:

$$a = (y_i - b \cdot \Sigma x_i - c \cdot \Sigma x_i^2)/N \tag{242}$$

$$b = \frac{Q_{xy} - c \cdot Q_{x^3}}{Q_{xx}} \tag{243}$$

$$c = \frac{Q_{xy} \cdot Q_{x^3} - Q_{x^2y} \cdot Q_{xx}}{Q_{x^3}^2 - Q_{xx} \cdot Q_{x^4}} \tag{244}$$

$$Q_{xx} = \Sigma x_i^2 - ((\Sigma x_i)^2/N) \tag{245}$$

$$Q_{xy} = \Sigma(x_i y_i) - ((\Sigma x_i) \cdot (\Sigma y_i)/N) \tag{246}$$

$$Q_{x^3} = \Sigma x_i^3 - ((\Sigma x_i) \cdot (\Sigma x_i^2)/N) \tag{247}$$

$$Q_{x^4} = \Sigma x_i^4 - ((\Sigma x_i^2)^2)/N \tag{248}$$

$$Q_{x^2y} = \Sigma(x_i^2 \cdot y_i) - ((\Sigma y_i) \cdot (\Sigma x_i^2)/N) \tag{249}$$

The residual standard deviation is then:

$$S_y = \sqrt{\frac{\sum_{i=1}^{n}(y_i - \bar{y}_i)^2}{N-3}} \quad \text{with } \hat{y}_i = a + bx_i + cx_i^2 \tag{250}$$

The measure for sensitivity results from the change in the measuring value when changing concentration. While in a linear calibration function sensitivity corresponds to the regression coefficient b, in a curved calibration function it depends on the respective concentration value and, therefore, corresponds to the first derivation of the calibration function:

$$E(x) = b + 2c \cdot x \tag{251}$$

It is recommended that the sensitivity in the middle of the working range be stated as a method parameter:

$$E(\tilde{x}) = b + 2c \cdot \tilde{x} \tag{252}$$

From this the method standard deviation

$$s_{x0} = s_y / E(\tilde{x}) \tag{253}$$

as well as the relative method standard deviation can be derived:

$$V_{x0} = s_{x0} \cdot 100\% / \tilde{x} \tag{254}$$

8.5.2
Testing of the Basic Calibration

If possible, one should work with a linear calibration function; a 2nd degree calibration function should only be used as an exception. In the simplest case, linearity testing can be carried out visually. For this, the calibration data as well as the calibration line are graphically illustrated, so obvious non-linearities can easily be identified. In case of doubt, linearity should be tested arithmetically by using, for example, the approximation test according to Mandel [22, 27]. For this test the calibration functions of the 1st and 2nd degree including the respective residual standard deviations S_y are used. From the residual standard deviations S_{y1} (of the 1st degree calibration function) and S_{y2} (of the 2nd degree calibration function), the difference between the variances DS^2 is calculated

$$DS^2 = (N-2) S_{y1}^2 - (N-3) S_{y2}^2 \tag{255}$$

with the degree of freedom $f = 1$

For the F-test the test quantity PW is calculated

$$PW = DS^2/s_{y2}^2 \qquad (256)$$

and compared with the value $F(f_1 = 1, f_2 = N - 3, P = 99\%)$ in the table. In case $PW \leq F$, a significantly better fit by a 2nd degree calibration function is not achieved; the calibration function is linear. In case $PW > F$, the individual steps of the analytical technique should first be investigated and improved if possible. If linearity cannot be achieved afterwards, sufficient linearity should be attempted by narrowing the working range.

As an alternative to the approximation test according to Mandel, residual analysis [28] can be used for linearity testing. The residuals d_i are the vertical distances of the measuring values from the regression curve.

$$d_i = y_i - \hat{y} \quad \text{for } i = 1, \ldots, n \qquad (257)$$

y_i Measuring value
$\hat{y}$ Estimated value corresponding to y_i (from the regression function)

For the graphical representation, it is practical to use the investigated working range as x-axis and the positively or negatively largest deviation as the intercept on the y-axis. If the mathematical model is correct, the residuals are normally distributed around the zero level.

8.5.3
Testing the Precision

8.5.3.1 Homogeneity of Variances

The linear regression calculation described above is based on a constant non-precision (variance of the measuring values) over the whole working range. Not only does an inhomogeneity of variances lead to a higher non-precision, it also results in a higher inaccuracy due to possible changes in the slope of the straight line.

For checking the homogeneity of variances the F-test is used, with which it is possible to prove whether the standard deviations from two different series of data are comparable. If this is the case, the data come from the same basic population. The two series of data do not necessarily have to have the same size. At first, the test quantity PW is constituted according to

$$PW = s_1^2/s_2^2 \quad \text{with: } s_1 > s_2 \qquad (258)$$

s_1, s_2 Standard deviations of the two series of data

8.5 Calibration of an Analytical Method (Basic Calibration)

The obtained value is then compared to the value $F(P, f_1, f_2)$ in the F-table, where

f_i Number of the degrees of freedom ($f = n - 1$)
n Number of measuring values

In case the F-Test indicates a significant difference of the variances, i.e. PW > F, one can either choose a narrower working range and again check the homogeneity of variances, apply a weighed regression [29], or carry out a so-called multiple curve fitting [30]. Setting a narrower working range is the recommended procedure.

8.5.3.2 Outlier Tests

From time to time, in practice, measuring values are obtained that deviate significantly from the others, without an obvious reason. However, as a matter of principle, calibration data have to be free of outliers. Outlier tests [31] aid in deciding if the value is part of a homogeneous data material. However, at first a suitable regression model has to be determined, because the validity of the chosen regression model is assumed when applying outlier tests.

Nalimov-Test

Given a series of measuring values with the characteristic standard data n, x, and s, the test quantity r^* is calculated by putting the value x^*, a suspected outlier, into Eq. (259):

$$r^* = \frac{|x^* - \bar{x}|}{s} \cdot \sqrt{\frac{n}{n-1}} \qquad (259)$$

The distinction is made by comparing r^* to r_i. The value r_i, which depends on the number n of individual measurements and on the statistical relevance P, may be taken from the r-table [19].

Example:
The following data with arithmetic mean and standard deviation are obtained as the measuring result of a chloride determination:

x_1 = 30.4 mg/L   Mean: $\bar{x}$ = 30.04 mg/L
x_2 = 30.0 mg/L   Standard deviation: $s = \pm 0.313$
x_3 = 30.5 mg/L
x_4 = 30.2 mg/L

x^* = 29.1 mg/L

For the Nalimov test, the degree of freedom is calculated by:

$$f = n - 2 \qquad (260)$$

With $n = 5$ individual data, the degree of freedom is $f = 3$. Substituting the values for $\bar{x}$ and s into Eq. (259), one obtains:

$$r^* = \frac{|29.1 - 30.04|}{0.313} \cdot \sqrt{\frac{5}{5-1}} \qquad (261)$$

$$= 3.358$$

The statistical factors r_i are taken from the r-table to decide if the suspected value is an outlier. For $f = 3$ one finds:

$r(95\%) = 1.757$
$r(99\%) = 1.918$
$r(99.9\%) = 1.982$

To determine if x^* represents an outlier, compare r^* with r_i.

$r(95\%) < r^* < r(99\%)$ probable outlier
$r(99\%) < r^* < r(99.9\%)$ significant outlier
$r^* > r(99.9\%)$ highly significant outlier

From the comparison of the value calculated for r^* with these statistical factors it is concluded that the suspected value is a highly significant outlier. Thus, arithmetic mean and standard deviation must be calculated again without the outlier:

Arithmetic mean $\bar{x} = 30.28$ mg/L
Standard deviation $s = 0.049$

The corrected values differ markedly from those that were distorted by the outlier.

Test according to Grubbs

The test according to Grubbs differs from that of Nalimov by the test quantity r_m, which is calculated with Eq. (262) as follows:

$$r_m = \frac{|x^* - \bar{x}|}{s} \qquad (262)$$

From the table "Comparison values $r_m(P)$ for the outlier test according to Grubbs" one selects the values $r_m(90\%)$, $r_m(95\%)$, and $r_m(99\%)$ for $f = n - 2$. It is then tested for the following criteria:

$r_m(90\%) < r_m < r_m(95\%)$ probable outlier
$r_m(95\%) < r_m < r_m(99\%)$ significant outlier
$r_m > r_m(99\%)$ highly significant outlier

8.5 Calibration of an Analytical Method (Basic Calibration)

If outliers are present that have been eliminated from the measuring series, $\bar{x}$ and s are again calculated.

The residual analysis mentioned above can also be used for identifying outliers in the calibration [32]. At first, the calibration line with the residual standard deviation S_{yA1} is calculated from all data pairs. The pre-selection of potential outliers is carried out arithmetically by determining the residuals. Any data pair with a conspicuously large residuum is a potential outlier. After eliminating the suspicious outlier pair (x_A, y_A) from the data collective, a new calibration line with the residual standard deviation S_{yA2} is calculated. The check can be carried out either with the F-test or the t-test. Both methods lead to mutually corresponding results.

If an outlier is statistically identified with the aid of the *F*- or *t*-test, it is absolutely necessary to look for the origin of the error and to eliminate it. The complete calibration has to then be repeated.

8.5.4 Calibration Methods

Calibration may be performed by one of the following four different methods with the aid of modern electronic data evaluation systems:
- Area normalization
- Internal standard
- External standard
- Standard addition

8.5.4.1 Area Normalization

The method of area normalization requires that all sample components eluted from the analytical separator column are detected. If response factors are not taken into account, this method can only be applied to the calibration of sample components that have the same response.

Area normalization is mainly used, therefore, in gas chromatographic analyses of hydrocarbons. Its application to ion chromatography is limited, as sample components only rarely exhibit the same response to the detection methods employed.

The calculation of the area for the unknown peak X in Fig. 8-5 is performed according to Eq. (263):

$$\% X = \frac{A_x \cdot 100}{A_x + A_y + A_z}$$

$$= \frac{A_x \cdot 100}{\sum_{i=1}^{n} A_i} \tag{263}$$

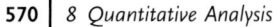

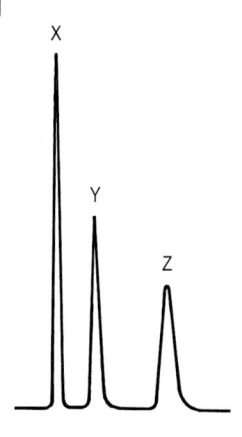

Fig. 8-5. Area normalization.

8.5.4.2 Internal Standard

In this method, a standard substance of known concentration that is not a sample constituent is added to the unknown sample. The sample composition is determined by comparing the peak areas of the standard and the sample component. This method does not require all constituents to be eluted and detected. Therefore, the method of internal standard is suited for samples in which not all of the components have to be determined. Moreover, analytical results based on internal standard calibration are used for control purposes. If the internal standard is added after all sample preparation steps, right before the measurement, the analytical results can be used to control the sample injection or even the detection system, whereas the addition of the standard prior to the sample preparation allows the specific control of the sample preparation steps.

The internal standard should meet a number of requirements:
- Chemical stability of the standard under the given chromatographic conditions, i.e., the standard itself must not cause a matrix effect
- Elution close to the analyte substance
- Baseline-resolved separation from both neighboring peaks
- The substance-specific correction factor is known or can be determined
- Similarity in the concentration and response to the analyte component
- High purity.

To calibrate according to the internal standard method, the analyte components must exist as reference compounds in sufficiently high purity. To determine the correction factors, a solution is prepared that contains known amounts of the analyte components and the internal standard. This solution is chromatographed and the correction factors based on the peak areas are calculated according to Eq. (264):

$$f_x = \frac{A_x \cdot W_{is}}{A_{is} \cdot W_x} \tag{264}$$

f_x Correction factor for component x
A_x Peak area of component x
W_x Concentration of component x
A_{is} Peak area of internal standard
W_{is} Concentration of internal standard

To plot the calibration curve, solutions with four different concentrations of the analyte components and a known concentration of the internal standard are chromatographed. The ratio of the peak area of the analyte component to that of the internal standard, A_x/A_{is}, is plotted as a function of the concentration W_x.

After the calibration, a known amount of an internal standard is added to a known sample amount and the sample is chromatographed again. The calculation of the composition is carried out with Eq. (265):

$$\% X = \frac{A_x \cdot f_x \cdot W_{is} \cdot 100}{A_{is} \cdot W_s} \tag{265}$$

With careful calibration based on peak areas, a relative precision of at most 0.1% can be achieved.

8.5.4.3 External Standard

The calibration method most often used in ion chromatography is the direct comparison of the peak area of a certain compound in an unknown sample with the peak area of a solution with a known content of the same compound. This method requires the injection of constant volumes under constant chromatographic conditions. Errors in sample delivery, however, are almost excluded when using a sample loop valve. The existence of reference compounds for all sample components to be analyzed is a prerequisite. In practice, several different standard solutions in the investigated concentration range are prepared and chromatographed [33]. When the resulting peak area is plotted versus the concentration of the standards, the compound-specific calibration function is obtained, according to Fig. 8-6.

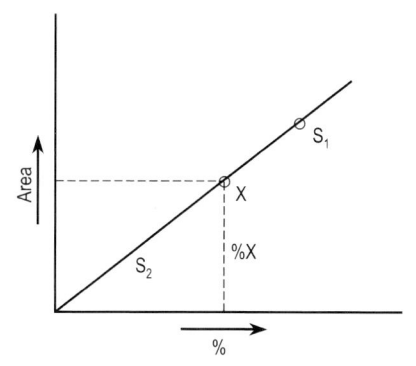

Fig. 8-6. External standard calibration.

The peak area of the analyte component should lie between those of the calibration standards, so that the concentration of the analyte component can be calculated via interpolation. Extrapolation is not recommended because the course of the function outside the investigated concentration range is unknown. The slope of the calibration line equals the compound-specific correction factor, which depends on the properties of both the compound itself and the detector. If possible errors in the sample delivery are excluded, a relative precision of 0.3% can be obtained in routine operation.

8.5.4.4 Standard Addition

The standard addition method [34] represents a combination of calibration with the aid of both external and internal standards. In ion chromatography it is used predominantly for the analysis of samples with difficult matrices. Matrix problems may lead to an increase in non-precision and/or express themselves as constant or proportional systematic deviations of the analytical results. Matrix influence can be identified via calculation of the recovery function. In constant systematic deviation, the error is independent of the analyte component. Such a deviation will cause a parallel shift of the calibration line. A possible origin for this deviation might be a co-detection of a matrix component. In proportional systematic deviations, the error depends on the concentration of the analyte component. This type of deviation results in a change of the slope of the calibration line. Deviations of this kind can be caused by individual sample preparation steps such as sample digestion and sample extraction, and also by matrix effects. Systematic deviations can be identified by standard addition and/or calculation of the recovery function.

The recovery or recovery rate is the ratio of the measured mean value under repeating conditions to the so-called true value of the analyte in the sample:

$$W = \bar{x}/x_R \cdot 100\% \tag{266}$$

W Recovery rate in %
$\bar{x}$ Measured mean value
x_R True value

The ideal value for W is 100%. With the aid of the recovery rate, the complete process can be assessed. If the true value is found, selectivity, accuracy, and robustness for this concentration level and this matrix under the given experimental conditions are proven. To identify a potential matrix influence, the sample matrix to be analyzed (which does not contain the analyte component) is divided into ten equally sized portions and spiked with concentrated standard solutions, so that the component concentrations in the spiked samples and in the aqueous calibration standards are the same. The spiked matrix samples are then analyzed with the corresponding analytical method. Ideally, the recovery function is a straight line with a residual standard deviation corresponding to the process standard deviation of the basic analytical method. In case a proportional systematic or constant systematic deviation is the result of the investigation of the matrix influence, the calibration function obtained with aqueous standards cannot be used for data evaluation; the standard addition method has to be applied.

After analyzing the analyte sample under suitable chromatographic conditions, a known amount of the component of interest is added and the sample is chromatographed again. In the calculation of the concentration, the volume change has to be taken into account. A peak that is not of interest serves as the internal standard to compensate for the dosing error. Calculation is performed according to Eq. (267):

$$Q_i = \frac{\Delta Q_i}{\frac{(A + \Delta A)_{i2}}{A_{i1}} \cdot \frac{A_{s1}}{A_{s2}} - 1} \tag{267}$$

Q_i Amount of the analyte ion i before addition
ΔQ_i Amount of the ion i added to the sample volume Q
$A_{i1}, (A + \Delta A)_{i2}$ Peak areas of the analyte ion i in the chromatograms 1 and 2
A_{s1}, A_{s2} Peak areas of the reference compound in the chromatograms 1 and 2

The concentration of the analyte ion may also be determined graphically. As seen in Fig. 8-7, the peak area of the compound of interest is plotted versus the added concentration. A linear correlation exists if the measuring quantity (electric conductance, UV absorption, etc.) is proportional to the analyte ion concentration. The concentration of the ion is obtained by extrapolating the straight line to the abscissa.

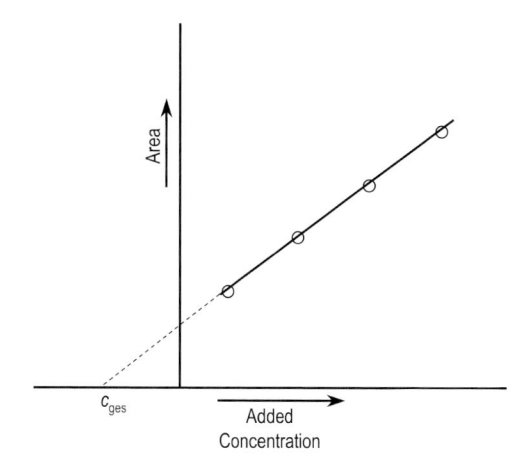

Fig. 8-7. Calibration via standard addition.

8.6
Detection Criteria, Limit of Detection, Limit of Determination

When determining very small amounts it is often very difficult to distinguish the measuring signal of a sample from the measuring signal of a blank or to quantitate the signal with sufficient precision. If the measuring value of a sample is only slightly above zero, two contradicting interpretations are possible:

a) The sample does not contain the analyte component; the measuring value is related only to the non-precision of the analytical method and, consequently, lays within the scatter.
b) The analyte component is indeed present in the sample; in this case, repeated analyses would result in a mean value with the questioned value within its scatter.

If the standard deviation is known, the probability for any measuring value to be either a blank value or a part of an existing compound amount can be calculated statistically. As a criteria for this decision the error probability is used:

a) The probability that the sample is in reality free of any component, although a positive measuring value is reported, is named α-error.
b) The probability that the sample does contain the component, although the analytical result states zero, is named β-error.

In analytical practice, a number of terms with different information content are discussed:
- Detection criteria
- Limit of detection
- Limit of determination
- Limit of coverage
- Critical value of the measuring quantity, etc.

Depending whether the analytical method requires a calibration or not, one distinguishes between two different procedures for determining the detection criteria and the limit of detection. Because ion chromatography is an analytical method requiring calibration, both parameters can be derived from the calibration data. When employing analytical methods that cannot be calibrated, the detection criteria or the limit of detection are determined via repeated analyses of blank samples.

8.6.1
Determination of Detection Criteria, Limit of Detection, and Limit of Determination

Funk et al. define the *detection criteria* L_c [26,35] as the measuring value, for which the β-error – with given α-error of 5% – is just 50% (see Fig. 8-8).

8.6 Detection Criteria, Limit of Detection, Limit of Determination

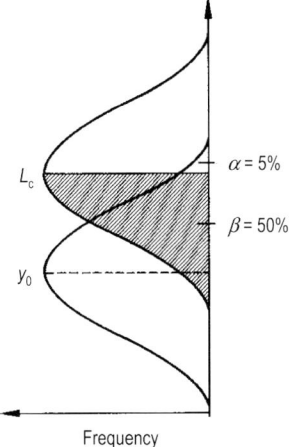

Fig. 8-8. Detection criteria. — α-error = 5%, β-error = 50%.

In the terminology of error probability this means:

If based on the detection criteria L_c the blank value is regarded to be exceeded, an error probability of only $\alpha = 5\%$ is against it. However, with repeated analyses of this sample 50% of all results will be below L_c ($\beta = 50\%$), i.e., the probability of not detecting this compound is 50%.

For the practical determination of the detection criteria, the standard deviation s_{y0} of the blank value y_0 or the standard deviation of the intercept a of the linear calibration function is used, where the standard deviation should be calculated from at least $n = 6$ measuring values. The detection criteria is then regarded as the upper limit of the blank value scatter

$$L_c = y_0 + t_\alpha\,(1 - \alpha, f) \cdot s_{y0} \tag{268}$$

or as the upper limit of the prognosis range of the intercept

$$L_c = a + t_\alpha\,(1 - \alpha, f) \cdot s_a \tag{269}$$

with

$$s_a = s_y \cdot \sqrt{\frac{1}{N} + 1 + \frac{\bar{x}^2}{\sum (x_i - \bar{x})^2}} \tag{270}$$

The term *limit of detection* XN [36, 37] is defined as the smallest detectable amount of a compound. A quantitative analysis with reported concentrations is only possible when the analytical result is equal or larger than the limit of detection, because the required significance level is achieved only for such an analytical result [38] (see Fig. 8-9). In contrast to the detection criteria the limit of detection is a quantitative statement of the amount of substance.

8 Quantitative Analysis

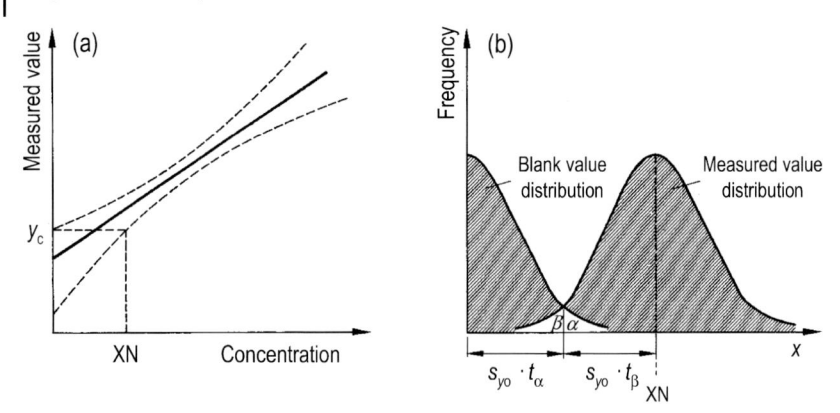

Fig. 8-9. Limit of detection.
a) Determined via the calibration function with the associated confidence interval
b) Determined via the scatter of the blank value. An important assumption: equal s!

The general formula for calculating the limit of detection via determination of the blank value scatter is:

$$XN = (t_\alpha + t_\beta) \cdot s_o \tag{271}$$

where the standard deviation s_o (in concentration units) should be calculated from at least $n = 6$ blank values (under repeating conditions).

The numerical value for the limit of detection XN, like that of the detection criteria L_c, depends on the chosen significance levels and the degrees of freedom.

For $\quad \alpha = \beta = 5\% \quad$ it follows
$\quad\quad t_\alpha = t_\beta = t$
$\Rightarrow \quad XN = x_o + 2 \cdot t(95\%, f) \cdot s_o$

Theoretically, x_o should be zero, which cannot be achieved in practice with analytical results $x \geq 0$.

For determining the limit of detection via the calibration function, the calibrations are carried out with increasingly smaller concentrations, until the calculated test quantity x_p corresponds approximately to the calibration concentration x_1 of the lower limit of the working range, i.e. the condition

$$0.5\, x_1 \leq x_p < x_1 \tag{272}$$

is fulfilled. In this case, x_p is declared to be the limit of detection:

$$XN = x_p$$

with

$$x_p = 2 \cdot s_{xo} \cdot t \cdot \sqrt{\frac{1}{N} + 1 + \frac{(y_p - \bar{y})^2}{b^2 \Sigma (x_i - \bar{x})^2}} \qquad (273)$$

Calculation of the limit of detection:

$$XN = 2 \cdot VB_x (y = y_c) \qquad (274)$$

$$XN = 2 \cdot \frac{s_y \cdot t_{f_1, P}}{b} \cdot \sqrt{\frac{1}{N_c} + \frac{1}{N_a} + \frac{(y_c - \bar{y})^2}{b^2 \Sigma (x_i - \bar{x})^2}} \qquad (275)$$

with

$$y_c = a + VB_y (x = 0) \qquad (276)$$

N_c Number of calibration standards

$$y_c = a + s_y \cdot t \cdot \sqrt{\frac{1}{N_c} + \frac{1}{N_a} + \frac{(\sigma - \bar{x})^2}{\Sigma (x_i - \bar{x})^2}} \qquad (277)$$

The term *limit of determination* XB is defined as the smallest concentration of a compound that can be determined with a given analysis precision, expressed as a relative confidence interval VB_{rel} [%]

$$VB_{rel} = VB(\hat{x}) / \hat{x} \cdot 100\% \qquad (278)$$

Hence the limit of determination depends on the largest random error that can still be tolerated in the analytical result statement.

Under the assumption of a constant standard deviation s_o for low concentrations, the confidence interval ($P = 95\%$) is given by

$$VB(\hat{x}) = t \cdot s_o \qquad (279)$$

For a maximal permissable relative confidence interval $VB_{rel,;>x}$ [%] the limit of determination is:

$$VB_{rel, max} = t \cdot s_o / XB \cdot 100\% \qquad (280)$$

converted to XB: $XB = t \cdot s_o / VB_{rel, max} \cdot 100\% \qquad (281)$

The statement of the limit of detection has always to be made in connection with the given $VB_{rel, max}$ value!

8.7
The System of Quality Control Cards

The quality control card [39] was introduced by Shewhart in 1931 and originally used for industrial product control. The principle of a quality control card is illustrated in Fig. 8-10. It is an optical representation of the "quality" based on
- Quality goal (set value of the analysis results for control samples) and
- Quality tolerance limits

With the tolerance limits one distinguishes between so-called *warning limits*, a single crossing of which is still tolerated, and so-called *control- or intervention limits*, the crossing of which results in immediate action.

In a so-called pre-period, one or several control sample(s) of every analysis series are analyzed together with the samples. For each control sample a precision check and, if possible, an accuracy check are carried out by comparison with given values. If the analysis quality is acceptable, warning and intervention limits are defined and the quality control card is drawn. In the subsequent control period, the compliance with the defined limits is monitored by measuring further control samples. The results from these control analyses or the statistical quantities derived from them are continuously registered in the quality control card. The abscissa relates to the continuing serial number, whereas control values are registered on the ordinate.

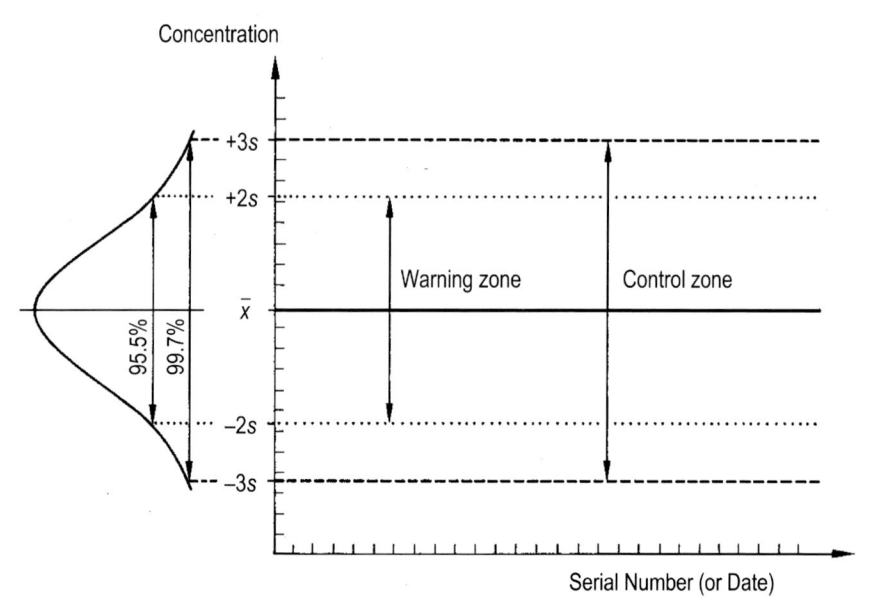

Fig. 8-10. Schematic representation of a quality control card

Thus, the quality control card aids in the fast recognition of errors because the given control criteria indicate an "out-of-control situation" for the monitored analytical process. With a quality control card each of the following statistical quantities can be monitored:
- Single measuring value, mean value
- Recovery rate
- Standard deviation
- Range

There are two important aspects in defining the limits of a quality control card:
- In case the analytical process changes significantly due to unknown influences, a statistical "out-of-control situation" should be indicated as quickly as possible.
- Low probability for a false alarm, i.e., a measuring value belongs to the same parent population, but is above the limit due to random variations. Setting the limits too narrowly creates the possibility of a false alarm.

An important parameter for evaluating quality control cards and "out-of-control situations" is the **A**verage **R**un **L**ength (ARL) which is defined as the average number of registrations in a quality control card for a single "out-of-control situation" [40]. A high ARL is required, should the analytical process run under control. The probability of a false alarm is then very low. On the other hand, a true "out-of-control situation" should be indicated as quickly as possible, i.e., the ARL has to be relatively small.

8.7.1
Types of Quality Control Cards and their Applications

In analytical quality control the following types of quality control cards – among others – are used:
- Shewhart control card as
  - Mean value control card (and blank value control card)
  - Recovery rate control card
- Span control card or standard deviation control card
- Cusum control card

Shewhart Control Card
With the Shewhart control card the precision of mean values from multiple analyses or even from individual results can be monitored. The decision as to whether a new value in the scope of control belongs to the same basic population, is carried out graphically based on how much the limits/lines are exceeded.

The application of quality control cards in quality control is based on the assumption that results are normally distributed [41]. Therefore, Shewhart developed a quality control card that shows the bell-shaped curve in a 90° turned form (see Fig. 8-10). For this a co-ordinate system is constructed, in which the ordinate represents the unit of the analytical result being subdivided accordingly. On the abscissa, time, serial number, day or similar units are registered in a chronological series. After a number of N days, or series, the mean value $\bar{x}$ as well as the standard deviation s are calculated and the Shewhart control card is drawn. The central line on the card represents the calculated mean value from the pre-period. The warning and control limits at $\bar{x} \pm 2s$ and $\bar{x} \pm 3s$ are plotted with the aid of the standard deviation.

The control and warning limits are the $\pm 2s$- and $\pm 3s$-ranges, respectively:

upper control limit	$KO = \bar{x} + 3s$
lower control limit	$KU = \bar{x} - 3s$
upper warning limit	$WO = \bar{x} + 2s$
lower warning limit	$WU = \bar{x} - 2s$

The area from the central line $+2s$ to the central line $-2s$ covers 95.5% of the area under the curve, i.e., the probability for a false alarm is only 4.5%. For this reason, one single crossing is still tolerated. The probability for a crossing of the $3s$-limits is 0.3%, i.e., if this happens, an "out-of-control situation" is more than likely.

Besides the indication of course errors the quality control card should also indicate systematic and tendentiously systematic errors. The following criteria for "out-of-control situations" are mentioned in literature:

- 1 value outside the control limit
- 7 subsequent values on one side of the central line
- 7 subsequent values with increasing tendency
- 7 subsequent values with decreasing tendency
- 2 of 3 subsequent values outside the warning limits and
- 10 of 11 subsequent values on one side of the central line

In analytical quality control, the Shewhart control card is applied in the form of three different control cards:

- Mean value control card
- Blank value control card and
- Recovery rate control card

The mean value control card corresponds to the original form of the Shewhart control card, although in analytical chemistry individual measuring values are prevailing. It is usually applied for monitoring the precision of an analytical process. For the mean value control card, the following can be used as control samples:

- Pure standard solutions
  Easy to prepare in any concentration
  Matrix effects are difficult to be identified
  When using an original standard systematic errors, which already occurred during calibration, cannot be identified

- Synthetic samples
    - Can be preparable on demand
    - Simulation of the sample matrix can be quite elaborate
    - Sample has to be stable over a long period of time
- Reference samples or
    - Samples which have been extensively analyzed with other analytical processes
    - Less expensive than certified reference standards
    - Matrix and analyte concentration are possibly not representative
- Certified reference standards
    - Independent sample with known target value
    - Matrix and analyte concentrations are possibly not representative
    - Well suited for monitoring analytical instruments

The blank value control card represents a special application form of the mean value control card and gives direction as to the reagents and the measuring system to be used. Because it would be too expensive to determine the blank value for every analysis, it is reasonable to carry out two blank value determinations for every analysis series, one at the beginning and one at the end of the analysis series. Such a procedure has the advantage of an additional drift control in addition to the precise determination of the blank value.

Mean value control cards, however, do not indicate any errors related to the sample matrix. It is possible to check the analytical process for matrix influences by determination of the recovery rate, although it only determines matrix related, proportional systematic deviations. Constant systematic deviations remain undiscovered. Thus, the determination of the recovery rate is only a limited measure for controlling accuracy. If certified reference standards are used, the target value (x_{set}) is known. Together with the measuring value x_{meas} the recovery rate can be determined according to:

$$WFR = (x_{meas}/x_{set}) \cdot 100\% \qquad (282)$$

The quality goal should be an average recovery rate of $\approx 100\%$, because only in this case can a systematic matrix influence be excluded. The sample should be spiked with an analyte concentration that corresponds at least to the concentration of the unspiked sample. However, the upper limit of the working range of the respective analytical process should not be exceeded. Moreover, it is important to add the spiked amount in the form of a solid or a concentrated solution. This is the only way to avoid any change in the sample matrix caused by larger volumes or additional errors such as dilution errors.

Regarding setup and decision criteria, the recovery rate control card corresponds to the mean value control card. The average recovery rate, RR, and its standard deviation s_{RR} are calculated as follows:

$$\overline{WFR} = \frac{1}{N} \cdot \sum_{i=1}^{N} WFR_i \qquad (283)$$

$$s_{WFR} = \sqrt{\frac{1}{N-1} \sum_{i=1}^{N} (WFR_i - \overline{WFR})^2} \qquad (284)$$

The warning and control limits are the $\pm 2s$- and $\pm 3s$-ranges, respectively:

upper control limit	$KO = \overline{RR} + 3s_{RR}$
lower control limit	$KU = \overline{RR} - 3s_{RR}$
upper warning limit	$WO = \overline{RR} + 2s_{RR}$
lower warning limit	$WU = \overline{RR} - 2s_{RR}$

In analogy to the mean value control card the average recovery rate represents the central line of the card.

Span Control Card

The Shewhart control card indicates how well the mean of the individual measuring values corresponds to the cumulative mean value. However, without allowing a statement on the scatter of the individual analysis results, the span control card serves precision control above all. The span is defined as the difference between the largest and the smallest measuring results in multiple analyses.

The span control card offers the following control possibilities:

1. Precision control
   A prerequisite for the use of the Shewhart control card is the constancy of the standard deviations. The span control card allows the disclosure of significant differences between the standard deviations of the sub-groups and, thus, represents a test for variance homogeneity.
2. Limited accuracy control – drift control
   Trends within a series as well as drift phenomena can be monitored with a span control card that has been modified to a difference control card.

For setting up a span control card the following quantities have to be known:
- Number n of replicate measurements per sub-group (n_i)
- Number N of the sub-groups
- R_i, span of the sub-group i
- $\bar{R}$, mean of the spans
- s^2, variance of the total measurement
- upper warning limit WO
- lower warning limit WU
- upper control limit KO
- lower control limit KU

At first, the spans, R_i, of all sub-groups are calculated and summed up to $\bar{R}$:

$$R_i = \text{largest value} - \text{smallest value of a sub-group } i \quad (285)$$
$$\text{from } n \text{ individual measurements}$$

$$\bar{R} = \Sigma R_i / N \quad (286)$$

$\bar{R}$ is the central line within the span control card. An "out-of-control situation" is indicated when:
- a span R_i is above the upper control limit
- a span R_i is below the lower control limit
- seven subsequent values show an increasing or a decreasing tendency
- seven subsequent values are above the average span $\bar{R}$

In analytical practice, the span control card will only be applied in its simplest form, e.g., with only one duplicate determination per analysis series, because the time effort for a larger number of parallel determinations is not acceptable.

Without doubt, the most common control card in quality control is the $\bar{x}$-R-Combination Card, which consists of a mean and a span control card. These are arranged in such a way that the mean and the span of a sub-group lay on top of each other (see Fig. 8-11). This procedure is applied to illustrate different types of changes at the same time on one card. The main advantage of this procedure is the ability to decide whether the scatter between the sub-groups is significantly larger than the scatter within one sub-group. However, as a prerequisite, the same control samples have to be used for the span and the mean control.

While for the span control card the absolute amount of the difference between the largest and the smallest measuring value is used for the check, the *difference control card* considers for control not only the absolute value of the difference but also the sign. The goal is the recognition of systematic errors, which occur first *during* an analysis series. For the difference control card a real sample of each series is analyzed twice. The differences $d_i = x_{i2} - x_{i1}$, where the first value always has to be subtracted from the second, are registered in a control card with the respective sign being taken into account. The central line of this card corresponds to the expected value of the difference, that is zero. Under optimum conditions, the difference (with the corresponding sign) should be scattered around the zero line. In case of a drift, x_2 will be always larger or smalller than x_1, which results in a one-sided position of the difference above or below the zero line.

Cusum Control Card

The Cusum control card was introduced by Page [42] in 1954 for industrial quality control. This card is a successive development of the mean value control card by Shewhart, where the individual control results are not registered, but rather the summation of the deviations of the individual results from a given value.

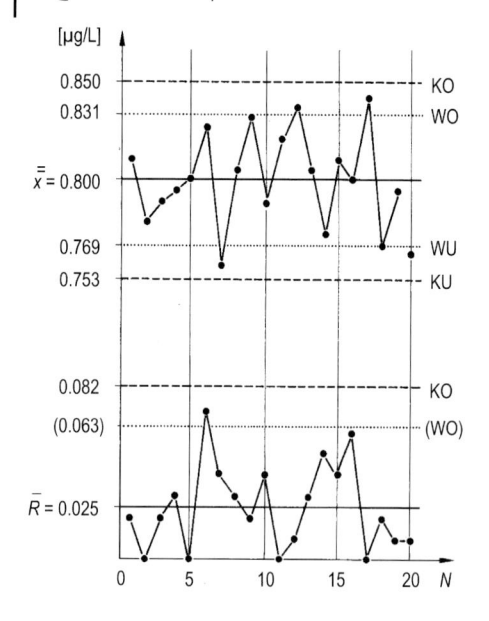

Fig. 8-11. Schematic representation of a $\bar{x}$-R-combination card.

Cumulative sum S:

$$S_1 = (x_1 - k)$$
$$S_2 = S_1 + (x_2 - k)$$
$$S_3 = S_2 + (x_3 - k) \qquad (287)$$
$$S_N = S_{N-1} + (x_n - k) = (\Sigma x_i) - nk$$

x_i Control analysis
k Reference value

Thus, every new registration contains the information about the current status and the previous analysis results obtained with this analytical method. In this way, changes which do not yet lead to "out-of-control situations" on the mean value control card can be more easily recognized, which is especially advantageous for processes with a large variance. Thus, the Cusum technique is suited for the following applications [43]:

- Recognition of systematic changes/shifts of the mean value during the process
- Determination of the magnitude with which the mean value has changed
- Determination of the moment when the change has happened
- Short-term predictions about the mean value to be

The Cusum control card should be dimensioned in such a way that it can react to the smallest changes of the mean value, which should be reflected by a significant change in the slope of the Cusum line. The slope of the line depends on the chosen reference value k and on the scaling of the y-axis (Cusum axis). If a certified standard is at hand, the certified concentration corresponds to the reference value k. Otherwise, the concentration of the control sample has to be determined. The estimation of the mean value should be based on at least 20 analyses.

If the card is well-dimensioned, changes in the slope can be visually identified in an easy way. In case the changes in the Cusum course are only minor, a visual interpretation poses difficulties. With the V-mask, a bilateral statistical test developed by Barnard [44] in 1959, positive and negative deviations from the mean value can be identified. The V-mask technique combines visual interpretations and objective test criteria. The V-mask is defined by the two parameters d and θ (see Fig. 8-12, top). The leading distance d corresponds to the distance between the tip of the V-mask and the last registration on the card. θ is the angle between the respective leg of the mask and the horizontal drawn through the mask tip. The V-mask is then positioned with a distance d from the last registration onto the Cusum card in such a way that the top (horizontal) points forward. With every new registration the mask is shifted, so that point E lays on the new Cusum value. This corresponds to the shift by one abscissa unit. An "out-of-control situation" (see Fig. 8-12, bottom) is indicated, if the Cusum line crosses one of the two V-mask legs. In practical application, the V-mask parameters d and θ are approximated according to a method described by Juran [45], which has a mathematical-statistical base. The parameters d and θ can be approximated as follows:

$$\theta = \arctan (D/2 \cdot w) \tag{288}$$

$$d = \frac{2s^2 \cdot \ln \alpha}{D^2} \tag{289}$$

- D Smallest deviation, which shall be detected with a certain probability
  $D = x_{max} - k$ or $D = x_{min} - k$ (in multiples of the standard deviation s)
- s Standard deviation of the reference value
- w Scaling factor
- α Probability of a 1st degree error
  (false alarm, although the process runs under control)
  ($\alpha = 0.0027$ corresponds to the 3s-control limit of the mean control card)

The advantages of the Cusum control card include the ability to visually recognize even small changes of the mean value in the form of an increase in the slope of the Cusum line. The moment at which this change in the analytical process happened can also be determined with high precision. The Cusum control card is predominantly used for monitoring processes with a large scatter. However, the calculation of the Cusum values, the scaling of the ordinate, and the calculation of the parameters for deciding on an "out-of-control situation" are more complicated as those for the mean value control card. For a much more detailed discussion of the quality control card system for internal quality control the interested reader is referred to the monography of Funk et al. [26].

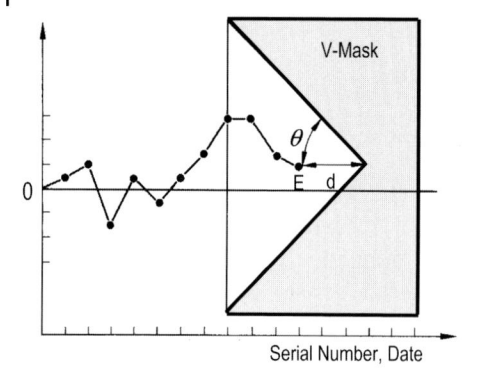

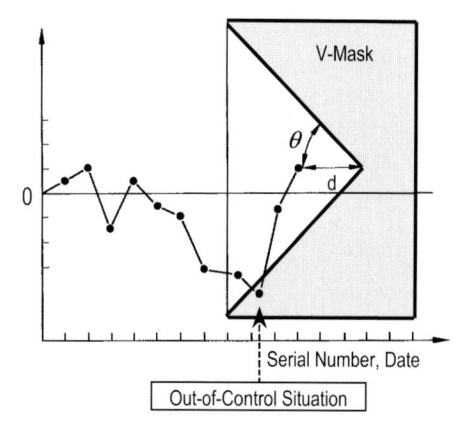

Fig. 8-12. Schematic representation of the V-mask for determining positive and negative deviations from the mean value in a Cusum control card.

9
Applications

The analytical method known as ion chromatography is suited for the analysis of a variety of inorganic and organic anions and cations. It is characterized by a high selectivities and sensitivity. The great number of applications portrayed below examplifies the versatility of this method.

Ion chromatography has become an indispensable tool for the analytical chemist, especially in the area of anion analysis. In many cases, this method has replaced conventional wet chemical methods such as titration, photometry, gravimetry, turbidimetry, and colorimetry, all of which are labor-intensive, time-consuming, and occasionally susceptible to interferences. Publications by Darimont [1] and Schwedt [2] have shown that results obtained by ion chromatographic methods are comparable to those of conventional analytical methods, thus completely dissolving the scepticism with which ion chromatography was initially met. In the field of cation analysis, it is attractive because of its simultaneous detection and sensitivity. It provides a welcome complement to atomic spectrometric methods such as AAS and ICP.

The emphasis of ion chromatographic applications is mostly on the following areas:

- Environmental analysis
- Power plant chemistry
- Semiconductor industry
- Household products and detergents
- Food and beverages
- Electroplating
- Metal processing
- Pharmaceuticals
- Biotechnology
- Mining
- Agriculture
- Pulp and paper industry

The separation and detection methods employed in ion chromatography, as well as a selection of ionic species that may be analyzed with this method, are summarized in Tables 9-1 and 9-2.

Table 9-1. Survey of separation techniques used in ion chromatography.

Separation technique	Mechanism	Resin functionality	Recommended Eluants	Species to be analyzed
HPIC Anions	Ion-exchange	$-NR_3^+$	Na_2CO_3/ $NaHCO_3$ $NaOH$	F^-, Cl^-, Br^-, I^-, SCN^-, CN^- $H_2PO_2^-$, HPO_3^{2-}, HPO_4^{2-}, $P_2O_7^{4-}$, $P_3O_{10}^{5-}$, NO_2^-, NO_3^-, S^{2-}, SO_3^{2-}, SO_4^{2-}, $S_2O_3^{2-}$, $S_2O_6^{2-}$, $S_2O_8^{2-}$, OCl^-, ClO_2^-, ClO_3^-, ClO_4^-, BrO_4^-, SeO_3^{2-}, SeO_4^{2-}, $HAsO_3^{2-}$, WO_4^{2-}, MoO_4^{2-}, CrO_4^{2-}, carbohydrates, peptides, proteins, etc.
HPIC Cations	Ion-exchange	$-SO_3^-$	MSA[a] H_2SO_4	Li^+, Na^+, NH_4^+, K^+, Rb^+, Cs^+, Mg^{2+}, Ca^{2+}, Sr^{2+}, Ba^{2+}, small aliphatic amines
		$-SO_3^-$/ $-NR_3^+$	PDCA[b] Oxalic acid	Fe^{2+}, Fe^{3+}, Cu^{2+}, Ni^{2+}, Zn^{2+}, Co^{2+}, Pb^{2+}, Mn^{2+}, Cd^{2+}, Al^{3+}, Ga^{3+}, V^{4+}, V^{5+}, UO_2^{2+}, lanthanides
HPICE	Ion-exclusion	$-SO_3^-$	HCl Octane-sulfonic acid Perfluoro-butyric acid	Aliphatic carboxylic acids, borate, silicate, carbonate, alcohols, aldehydes
MPIC Anions	Ion-pair formation	neutral	NH_4OH TMAOH[c] TPAOH[d] TBAOH[e]	In addition to the anions listed under HPIC: anionic surfactants, metal-cyanide complexes, aromatic carboxylic acids
MPIC Cations	Ion-pair formation	neutral	HCl Hexane-sulfonic acid Octane-sulfonic acid	In addition to the cations listed under HPIC: alkylamines, alkanolamines, quaternary ammonium compounds, cationic surfactants, sulfonium compounds, phosphonium compounds

a) MSA   Methanesulfonic acid
b) PDCA Pyridine-2,6-dicarboxylic acid
c) TMAOH Tetramethylammonium hydroxide
d) TPAOH Tetrapropylammonium hydroxide
e) TBAOH Tetrabutylammonium hydroxide

9.1
Ion Chromatography in Environmental Analysis

Environmental analysis is one of the most important fields of application in ion chromatography; it is divided into water-, soil-, and air hygiene.

The main focus of applications in environmental analytical chemistry is the qualitative and quantitative analysis of anions and cations in all kinds of water [3-8]. For example, the concentration of the anions chloride, nitrite, bromide, nitrate, orthophosphate, and sulfate in water determines its quality. Using ion chromatography, these anions can all be separated and determined in less than ten minutes. The high sensitivity of this method (detection limit with a *direct injection* of 50 µL sample: ~ 10 µg/L) and the potential for automation were two

Table 9-2. Survey of the detection methods used in ion chromatography.

Detection method	Principle	Applications
Conductivity	electrical conductivity	Anions and cations with pK_a or $pK_b < 7$
Amperometry	Oxidation or reduction on Ag-/Pt-/Au-/Glassy Carbon- and Carbon Paste electrodes	Anions and cations with pK_a or $pK_b > 7$
UV/Vis detection with or without post-column derivatization	UV/Vis light absorption	UV-active anions and cations, transition metals after reaction with PAR, aluminum after reaction with Tiron, lanthanides after reaction with Arsenazo I, polyvalent anions after reaction with iron(III), silicate and orthophosphate after reaction with molybdate
Fluorescence in combination with post-column derivatization	Excitation and emission	Ammonium, amino acids, and primary amines after reaction with OPA
Refractive index	Change in refractive index	Anions and cations at higher concentrations
Radioactivity	Szintillation measurement	Radiostrontium analysis
ICP-OES, ICP-MS	Atomic emission	Hyphenation technique for selective and sensitive transition metal analysis
MS	Electrospray ionization	Hyphenation technique for structural elucidation of organic anions and cations

features that helped ion chromatography rapidly become a widely used analytical tool. Refer to Fig. 9-1 for a typical example of the separation of anionic constituents in drinking water on two different stationary phases. Chromatogram A represents the separation on a conventional latexed anion exchanger with a sodium carbonate/sodium bicarbonate eluant, on which the three major components are separated with high resolution within eight minutes. However, the evaluation of the fluoride signal poses a problem, because fluoride elutes close to the system void on this stationary phase and, consequently, cannot be integrated with high precision. Under these chromatographic conditions, reliable fluoride data can only be generated for fluorinated drinking water samples containing up to 1.5 mg/L fluoride, as the measuring error decreases significantly with increasing fluoride content. Nevertheless, unassailable identification of compounds eluting close to the system void is not possible. For this reason, the IonPac AS4A-SC used in this separation is only suitable for analyzing the major components chloride, nitrate, and sulfate. A significantly better resolution between fluoride and the negative water dip is obtained on IonPac AS14 (see Section 3.4.1) using the same eluant mixture with different concentrations of the two components. Compared with IonPac AS4A-SC, IonPac AS14 has a longer analysis time and a column-specific lower chromatographic efficiency that results in a lower sensitivity, both of which are obviously disadvantageous. Chromatogram B in Fig. 9-1

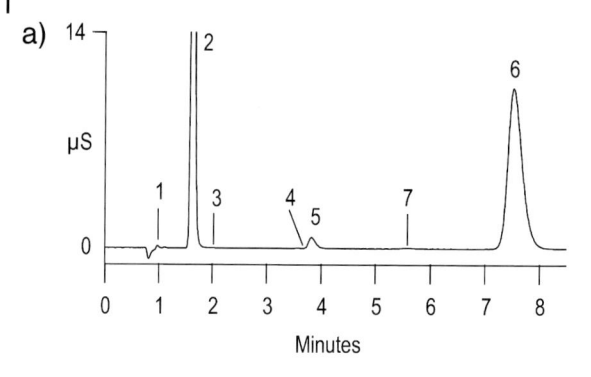

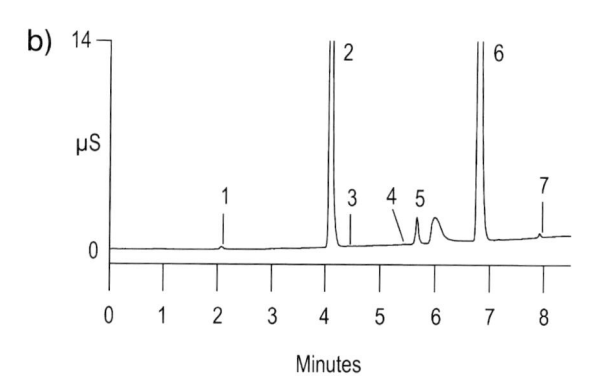

Fig. 9-1. Anion analysis of a drinking water. – Separator columns: (a) IonPac AS4A-SC, (b) IonPac AS17; eluants: (a) 1.7 mmol/L NaHCO$_3$ + 1.8 mmol/L Na$_2$CO$_3$, (b) 1 mmol/L KOH isocratically for 1.5 min, then to 20 mmol/L in 3.5 min and from 20 mmol/L to 40 mmol/L in 2 min; flow rates: 2 mL/min; detection: suppressed conductivity; injection: 25 µL drinking water (undiluted) from Sunnyvale, USA; solute concentrations: (a) 0.06 mg/L fluoride (1), 24.1 mg/L chloride (2), 0.01 mg/L nitrite (3), 0.02 mg/L bromide (4), 1.48 mg/L nitrate (5), 31.4 mg/L sulfate (6), and 0.29 mg/L orthophosphate (7), (b) 0.05 mg/L fluoride (1), 21.5 mg/L chloride (2), 0.015 mg/L nitrite (3), 0.03 mg/L bromide (4), 1.41 mg/L nitrate (5), 29.7 mg/L sulfate (6), and 0.48 mg/L orthophosphate (7).

is obtained with the innovative IonPac AS17 hydroxide-selective anion exchanger using a potassium hydroxide gradient. This chromatogram fascinates with its excellent peak symmetry. The significantly higher sensitivity results from the lower background conductivity, which can be lower than 1 µS/cm when using high purity hydroxide eluants. In addition, the half widths of the signals are significantly smaller when applying a gradient technique as compared to an isocratic run. This sensitivity increase has a favorable effect on the analysis of orthophosphate, which can be determined without any problem despite the higher retention time. In contrast to separations using carbonate-based eluants, an additional peak is observed in the chromatogram when hydroxide eluants are employed; this can be attributed to bicarbonate present in the sample. Due to the high pH of the mobile phase, bicarbonate elutes as divalent carbonate between nitrate and sulfate. A determination of carbonate under these chromatographic conditions is only possible using an Eluant Generator™ (see Section 3.9) in combination with a continuously regenerated trap column (CR-ATC). Even traces of carbonate in the mobile phase must be excluded for quantitative carbonate analysis.

In addition to mineral acids, it is also possible to determine salts from weak inorganic acids via ion chromatography. This includes, for example, orthosilicate [9, 10] which can be separated by both anion exchange and ion-exclusion chromatography. In both cases, photometric detection after post-column reaction with sodium molybdate is employed for an extremely selective identification. Both separation modes also allow the separation of orthophosphate, which may also be determined with photometric detection. When the effluent of an anion exchanger is derivatized after passing the suppressor system and conductivity cell, a simultaneous analysis of strong and weak inorganic acids becomes possible. A representative standard chromatogram is illustrated in Fig. 9-2. A combination of conductivity detection and post-column derivatization is also feasible when applying ion-exclusion chromatography if, for example, weak inorganic acids such as orthophosphate, silicate, and carbonate are to be determined together with aliphatic monocarboxylic acids in strongly contaminated water samples.

In the past, drinking water disinfection was carried out by chlorination until it was found in the 1970s that carcinogenic compounds such as trihalomethanes are formed during this process. Therefore, people searched for alternative disinfection techniques to minimize the amount of compounds that were supposed to be harmful. The most promising alternative to chlorination is ozonation. If the raw water being processed for drinking water contains bromide, the latter will be oxidized to bromate by ozone. The concentration of the bromate formed depends on the following parameters [11, 12]:

- Bromide concentration in the raw water
- Ozone dose
- Ozonation time
- Water temperature
- pH of the water

Two mechanisms are discussed for bromate formation during ozonation. On one hand, bromate can be formed as an intermediate product of the reaction between ozone and hypobromite [13, 14]:

$$Br^- + O_3 + H_2O \rightarrow HOBr + O_2 + OH^- \qquad (290)$$

$$HOBr + H_2O \rightarrow H_3O^+ + OBr^- \qquad (291)$$

$$OBr^- + 2 O_3 \rightarrow BrO_3^- + 2 O_2 \qquad (292)$$

$$HOBr + O_3 \rightarrow \text{no reaction}$$

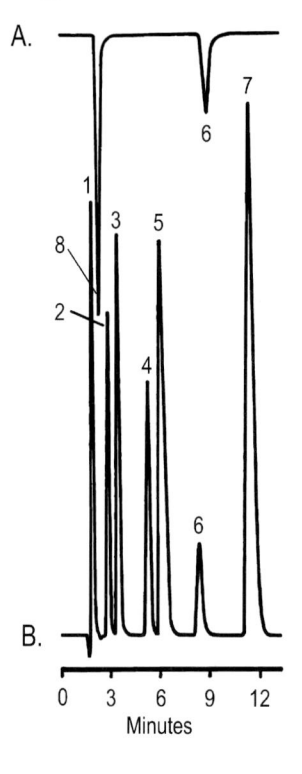

Fig. 9-2. Simultaneous analysis of weak and strong inorganic acids. — Separator column: IonPac AS4A-SC; eluant: 1.7 mmol/L NaHCO$_3$ + 1.8 mmol/L Na$_2$CO$_3$; flow rate: 1 mL/min; detection: (A) suppressed conductivity, (B) photometry at 410 nm after post-column reaction with sodium molybdate; injection volume: 50 µL; solute concentrations: 3 mg/L fluoride (1), 4 mg/L chloride (2), 10 mg/L each of nitrite (3) and bromide (4), 20 mg/L nitrate (5), 10 mg/L orthophosphate (6), 25 mg/L sulfate (7), and 27 mg/L orthosilicate (8).

However, at low pH, hypobromite is converted to hypobromous acid, which is not oxidized to bromate by ozone. Thus, in the pH range (pH 6.5−9.5) relevant for drinking water processing, only a small amount of hypobromous acid (pK ≈ 9) is present as hypobromite and able to react with molecular ozone. According to this reaction mechanism, bromate formation very much depends on the pH during the ozonation process, i.e., the lower the pH, the less bromate is formed (Fig. 9-3).

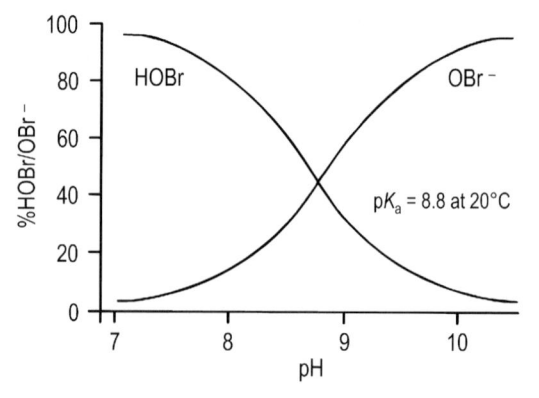

Fig. 9-3. Equilibrium between hypobromous acid and hypobromite in aqueous solution.

Alternatively, von Gunten et al. [15, 16] postulated the formation of hydroxide radicals during the decomposition of ozone in aqueous solutions that are able to oxidize both hypobromous acid and hypobromite to bromate. Even though the transformation of ozone into hydroxide radicals is faster at higher pH, according to this mechanism bromate formation is less pH dependent than the reaction with molecular ozone.

Toxicological studies [17] indicate that bromate is suspected to be carcinogenic even in the low µg/L range. Various health and environmental protection agencies have been discussing concentration limits for bromate in drinking water for quite some time. The World Health Organization (WHO), for example, recommends in their directive a concentration limit of 25 µg/L [18], while the American Environmental Protection Agency (EPA) and the European Union currently endorse a limit of 10 µg/L [19]. In connection to this value, a minimum detection limit of 2.5 µg/L is recommended for bromate analyses.

A number of different ion chromatographic techniques exist for bromate analysis. In method 300.0 (Part B) [20], the U.S. EPA describes the separation of oxyhalides – chlorite, chlorate, and bromate – in the presence of mineral acids on IonPac AS9-SC with suppressed conductivity detection. On this stationary phase, adequate separation between bromate and chloride is achieved with the standard eluant mixture of sodium carbonate and sodium bicarbonate if the chloride excess is less than 100-fold. Because in real-world samples the chloride content can easily be 50-100 mg/L, bromate quantification in the concentration range between 1 µg/L and 10 µg/L is impossible under standard conditions. Kuo et al. [21] improved the separation by treating the sample with a cation exchanger in the silver form, through which the chloride content is decreased to a value determined by the solubility of silver chloride in solution ($\sim$ 0.4 mg/L chloride). With this sample pre-treatment, the minimum detection limit for bromate could be lowered to 10 µg/L. Hautman et al. [22, 23] achieved a further sensitivity increase by using a borate eluant, with which the separation between bromate and chloride could be improved. With an injection volume of 200 µL, the minimum detection limit for bromate (based on a signal-to-noise ratio of 3:1) is 5.9 µg/L. This represents the limit of the performance of this technique. It is not possible to obtain a further sensitivity increase by injecting larger volumes without sacrificing chromatographic efficiency. A big disadvantage of this technique is the fact that the typical bromate content in ozonated drinking water samples is in the range of the minimum detection limit or even below. To detect these low concentrations, pre-concentration is unavoidable [24, 25].

Figure 9-4 schematically depicts the required system setup for pre-concentration (according to ISO/DIS Draft 15061). At first, the sample to be analyzed is sparged with inert gas to remove reactive gases such as ozone or chlorine dioxide. Then, ethylenediamine is added to convert existing hypobromite into the corresponding bromoamines, by which a further reaction to bromate is prevented. The sample thus pretreated is then directed through a cartridge containing a cation exchanger in the silver form (OnGuard-Ag or similar products). This procedure to lower the chloride content in the sample is essential, because even with weak borate eluant a separation between bromate and chloride is not possible without lowering the chloride content. Moreover, lowering the chloride content allows the concentrated volume to be larger, consequently increasing sensitivity. However, these cartridges release silver ions, which may accumulate in the concentrator column and the subsequent analytical separator, resulting in a significant loss of separation power. Therefore, a second cartridge containing a cation exchanger in the hydrogen form is used to remove silver ions. For safety reasons, another special trap column with similar properties — MetPac CC-1 — is placed between the injection valve and the concentrator column. The MetPac CC-1 removes silver ions as well as any cations which may lead to a malfunction of the analytical system.

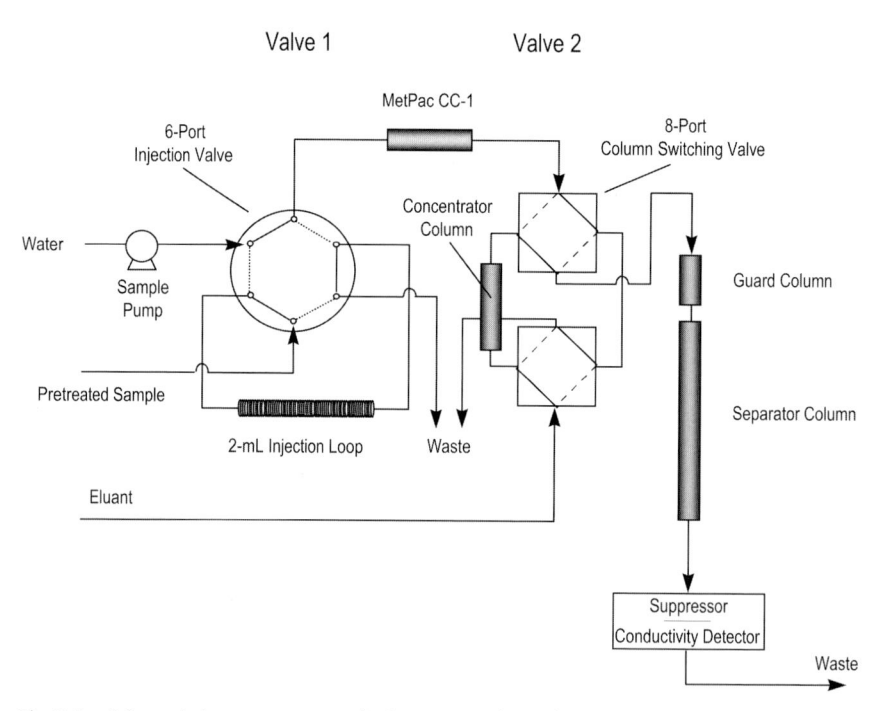

Fig. 9-4. Schematic instrument setup for bromate analysis after pre-concentration (according to ISO/DIS Draft 15061).

The respective analytical procedure consists of three individual steps:
- Filling of the injection loop
- Transfer of the sample onto the concentrator column
- Separation of bromate on the analytical separator

Originally, Joyce and Weinberg [24, 25] used a borate eluant (40 mmol/L boric acid + 20 mmol/L NaOH) for separating bromate. After the elution of chloride, they switched to a more concentrated eluant (0.25 mol/L boric acid + 0.1 mol/L NaOH) to elute the remaining anions off the separator column. Instead of a borate eluant, a dilute sodium bicarbonate solution can also be used as an eluant [26]; rinsing is then carried out with a mixture of sodium carbonate and sodium bicarbonate. Figure 9-5 shows an example chromatogram of a round robin sample containing 4 µg/L bromate that was obtained under the chromatographic conditions described above. With this method the minimum detection limit for bromate could be lowered to 1 µg/L.

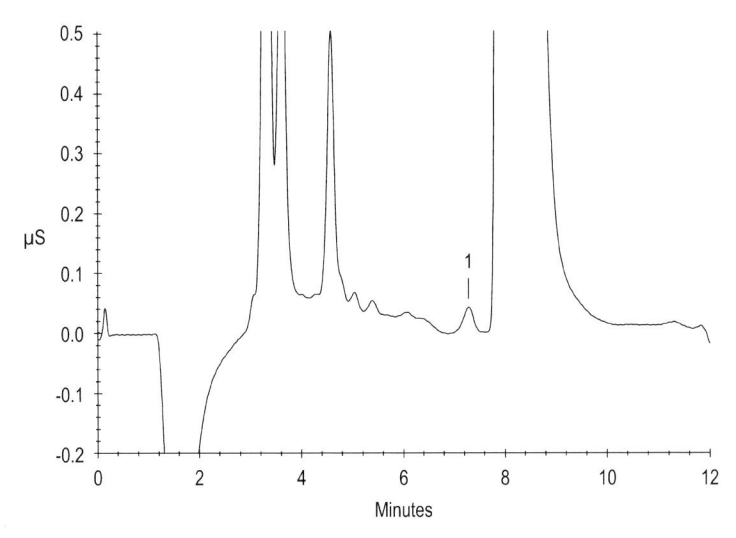

Fig. 9-5. Trace analysis of bromate in drinking water after pre-concentration according to ISO/DIS Draft 15061. – Separator column: IonPac AS9-SC; eluant 1: 0.7 mmol/L NaHCO$_3$, eluant 2: 5 mmol/L NaHCO$_3$ + 5 mmol/L Na$_2$CO$_3$; flow rate: 1.5 mL/min; detection: suppressed conductivity; injection volume: 2 mL; pre-concentration with IonPac AG9-SC; transfer solution: de-ionized water with a flow rate of 2.5 mL/min; sample: drinking water with 4 µg/L bromate (1).

As an alternative to conductivity detection, Heitkemper et al. [27] and Seubert et al. [28] successfully employed ICP-MS as a detection method for bromate analysis and achieved a minimum detection limit of 0.8 µg/L. Regarding sample preparation, this detection method does not require a reduction of the chloride content prior to the ion chromatographic separation because of its high selectivity for bromine-containing compounds. However, this method is also not free of interferences. Tribromoacetic acid, for example, interferes and thus has to be removed completely prior to analysis.

At present, a number of derivatization methods for the trace analysis of bromate are discussed [29-31] that enable the simultaneous analysis of oxyhalides and mineral acids. The method introduced by Weinberg et al. [29] is based on the post-column derivatization of bromate with bromide under acidic conditions to yield tribromide, Br_3^-, which can be detected by UV at 267 nm. The reaction scheme was already described by Equations (198) and (199) in Section 7.2.1. In addition to bromate, this method can also be used for determining other oxyhalides such as iodate and chlorite. A characteristic of this method is the *in situ* generation of the reducing agent to transform bromate into tribromide — a combination of hydrobromic acid, HBr, and nitrous acid, HNO_2 — by means of a membrane suppressor with cation exchange functional groups. When injecting large volumes (500 µL) a minimum detection limit of 0.5 µg/L bromate can be achieved without any problem.

Following the proposed method of Weinberg et al., von Gunten et al. [30] developed a derivatization technique for the determination of bromate, iodate, chlorite, and nitrite, all of which specifically react with iodide in acidic solution. The reaction scheme for bromate is as follows:

$$BrO_3^- + 3I^- + 3H^+ \xrightarrow{Mo(VI)} 3HOI + Br^- \quad (293)$$

$$3HOI + 3I^- + 3H^+ \rightarrow 3I_2 + 3H_2O \quad (294)$$

$$3I_2 + 3I^- \rightarrow 3I_3^- \quad (295)$$

According to this scheme, triiodide, I_3^-, is formed in excess of iodide, which can be detected very sensitively due to its large molar extinction coefficient. The absorption spectrum of the triiodide anion shows two maxima at 288 nm (ε = 38,200 L mol^{-1} cm^{-1}) and 352 nm (ε = 26,400 L mol^{-1} cm^{-1}). The second maximum is suited as a measuring wavelength, because fewer matrix interferences are to be expected at higher wavelengths. The derivatization reagent consists of a 0.26 mol/L KI solution containing catalytic amounts of Mo(VI) (43 µmol $(NH_4)_6Mo_7O_{24}$). Because iodide in the derivatization reagent is oxidized by oxygen at low pH, reagent acidification — as already described by Weinberg et al. [29] — is also carried out with a membrane-based suppressor system that is continuously regenerated with sulfuric acid.

The biggest advantage of the derivatization techniques developed by Weinberg et al. and von Gunten et al. is the ability to combine it with conductivity detection for determining the major constituents — chloride, nitrate, and sulfate. Figure 9-6 shows a chromatogram of a drinking water sample spiked with 2 µg/L bromate that was obtained on IonPac AS9-HC. The upper chromatogram results from conductivity detection of the matrix anions — chloride, bromide, nitrate, orthophosphate, and sulfate. Bromide with a concentration of 39 µg/L is well separated from nitrate. The lower chromatogram represents the UV detection with three easily detectable peaks quantified as 2 µg/L iodate, 2 µg/L bromate, and 1.6 µg/L nitrite. Bromate and nitrite are not seen in the upper chromatogram because they lie under the large chloride peak. To determine these anions via conductivity detection, chloride had to be removed with a cation exchanger in the silver form. Because this kind of sample preparation also removes bromide, two separate chromatographic runs would be necessary to analyze bromate and bromide. Regardless of this, the minimum detection limit for bromate of 0.1 µg/L with this post-column derivatization and UV detection can only be achieved with conductivity detection via pre-concentration.

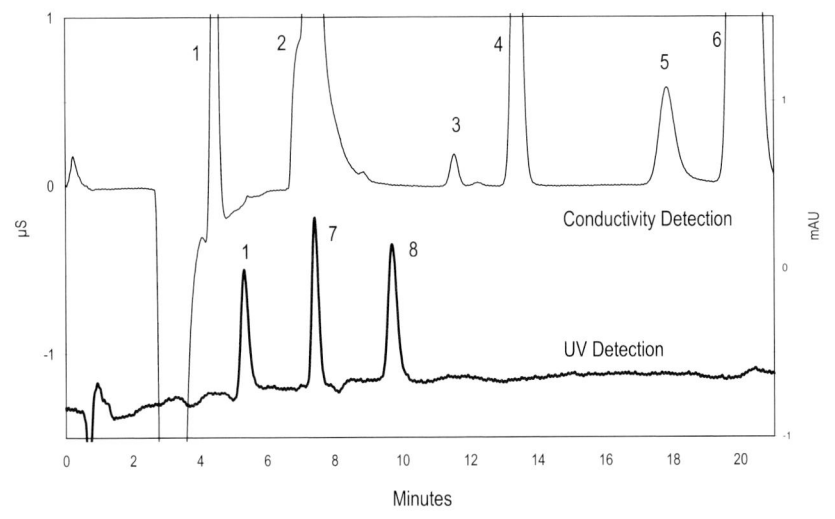

Fig. 9-6. Trace analysis of bromate in drinking water using conductivity detection in combination with post-column derivatization via triiodide. — Separator column: IonPac AS9-HC; eluant: 9 mmol/L Na_2CO_3; flow rate: 1 mL/min; detection: suppressed conductivity, UV detection (352 nm); derivatization reagent: 0.26 mol/L KI + 43 µmol/L $(NH_4)_6Mo_7O_{24}$; suppressor for acidification: ASRS-I; regenerant: 0.15 mol/L H_2SO_4; injection: 500 µL drinking water (undiluted) spiked with 2 µg/L bromate; analytes: (1) 2 µg/L iodate, (2) chloride, (3) 39 µg/L bromide, (4) nitrate, (5) orthophosphate, (6) sulfate, (7) 2 µg/L bromate, and (8) 1.6 µg/L nitrite; (taken from [30]).

A similarly high sensitivity for bromate can also be achieved with a derivatization technique introduced by Wagner et al. [31] that is carried out with o-dianisidine (ODA). This technique, known today as U.S. EPA Method 317.0, can be directly combined with the American EPA Method 300.1 [32]. In Part B this method describes the separation of bromate, bromide, chloride, and chlorite on an IonPac AS9-HC high-capacity anion exchanger. The method described by Wagner et al. (see also Section 7.2.1) is based on the reaction of bromate with o-dianisidine in strongly acidic solution with subsequent photometry at 450 nm. The authors report the limit of determination to be 0.5 µg/L. Post-column derivatization with ODA is only interfered with by large amounts of chlorite (> 200 µg/L), which also reacts with o-dianisidine, eluting ahead of bromate. Interferences by chlorite only occur when drinking water disinfection is carried out with a mixture of chlorine dioxide and ozone.

A number of procedures were investigated in an effort to find an approach that preferentially removes chlorite without adversely affecting bromate levels. Of these, the procedure that employs Fe(II) in acidic solution was found to be the most effective [33]. Removing residual chlorite from drinking water using ferrous iron under slightly acidic (pH 5-6.5) conditions has been extensively studied [34]. The molar stoichiometry, based on Eq. (296), predicts that 3.3 mg of Fe(II) would be required to completely reduce 1.0 mg ClO_2^- [35].

$$4Fe^{2+} + ClO_2^- + 10H_2O \rightarrow 4Fe(OH)_3 \text{ (s)} + Cl^- + 8H^+ \qquad (296)$$

Because elevated levels of iron were projected to pose potential fouling problems with the IonPac AS9-HC column and the membrane suppressor system, efforts were directed towards finding an acceptable tool for removing iron from slightly acidic solutions. Solid-phase extraction cartridges containing a cation exchange resin in the hydrogen form have been successfully used to remove cationic species such as Fe(III) from some sample matrices [36]. Thus, it was necessary to determine if Fe(II) could be removed in a similar way. When adding Fe(II) to drinking water containing chlorite, a portion of the Fe(II) is oxidized to $Fe(OH)_3$, while the excess remains as Fe(II). Wagner et al. [33] found out that both the particulate filter and the solid-phase extraction cartridge were necessary to reduce the iron to acceptable levels and thereby prevent fouling of the IC system. They also proved that no loss of bromate was recorded over a period of 36 hours when treating drinking water with Fe(II). Thus, U.S. EPA Method 317.0 provides a rugged direct injection analysis for determining trace bromate in all drinking water matrices.

The simultaneous analysis of alkali- and alkaline-earth metals is another important ion chromatographic application in the field of drinking water and surface water analysis. A corresponding chromatogram in Fig. 9-7 shows the separation of sodium, ammonium, potassium, magnesium, and calcium in a drinking water sample, which nowadays can be obtained on an IonPac CS12 weak acid cation exchanger (see Section 4.1.1) in about 12 minutes.

The analysis of anions and cations in snow and ice core samples [37-40], as well as in rain water [41], ground water, and swimming pool water, is equally simple. The very low electrolyte content in snow and ice core samples sometimes requires the injection of large sample volumes (200 µL). Special caution is necessary during sampling. To avoid contaminations during the transport between the sampling location and the laboratory, it is recommended that the samples be kept in airtight polyethylene containers, which have previously been rinsed with dilute hydrochloric acid and also rinsed several times with high-purity deionized water. As a further safety measure, safety gloves and face masks should be worn during sampling. The samples should be transported in a frozen state, kept at $-18\,°C$ until measurement, and thawed at room temperature right before the ion chromatographic analysis. The concentration range of anions and cations in snow samples from the Austrian Alps measured by Nickus et al. [38], for example, extends over three orders of magnitude. While the nitrate content during snow melting can be in the lower mg/L level, the concentrations of alkali- and alkaline-earth metals are usually in the lower µg/L level. A typical anion and cation profile is illustrated in Fig. 9-8.

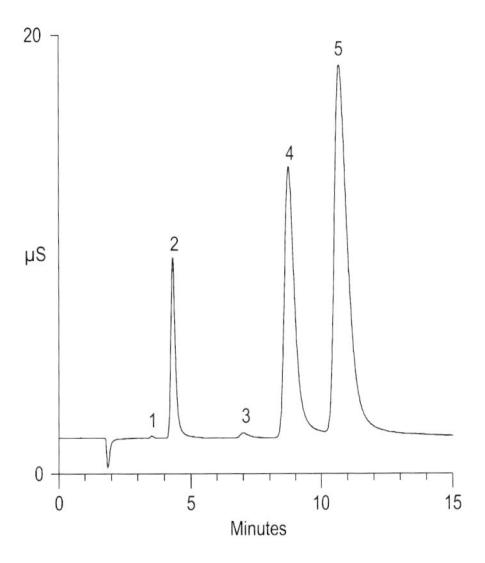

Fig. 9-7. Simultaneous analysis of alkali- and alkaline-earth metals in drinking water. – Separator column: IonPac CS12; eluant: 9 mmol/L H_2SO_4; flow rate: 1 mL/min; detection: suppressed conductivity; injection volume: 25 µL potable water from Idstein (undiluted, acidified to pH 3); peaks: (1) lithium, (2) sodium, (3) potassium, (4) magnesium, and (5) calcium.

Apart from the obligatory membrane filtration (0.45 µm), the only sample preparation in the analysis of rain water, ground water, and swimming pool water is a dilution with de-ionized water. Figure 9-9 shows a cation chromatogram of a rain water sample [42] that applies non-suppressed conductivity detection; the sample was injected without any dilution because of its low electrolyte content. Metrosep Cation 1-2 served as the separator column and the eluant was a mixture of tartaric acid and pyridine-2,6-dicarboxylic acid.

Ground water – especially from wood-rich areas – should be injected through an OnGuard-P (Dionex) extraction cartridge to remove humic acids. This cartridge contains a polyvinylpyrrolidone resin (PVP) which selectively retains humic acids. As an example, Fig. 9-10 shows a chromatogram of a correspondingly diluted sample from a sampling station on the Main river (Germany).

The detection of ammonium perchlorate, an important constituent of solid rocket propellants, in the ground waters of some U.S. federal states is alarming. Significant amounts have been measured in areas where rocket fuel, ammunition, or pyrotechnic articles are developed, tested, or manufactured. Even in low concentrations, perchlorate represents a health risk for human beings, because it affects hormone production in the thyroid. Initial investigations by the U.S. EPA indicate a health risk at perchlorate concentrations above 4-18 µg/L. In California, perchlorate has been detected in more than 100 drinking water wells; 20 of them had to be closed because they exceed the above-mentioned limit. Trace analysis of perchlorate is a difficult analytical task. Ion chromatogra-

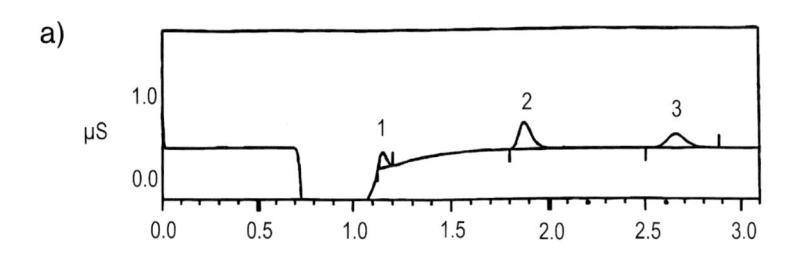

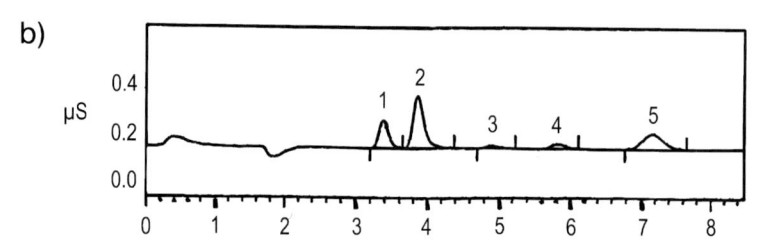

Fig. 9-8. Anion and cation profile of a snow sample from the Austrian Alps. — (a) Separator column: IonPac AS4A-SC; eluant: 4 mmol/L Na_2CO_3 + 1.5 mmol/L $NaHCO_3$; flow rate: 2 mL/min; detection: suppressed conductivity; injection volume: 200 µL; solute concentrations: 16.5 µg/L chloride (1), 65.1 µg/L nitrate (2), and 36.7 µg/L sulfate (3); (b) separator column: IonPac CS12; eluant: 21 mmol/L methanesulfonic acid; flow rate: 1 mL/min; detection: suppressed conductivity; injection volume: 200 µL; solute concentrations: 4.9 µg/L sodium (1), 16.5 µg/L ammonium (2), 2.3 µg/L potassium (3), 0.1 µg/L magnesium (4), and 4.2 µg/L calcium (5); (taken from [38]).

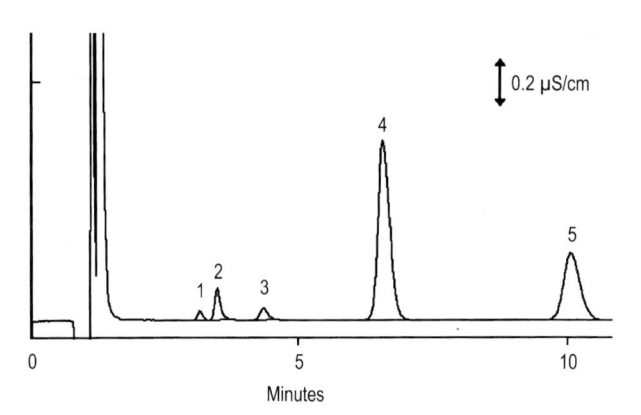

Fig. 9-9. Cation analysis of a rain water sample. — Separator column: Metrosep Cation 1-2; eluant: 4 mmol/L tartaric acid + 1 mmol/L pyridine-2,6-dicarboxylic acid; flow rate: 1 mL/min; detection: non-suppressed conductivity; injection volume: 10 µL; solute concentrations: 0.07 mg/L sodium (1), 0.26 mg/L ammonium (2), 0.32 mg/L potassium (3), 4.18 mg/L magnesium (4), and 1.04 mg/L calcium (5); (taken from [42]).

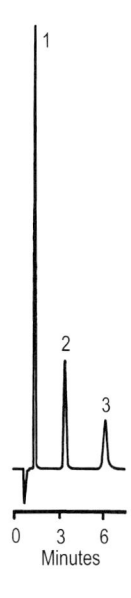

Fig. 9-10. Anion analysis of a ground water. — Separator column: IonPac AS4A; eluent: 1.7 mmol/L NaHCO$_3$ + 1.8 mmol/L Na$_2$CO$_3$; flow rate: 2 mL/min; detection: suppressed conductivity; injection: 50 μL ground water (1:10 diluted); solute concentrations: 32 mg/L chloride (1), 38 mg/L nitrate (2), and 27 mg/L sulfate (3).

phy is probably the only method that allows such low perchlorate concentrations to be determined. Large polarizable anions such as perchlorate are strongly retained on conventional anion exchangers, so they often elute with a marked tailing [43]. In the past, organic solvents such as methanol or organic modifiers such as *p*-cyanophenol were added to the mobile phase to reduce this tailing. When using modern hydroxide-selective stationary phases such as IonPac AS11 and AS16, one does not have to add any organic modifiers, because adsorption phenomena are considerably suppressed due to the extreme hydrophilicity of the ion-exchange functional groups [44, 45]. As can be seen from the chromatogram of a drinking water sample spiked with perchlorate in Fig. 9-11, perchlorate elutes from IonPac AS16 within ten minutes with a purely aqueous eluant (0.05 mol/L NaOH). The perchlorate peak exhibits excellent symmetry and is separated from other anionic sample constituents. Large sample volumes have to be injected to obtain detection limits in the sub-μg/L range. Plotting peak

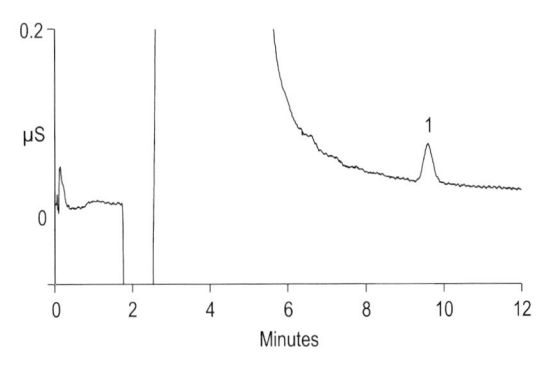

Fig. 9-11. Trace analysis of perchlorate in a drinking water matrix. — Separator column: IonPac AS16; eluent: 0.05 mol/L NaOH; flow rate: 1.5 mL/min; detection: suppressed conductivity; injection volume: 1000 μL; sample: drinking water from Sunnyvale (USA) spiked with 6 μg/L perchlorate (1).

areas as a function of the perchlorate concentration, linear regression in the investigated concentration range between 2.5 and 100 µg/L yields a regression coefficient of $r^2 = 0.9996$. According to the EPA directive [46], the minimum detection limit was determined to be 0.3 µg/L.

Apart from the analysis of the main components — chloride, nitrate, and sulfate — a peculiarity of swimming pool water analysis is the determination of the two chlorine species — chlorite, ClO_2^-, and chlorate, ClO_3^-. These are disproportionation products of chlorine dioxide, which some countries use to fumigate drinking- and swimming pool water for disinfection purposes [47, 48]. Chlorine dioxide easily dissolves in water, but decomposes slowly by formation of chlorous acid and chloric acid:

$$2ClO_2 + H_2O \rightarrow HClO_2 + HClO_3 \qquad (297)$$

Chlorous acid rapidly decomposes to yield hydrochloric acid and chloric acid as disproportionation products:

$$6ClO_2 + 3H_2O \rightarrow HCl + 5HClO_3 \qquad (298)$$

Chlorous acid is retained only in alkaline solution, because the corresponding chlorite is more stable than the corresponding acid:

$$2ClO_2 + 2OH^- \rightarrow ClO_2^- + ClO_3^- + H_2O \qquad (299)$$

For water treatment purposes, only small amounts of chlorine dioxide are necessary, so the reaction products chlorite and chlorate are also present only at low concentrations. If chlorination is carried out with chlorine gas, hypochlorous acid is rapidly formed, which acts as a biozide.

$$Cl_2 + H_2O \rightarrow HOCl + H^+ + Cl^- \qquad (300)$$

During the decomposition of HOCl, chlorate can also be formed:

$$3HOCl \rightarrow 3H^+ + 2Cl^- + ClO_3^- \qquad (301)$$

Both chlorite and chlorate pose potential health risks. However, the World Health Organization (WHO) suggested only a threshold of 200 µg/L for chlorite. At present, no threshold exists for chlorate as sufficient information about its toxicology is not available. Therefore, the WHO recommends keeping the chlorate content as low as possible until secured findings are submitted. Chlorite and chlorate can be separated from mineral acids on IonPac AS9(-SC). Figure 9-12 shows the chromatogram of a test mixture adjusted to the matrix of such water; with this chromatogram the applied chromatographic method can be evaluated for separation efficiency and sensitivity.

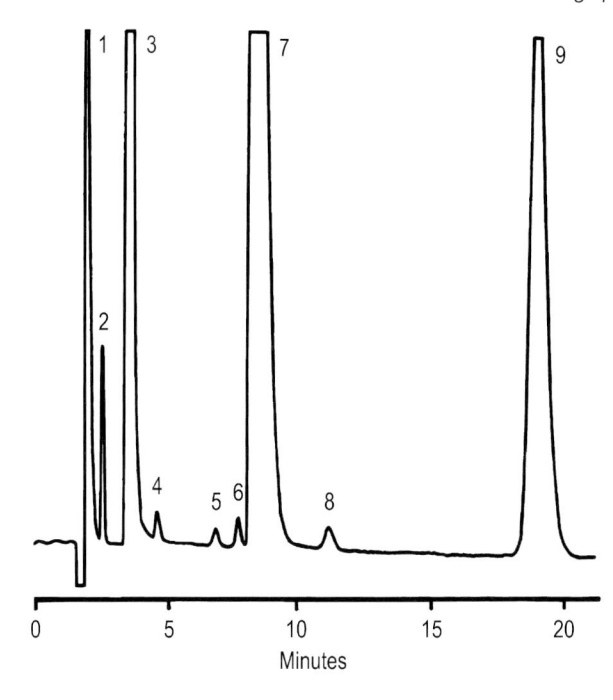

Fig. 9-12. Separation of chlorite and chlorate from mineral acids. — Separator column: IonPac AS9 (−SC); eluant: 0.75 mmol/L $NaHCO_3$ + 2 mmol/L Na_2CO_3; flow rate: 1 mL/min; detection: suppressed conductivity; injection: 50 µL; solute concentrations: 5.6 mg/L fluoride (1), 0.5 mg/L chlorite (2), 45.4 mg/L chloride (3), 0.06 mg/L nitrite (4), 0.08 mg/L bromide (5), 0.14 mg/L chlorate (6), 42.1 mg/L nitrate (7), 0.17 mg/L orthophosphate (8), and 5 mg/L sulfate (9).

The selectivity of IonPac AS12A has also been optimized for the separation of oxyhalides and mineral acids. In this context, the chromatogram of a swimming pool water sample in Fig. 9-13 is of interest. This swimming pool water sample,

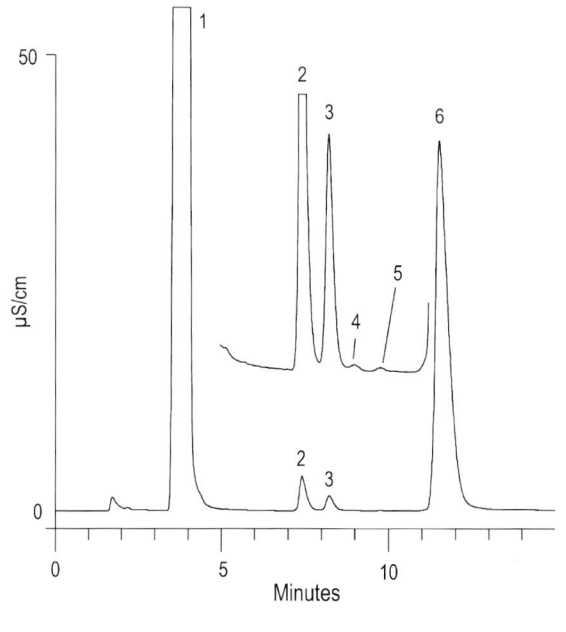

Fig. 9-13. Analysis of a swimming pool water sample. — Separator column: IonPac AS12A; eluant: 0.3 mmol/L $NaHCO_3$ + 2.7 mmol/L Na_2CO_3; flow rate: 1.5 mL/min; detection: suppressed conductivity; injection volume: 50 µL; solute concentrations: 635 mg/L chloride (1), 8.4 mg/L bromide (2), 10 mg/L chlorate (3), <0.04 mg/L nitrate (4), 0.14 mg/L orthophosphate (5), and 105 mg/L sulfate (6).

which was analyzed with a conventional IonPac AS4A-SC anion exchanger, showed an unusually high nitrate concentration. Only with the use of IonPac AS12A was it found that the supposed nitrate peak is actually chlorate, which co-elutes with nitrate on IonPac AS4A-SC. Only trace amounts of nitrate could be detected in the sample. Once more, this case shows that uncritical selection of stationary phases can lead to erroneous results.

Investigations by Hautman et al. [23] with IonPac AS9-SC and several different eluant mixtures revealed that conductivity detection, even in combination with a suppressor system, is often not sensitive enough. As described above, Salhi and von Gunten [30] developed an analytical method for iodate, bromate, and chlorite based on post-column derivatization with iodide, which is oxidized to iodine yielding triiodide in excess of iodide. Triiodide, in turn, can be photometrically detected at 288 nm due to its high extinction coefficient [49]. Chlorate cannot be detected under these conditions, because it oxidizes iodide only very slowly. A higher reaction speed can only be achieved in a strongly acidic medium (6 mol/L HCl), which is not suited for post-column derivatization. On the other hand, it has long been known that osmium tetroxide accelerates the oxidation of iodide by chlorate [50]. Nowak and von Gunten [51] finally succeeded to modify the triiodide method by using an osmium tetroxide catalyst, so that the oxidation of iodide to iodine by chlorate can be carried out at pH 3 with sufficient speed. The following reactions occur:

$$6I^- + ClO_3^- + 6H^+ \rightarrow 3I_2 + Cl^- + 3H_2O \tag{302}$$

$$I_2 + I^- \rightarrow I_3^- \tag{303}$$

If some osmium tetroxide is added to the derivatization reagent, the reaction yield is only 10% within a time frame of about one minute, which is suitable for post-column derivatization. Lowering the pH enhances the reaction speed between chlorate and iodide, but also the side reaction between osmium tetroxide and iodide. However, when using osmate, OsO_4^{2-}, the reduced form of osmium tetroxide, complete reaction within one minute is achieved. Thus, osmate is a better catalyst than osmium tetroxide. Reduction of the volatile and poisonous (!) osmium tetroxide to non-volatile osmate is carried out with ethanol in an alkaline medium. The schematic instrument setup for this method is depicted in Fig. 9-14. The only difference in comparison with a conventional post-column derivatization is an additional manifold, placed between the suppressor system and the reaction coil, for rinsing the reaction coil with 0.1 mol/L H_2O_2 solution. This rinse has to be performed every working day, as osmate (Os^{VI}) slowly disproportionates to yield Os^{VII} and Os^V. While Os^{VII} is soluble, Os^V forms a black precipitate, which leads to an enhanced background absorption and to a lower response factor for chlorate. However, in the presence of H_2O_2 the black precipitate immediately dissolves. As an example, Fig. 9-15 shows the chromatogram of a drinking water obtained on an IonPac AS9-HC high-capacity anion exchanger under these conditions. An alkaline borate solution was used as an eluant in this case; the sodium carbonate eluant that is normally used is not suitable, because a negative peak is observed where chlorite elutes. Acidification of the column effluent is carried out with the aid of a membrane suppressor and chemical regeneration. A small amount of sodium formate is added to the eluant to adjust the column effluent pH after passing the suppressor system to 3.3, an optimal value for the post-column reaction. As can be seen from the chromatogram in Fig. 9-15, chlorite, bromate, and nitrite can be determined with this method, in addition to chlorate. The minimum detection limit for chlorate is 0.4 µg/L.

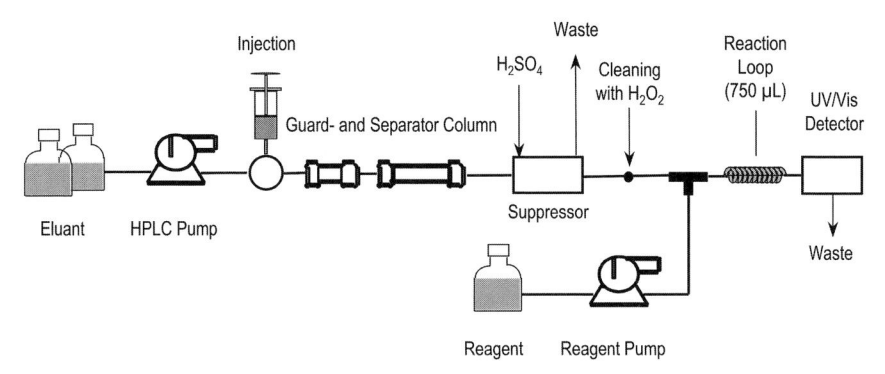

Fig. 9-14. Schematic setup of an ion chromatographic system for trace analysis of chlorate via reaction to triiodide.

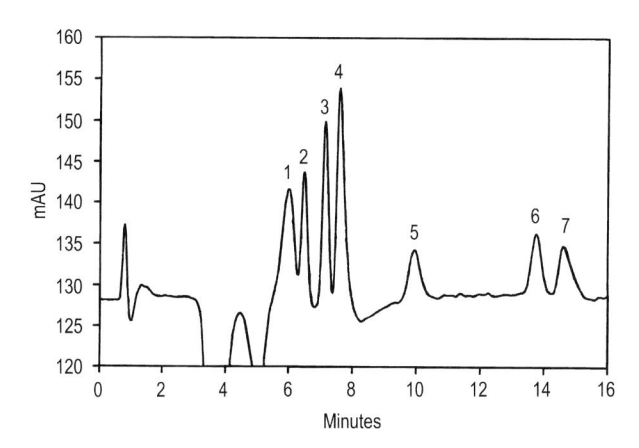

Fig. 9-15. Trace analysis of chlorate in drinking water via osmate catalyzed reaction to triiodide. – Separator column: IonPac AS9-HC; eluant: 50 mmol/L $Na_2B_4O_7$ + 6 mmol/L sodium formate; flow rate: 1 mL/min; detection: photometry at 288 nm via reaction with iodide to triiodide; suppressor: ASRS-1; derivatization reagent: 0.1 mol/L KI/0.1 mmol/L OsO_4 + 5 mmol/L NaOH – ethanol (95:5 v/v); reagent flow rate: 0.2 mL/min; injection volume: 750 µL; sample: drinking water from Dübendorf (Switzerland); Peaks: (1) system peak, (2) 27 µg/L chlorite, (3) 26 µg/L bromate, (4) 10 mg/L chloride, (5) 92 µg/L nitrite, (6) 33 µg/L chlorate, and (7) 6.5 mg/L nitrate.

Hypochlorous acid, which is formed during disinfection of drinking water with chlorine dioxide, can also react with humic acids that naturally occur in water. In addition to trihalomethanes, haloacetic acids are formed, which are carcinogenic even at low concentrations. The current method for determining haloacetic acids via liquid-liquid extraction with subsequent gas chromatographic analysis utilizing an ECD detector or a mass spectrometer [52] is very sensitive (minimum detection limit: ≈ 0,4 µg/L) but time-consuming. Because haloacetic acids have become of environmental interest only very recently, very few references exist regarding their analysis by liquid chromatography. Bächmann et al. [53] used anion exchange chromatography for analyzing mono- and di-chloroacetic acid in rain samples. The same method was used by Nair et al. [54] for separating the three chloroacetic acids as well as mono- and di-bromoacetic acid under isocratic conditions. Nair et al. used a packed bed suppressor based on two cartridges that are alternated between separator column and conductivity cell after every chromatographic run. Because the affinities of haloacetic acids to the stationary phase of an anion exchanger strongly increase with increasing degree of halogenation, higher haloacetic acids are characterized by extreme peak widths. Thus, the application of this method for real-world water samples with significant amounts of chloride and sulfate is practically excluded. A very detailed study on the ion chromatographic analysis of haloacetic acids was recently published by Sarzanini et al. [55]. They investigated a number of ion-pair- and anion exchange chromatographic techniques for their applicability to separating real-world samples containing di- and tri-chloroacetic acid as well as the three bromoacetic acids. In the course of their ion-pair chromatographic studies on chemically bonded silica with cetyltrimethylammonium salts as an ion-pair reagent, minimum detection limits between 50 and 400 µg/L were obtained after careful optimization of the eluant composition. The authors utilized UV detection at 210 nm to minimize interferences by chloride. Even though the separation efficiency and sensitivity of this technique could be improved by applying gradient elution, UV detection is also limited in that the simultaneous analysis of mineral acids is not possible under these conditions. According to Sarzanini et al., similar detection limits were also achieved with anion exchange chromatography on IonPac AS9-SC with a carbonate/bicarbonate mixture as an eluant. Hydroxide-selective stationary phases such as IonPac AS11 were also evaluated, but satisfying results could not be obtained due to separation problems between chloride and monobromoacetic acid as well as between nitrate and dichloroacetic acid. According to the latest findings, however, those stationary phases offer the ability to separate all haloacetic acids in the same run by applying a gradient elution technique [56]. As can be seen from the chromatogram A in Fig. 9-16, all haloacetic acids can be separated to baseline on IonPac AS16 within 25 minutes. With the exception of monobromoacetic acid, which co-elutes with chloride under these chromatographic conditions, this analysis is principally feasible for a drinking water matrix (chromatogram B in Fig. 9-16).

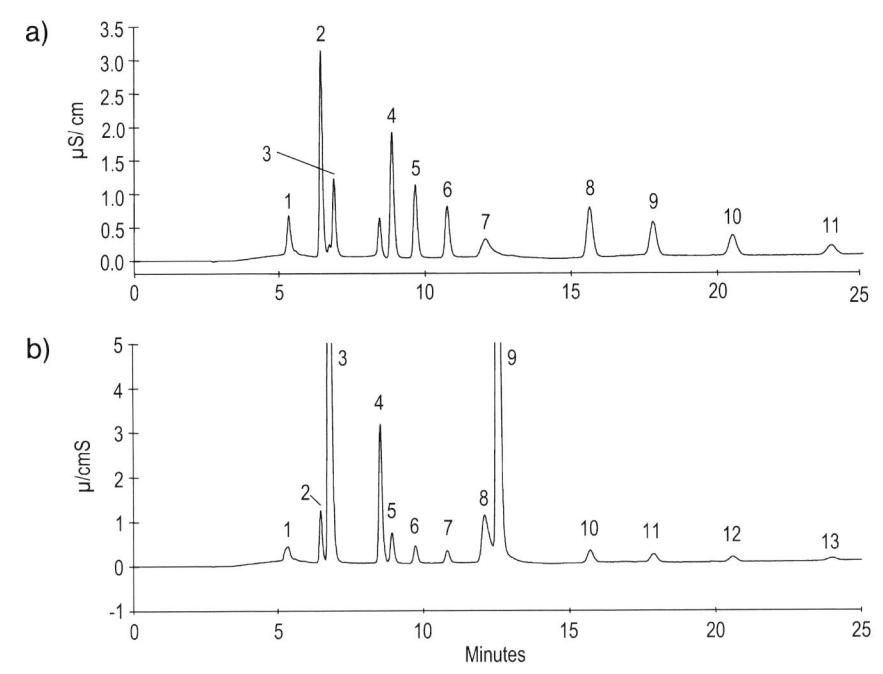

Fig. 9-16. Separation of haloacetic acids on a hydroxide-selective anion exchanger using a gradient elution technique. – Separator column: IonPac AS16; eluant: KOH with Eluant Generator™; gradient: linear, 3 mmol/L to 60 mmol/L in 25 min; flow rate: 1 mL/min; detection: suppressed conductivity; injection volume: 25 µL; solute concentrations: (a) 5 mg/L each of monochloroacetic acid (2), monobromoacetic acid (3), dichloroacetic acid (4), bromochloroacetic acid (5), dibromoacetic acid (6), trichloroacetic acid (8), bromodichloroacetic acid (9), dibromochloroacetic acid (10), and tribromoacetic acid (11), peak (1) acetic acid and peak (7) carbonate are impurities in the standard solution, (b) 3.9 mg/L monochloroacetic acid (2), 12.2 mg/L chloride (3), 8.1 mg/L nitrate (4), 3.9 mg/L each of dichloroacetic acid (5) and bromochloroacetic acid (6), 4 mg/L dibromoacetic acid (7), 19.3 mg/L sulfate (9), 3.7 mg/L trichloroacetic acid (10), 3.5 mg/L each of bromodichloroacetic acid (11) and dibromochloroacetic acid (12), and 3.4 mg/L tribromoacetic acid (13), peak (1) acetic acid and peak (8) carbonate are impurities in the sample; (taken from [56]).

Because modern anion exchangers are solvent-compatible, solvents can also be used to alter selectivity. Figure 9-17 illustrates the gradient elution of chloro- and bromo-acetic acids on IonPac AS11 with a constant content of methanol in the mobile phase. Even though separations of this kind are promising, it is questionable whether the necessary sensitivity for water analysis can be achieved with suppressed conductivity detection. Thus, the method of IC-MS via an electrospray interface introduced by Hashimoto and Otsuki [57] creates a lot of interest. With this method the authors analyzed nine different chloro- and bromo-acetic acids at trace levels. However, the method requires analyte extraction with methyl-*tert*-butyl ether. Based on a 200-mL water sample, detection limits in the lower ng/L range can be obtained. The fact that IC-MS has recently been evaluated for the trace analysis of bromate [58] indicates a clear trend towards mass-selective detection in ion chromatographic trace analysis.

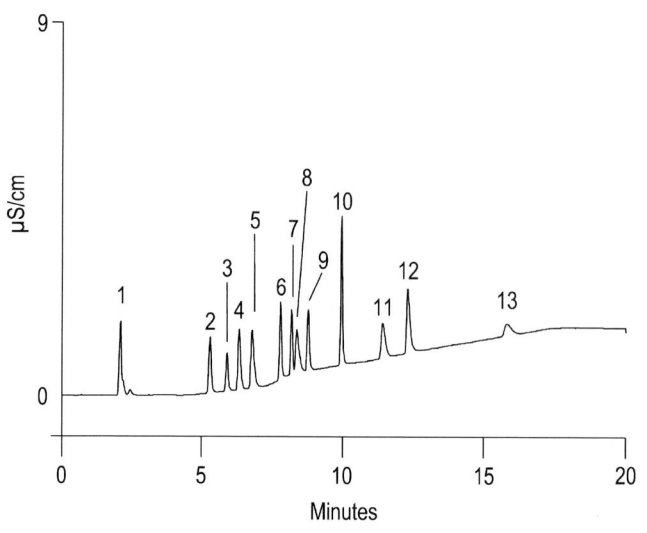

Fig. 9-17. Influence of methanol on the separation of haloacetic acids on a hydroxide-selective anion exchanger using a gradient elution technique. – Separator column: IonPac AS11 (2-mm); eluant: NaOH – methanol (85:15 v/v); gradient: linear, 0.5 mmol/L NaOH for 2 min isocratic, then to 4.25 mmol/L in 3 min, then to 37.5 mmol/L in 10 min; flow rate: 0.5 mL/min; detection: suppressed conductivity; solute concentrations: 1 mg/L fluoride (1), 5 mg/L monochloroacetic acid (2), 10 mg/L monobromoacetic acid (3), 2 mg/L chloride (4), 5 mg/L nitrite (5), 5 mg/L dichloroacetic acid (6), 3 mg/L bromide (7), 3 mg/L nitrate (8), 10 mg/L dibromoacetic acid (9), 10 mg/L sulfate (10), 10 mg/L trichloroacetic acid (11), 10 mg/L orthophosphate (12), and 20 mg/L tribromoacetic acid (13).

Surface waters are mostly analyzed using ion exchangers with a slightly higher ion-exchange capacity that compensates for the occasionally high electrolyte concentration in such samples. High concentration differences between major and minor components sometimes also require the capability of injecting such samples undiluted without overloading the separator column. This is illustrated in Fig. 9-18 with a chromatogram of a Rhine water sample in which small quantities of orthophosphate and bromide could be detected in addition to the main components chloride, nitrate, and sulfate.

Iodide can be detected very selectively by UV at 227 nm. Although iodide belongs to the class of polarizable anions and, thus, exhibits a high affinity to the stationary phase of an anion exchanger, it elutes fairly rapidly and with high peak symmetry from IonPac AS9(-SC). With a pure carbonate eluant and a flow rate of 2 mL/min the retention time of iodide is around ten minutes (Fig. 9-19). If conductivity and UV detection are applied simultaneously, mineral acids can be determined in the same run.

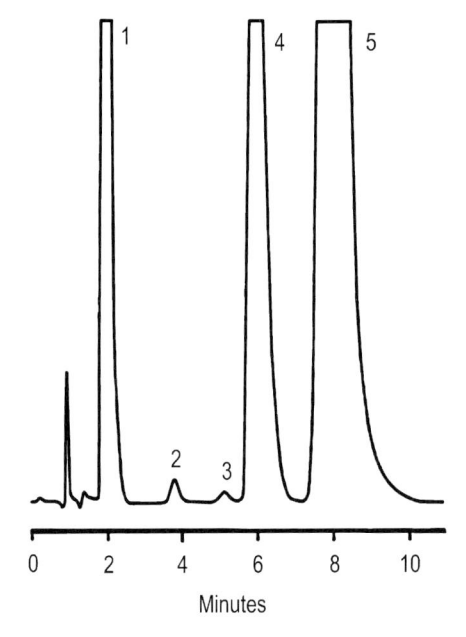

Fig. 9-18. Anion analysis of a Rhine water sample. — Separator column: IonPac AS3; eluant: 2.8 mmol/L $NaHCO_3$ + 2.2 mmol/L Na_2CO_3; flow rate: 2.3 mL/min; detection: suppressed conductivity; injection: 50 µL Rhine water (Düsseldorf); peaks: (1) chloride, (2) 1.2 mg/L orthophosphate, (3) 0.2 mg/L bromide, (4) nitrate, and (5) sulfate.

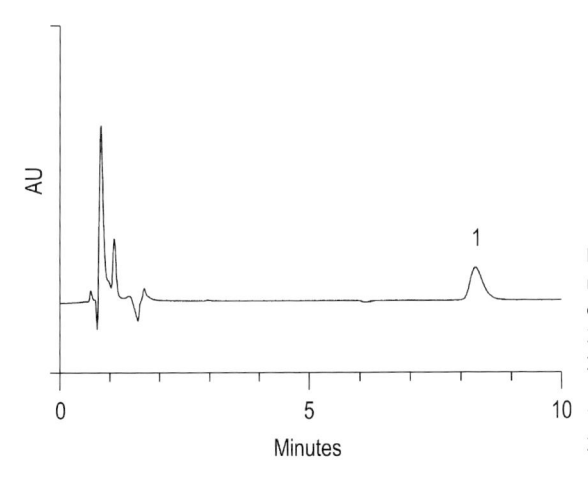

Fig. 9-19. Analysis of iodide in natural waters. — Separator column: IonPac AS9-SC; eluant: 3 mmol/L Na_2CO_3; flow rate: 2 mL/min; detection: UV (227 nm); injection volume: 100 µL; solute concentration: 268 µg/L iodide (1).

The determination of the three nitrogen compounds — nitrite, nitrate, and ammonium — is of special interest in the field of wastewater analysis. Because wastewater samples usually show a high chloride content, nitrite, which elutes immediately after chloride, can only be determined unequivocally by means of UV detection. Figure 9-20 shows the chromatogram of a wastewater sample rich in chloride that had to be injected undiluted owing to its low nitrite content. Although an IonPac AS9-HC high-capacity anion exchanger was used as a stationary phase, a quantitative evaluation of the nitrite signal via peak area is not possible with conductivity detection (chromatogram A). In contrast, UV detection at 214 nm allows an unequivocal evaluation of the nitrite signal even though the anion exchanger is overloaded by the predominance of chloride. The later eluting nitrate, which is present in much higher concentration, may also be detected at this wavelength owing to its optical absorption properties. The negative peak in front of nitrite is caused by the UV-transparent chloride which lowers the absorption of the column effluent for a short time due to its high concentration. The analyzable maximum concentration differences between chloride and nitrite on various anion exchangers with suppressed conductivity or UV detection are summarized in Table 9-3.

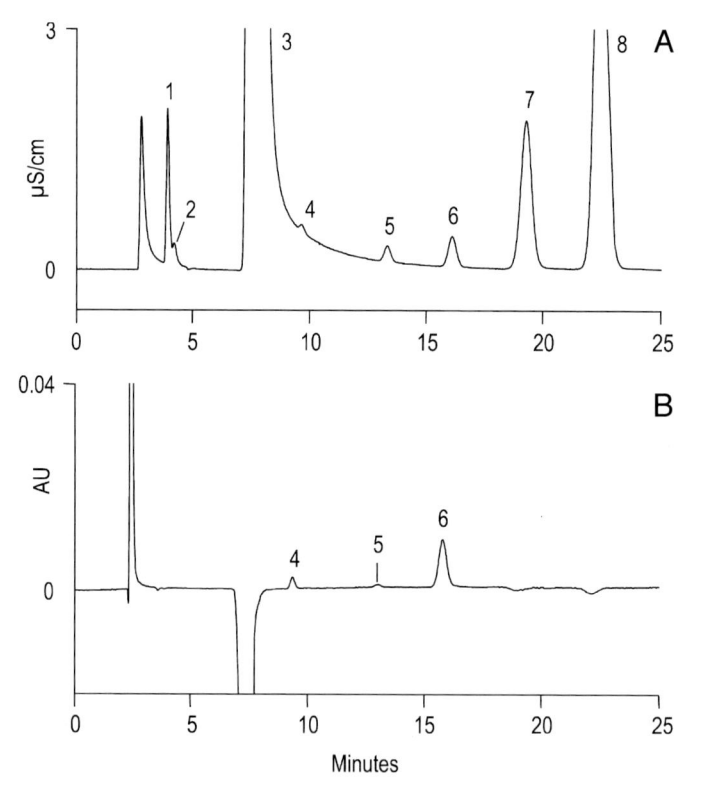

Fig. 9-20. Analysis of nitrite and nitrate in strongly contaminated wastewater using conductivity and UV detection. − Separator column: IonPac AS9-HC; eluant: 9 mmol/L Na_2CO_3; flow rate: 1 mL/min; detection: (A) suppressed conductivity, (B) UV (214 nm); injection volume: 25 µL; solute concentrations: 1.14 mg/L fluoride (1), acetate (2), 2 g/L chloride (3), 0.14 mg/L nitrite (4), 0.82 mg/L bromide (5), 1.5 mg/L nitrate (6), 21.1 mg/L orthophosphate (7), and 30.3 mg/L sulfate (8).

Table 9-3. Analyzable maximum concentration differences between chloride and nitrite in dependence on the type of separator column and detection mode.

Separator/Detection Mode	Cl : NO$_2$	Cl : NO$_2$–N
IonPac AS14/Conductivity	100 : 0.1 mg/L (1000 : 1)	100 : 0.034 mg/L (3285 : 1)
IonPac AS14/UV (214 nm)	500 : 0.03 mg/L (16667 : 1)	500 : 0.009 mg/L (54761 : 1)
IonPac AS9-HC/Conductivity	2000 : 0.2 mg/L (10000 :1)	2000 : 0.068 mg/L (32850 : 1)
IonPac AS9-HC/UV (214 nm)	5000 : 0.045 mg/L (111000 : 1)	5000 : 0.014 mg/L (365000 : 1)

Apart from UV detection there are two other alternative detection methods for trace analysis of nitrite. A very sensitive detection method for nitrite is DC amperometry at +1.1 V on a Carbon Paste electrode. In this case, amperometry and suppressed conductivity detection can be applied simultaneously by directing the conductivity cell effluent through the amperometric cell. Figure 9-21 exemplifies this with a surface water sample which was injected undiluted [59]. As can be seen from this chromatogram, nitrite concentrations in the lowest µg/L range can be detected without any problem.

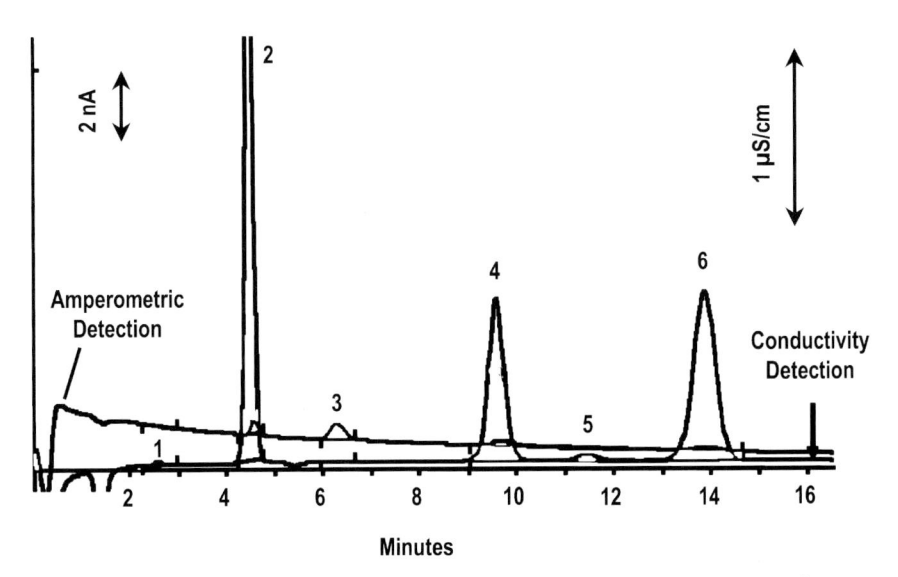

Fig. 9-21. Amperometric detection of nitrite at trace level. — Separator column: Metrosep Anion Dual 2; eluant: 2 mmol/L NaHCO$_3$ + 1.5 mmol/L Na$_2$CO$_3$; flow rate: 0.8 mL/min; detection: suppressed conductivity and DC amperometry on a carbon paste electrode; oxidation potential: +1.1 V; injection volume: 20 µL; solute concentrations: 0.018 mg/L fluoride (1), 5.53 mg/L chloride (2), 0.0036 mg/L nitrite (3), 4.23 mg/L nitrate (4), 0.41 mg/L orthophosphate (5), and 4.86 mg/L sulfate (6); (taken from [59]).

The second alternative for detecting traces of nitrite is the detection via formation of triiodide, which was discussed above for bromate analysis [30]. It is described in a similar form by Miura et al. [60]. The reaction mechanism for nitrite is as follows:

$$2NO_2^- + 4H^+ + 2I^- \rightarrow 2NO + I_2 + 2H_2O \qquad (304)$$

In contrast to von Gunten et al., Miura et al. acidify the derivatization reagent (KI) directly with nitric acid, although it is known that iodide is oxidized by oxygen at low pH. This leads to a relatively high background absorption and, consequently, to a lower sensitivity. The proof for this hypothesis is the minimum detection limit for nitrite of 4.6 µg/L calculated by Miura et al., which is about one order of magnitude higher than that of Salhi and von Gunten [30].

Unlike nitrite, the determination of ammonium, even in high excess of sodium, is possible with conductivity detection, because high resolution between sodium, ammonium, and potassium is obtained with special cation exchangers [61]. Figure 9-22 illustrates the separation between sodium and ammonium with a concentration ratio of 4000:1, obtained on a weak acid cation exchanger modified with carboxylate-, phosphonate- and crown ether groups. The higher selectivity of this stationary phase towards ammonium and potassium ions can be attributed to 18-crown-6, which was immobilized on the stationary phase surface [62]. Optimum separations are obtained at an elevated temperature of 40 °C, which predominantly affects the retention of potassium, which elutes last. However, it is not so much the column temperature itself that is important, but the constancy of the temperature during the chromatographic run. Resolution between sodium and ammonium can even be improved by lowering the acid concentration at the beginning of the analysis, so that even more disparate concentration differences can be quantified. To keep the total analysis time within an acceptable time frame, the acid concentration is increased step-wise during the chromatographic run. For quantifying sodium and ammonium with concentration ratios of less than 1000:1, conventional weak acid cation exchangers such as IonPac CS12A can be combined with an IonPac CG15 pre-column as an alternative. One advantage of such a column combination is that elevated column temperatures and organic solvents in the mobile phase are not necessary.

In addition to municipal wastewaters, industrial wastewaters may also be analyzed by ion chromatography, although the analytical problems may strongly differ. An unusual example in anion analysis is the separation of monoisopropyl sulfate and inorganic sulfate in the wastewater that results from isopropanol synthesis is shown in Fig. 9-23. In analogy to the retention behavior of inorganic and organic phosphates (see Fig. 3-130 in Section 3.8.1), sulfates exhibit also a much shorter retention for the organic component.

At first glance, the separation of transition metals shown in Fig. 9-24 merely illustrates the applicability of ion chromatography to a class of compounds which are normally analyzed using atom spectrometric methods. However, in view of

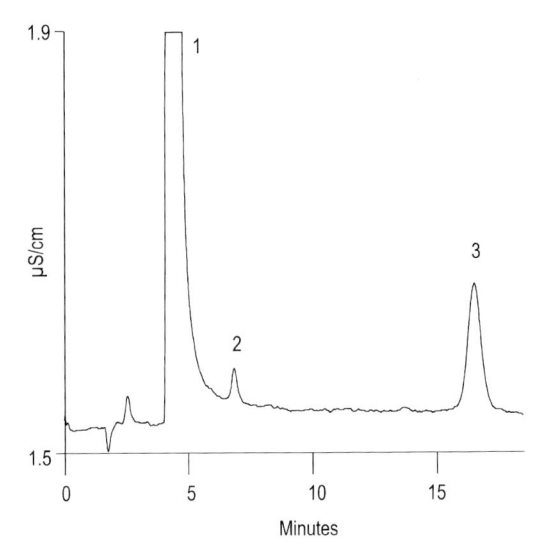

Fig. 9-22. Ammonium analysis at high excess of sodium. — Separator column: IonPac CS15; column temperature: 40 °C; eluant: 5 mmol/L H_2SO_4 — acetonitrile (91:9 v/v); flow rate: 1.2 mL/min; detection: suppressed conductivity; injection volume: 25 µL; solute concentrations: 100 mg/L sodium (1), 0.025 mg/L ammonium (2), and 0.025 mg/L calcium (3).

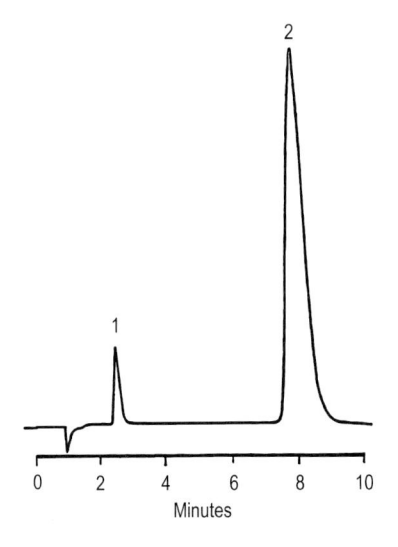

Fig. 9-23. Anion analysis of a wastewater resulting from isopropanol synthesis. — Separator column: IonPac AS3; eluant and flow rate: see Fig. 9-18; detection: suppressed conductivity; injection: 50 µL wastewater (1:5000 diluted); peaks: (1) monoisopropyl sulfate and (2) sulfate.

matrix problems, these methods are subject to limitations when the analyte sample, as in the present case, is an aqueous eluate from a technical process in which small amounts of nickel are to be determined in presence of iron as a main component. This poses no difficulty in ion chromatography because the nickel determination is not interfered with by high iron concentrations when oxalic acid is applied as a complexing agent. As can be seen from Fig. 9-24, other transition metals such as copper, cadmium, cobalt, and zinc can also be detected under these chromatographic conditions.

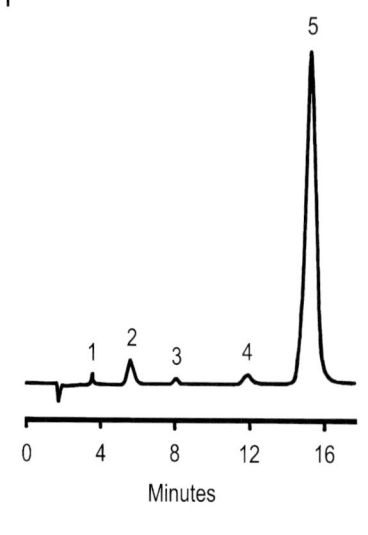

Fig. 9-24. Transition metal analysis in an industrial wastewater. — Separator column: IonPac CS5A; eluant: 0.05 mol/L oxalic acid + 0.095 mol/L LiOH; flow rate: 1 mL/min; detection: photometry at 520 nm after post-column reaction with PAR; injection: 50 µL wastewater (1:50 diluted); solute concentrations: 1.2 mg/L copper (1), 70 mg/L cadmium (2), 0.8 mg/L cobalt (3), 5.2 mg/L zinc (4), and 96 mg/L nickel (5).

In principle, care must be taken in the analysis of wastewater samples so that the ion exchanger is not contaminated by organic materials such as fats, oils, surfactants, etc. They can be removed from the sample via solid-phase extraction on suitable resins (e.g., OnGuard cartridges). Further details regarding the subject of sample preparation may be found in Section 9.11.

In comparison to municipal and industrial wastewaters, landfill leachates and seepage waters are significantly more contaminated. In addition to high concentrations of inorganic anions and cations they often contain significant amounts of short-chain aliphatic carboxylic acids such as acetic acid and butyric acid. In particular cases, the contribution of those organic acids to the total amount of dissolved organic carbon (DOC) can be up to 90% or even higher. Thus far, aliphatic carboxylic acids in landfill leachates have been determined almost exclusively by means of gas chromatography [63], even though sample preparation (extraction and derivatization) is labor-intensive, the column material is not suitable for separating carboxylic acids, and thermolabile components may degrade. However, di-, tri-, and hydroxy-carboxylic acids are responsible for metal releasing processes and are also important as pilot parameters (for example lactic acid) for a number of biotechnological waste treatment processes. In contrast, mono- and multi-functional acids can be determined simultaneously by ion-exclusion chromatography. The method introduced by Fischer [64] offers the additional advantage of a simple sample preparation, as membrane filtrated samples can be injected directly after removing humic acids with a polyvinylpyrrolidone extraction cartridge (OnGuard-P). The chromatograms of a seepage water from a sewage sludge dump obtained with two different stationary phases — Polyspher OA-HY from Merck (Darmstadt, Germany) and IonPac ICE-AS6 from Dionex (Sunnyvale, USA) — are compared in Fig. 9-25. The respective organic acid concentrations are summarized in Table 9-4. By comparing the retention times with those of reference compounds, 18 components could be identified. However, even under optimized chromatographic conditions not all components can be separated to baseline. Mono- and di-carboxylic acids exhibit higher retention on IonPac ICE-AS6 than on Polyspher OA-HY, so that short-chain hydroxy- and keto-carboxylic acids are better separated due to a stronger temperature dependence of the retention. IonPac ICE-AS6 also allows a higher flexibility of the chromatographic conditions. Thus, complementary chromatographic information is obtained with the two separator columns. Both chromatograms indicate the material diversity of aliphatic carboxylic acids in these matrices, which is significantly larger than existing studies reveal. Monocarboxylic acids dominate in terms of amount, but higher concentrations of dicarboxylic acids as well as the presence of lactic acid and succinic acid emphasize that the analytical neglect of multifunctional acids is not justified. Especially when investigating the dumping of transition metal-containing waste, this class of compounds should not be overlooked, due to its potential mobilization effects.

Seawater represents the most difficult water matrix in the analysis of trace ionic constituents, because chloride, sulfate, and sodium are the primary ions and they are present at extremely high concentrations. Trace ammonium, for example, is difficult to quantify, because sodium and ammonium elute in close proximity to each other on weak acid cation exchangers. To overcome this problem, Rey et al. developed an isocratic column-switching method, which greatly increases the resolution between sodium and ammonium ions, and allows for the determination of sodium-to-ammonium concentration ratios in the order of 20,000:1. This method requires two columns containing different functional groups [65]. An innovative column-switching technique for determining ammonium in seawater using common cation exchangers was developed by Huang et al. [66]. This two-stage procedure consists of matrix elimination followed by IC analysis of ammonium. In the ammonium analysis, from 5.0 to 6.1 min, the conductivity cell effluent is directed to an IonPac CG12A concentrator column (see Fig. 9-26a). Because the sulfuric acid eluant is converted to water in the suppressor, the effluent is collected onto the concentrator column. During the remaining analysis time and by switching the four-way valve V2 (Fig. 9-26b), the major part of the matrix and the strongly retained cations in the sample are eluted to waste. After all the sample cations are eluted from the separator column, ammonium is eluted from the concentrator column (Fig. 9-26c). The valve configuration is schematically depicted in Fig. 9-26. The corresponding program for the two-stage procedure is given in Table 9-5. Figure 9-27 shows a chromatogram of a 20-fold diluted seawater sample obtained under these conditions. The authors report spike recoveries between 94% and 104% and RSD values around 3% for ammonium levels between 19 µg/L and 45 µg/L.

A more exotic application is the determination of beryllium in seawater described by Bashir et al. [67]. Beryllium is one of the most toxic non-radioactive elements to be found at trace levels in natural waters. It acts as an insidious

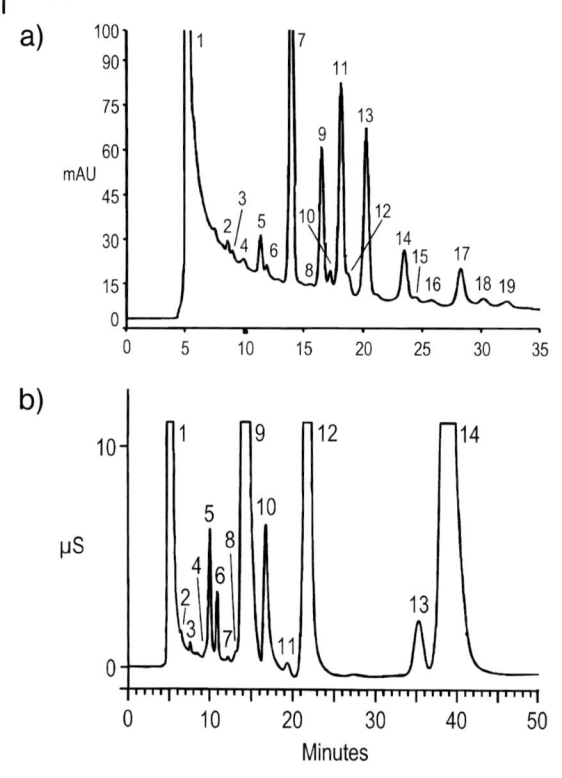

Fig. 9-25. Separation of aliphatic carboxylic acids in the seepage water of a sewage sludge dump. — Separator columns: (a) Polyspher OA-HY (Merck), (b) IonPac ICE-AS6 (Dionex); column temperatures: (a) 45 °C, (b) 60 °C; eluants: (a) 5 mmol/L H_2SO_4, (b) 0.4 mmol/L perfluorobutyric acid; flow rates: (a) 0.5 mL/min, (b) 0.5 mL/min; detection: (a) UV (219 nm), (b) suppressed conductivity; injection volumes: (a) 20 µL, (b) 25 µL; sample preparation: OnGuard-P; peaks: (a) (1) strong acids, (2) pyruvic acid, (3) glyoxylic acid, (4) glycerolic acid, (5) lactic acid/succinic acid, (6) glycolic acid, (7) acetic acid/glutaric acid, (8) adipic acid (9) propionic acid, (10)/(11) unknown, (12) isobutyric acid, (13) n-butyric acid, (14) isovaleric acid, (15)/(16) unknown, (17) valeric acid, (18)/(19) unknown, (b) (1) strong acids, (2) pyruvic acid, (3) unknown, (4) glycolic acid, (5) formic acid, (6) lactic acid, (7) unknown, (8) succinic acid, (9) acetic acid, (10) unknown, (11) glutaric acid, (12) propionic acid, (13) isobutyric acid, and (14) n-butyric acid; (taken from [64]).

carcinogenic poison, affecting cellular membranes and binding specific regulatory proteins in cells. The monitoring of trace levels of Be(II) in natural waters is of interest as it indicates the extent of environmental pollution from anthropogenic sources such as nuclear, aeronautical, and metallurgical industries. Concentrations of Be(II) in natural waters range from 0.1 to 500 µg/L [68]. Of all the spectrometric analytical techniques, inductively coupled plasma mass spectrometry (ICP-MS) [68] is the most ideal. The combination of a relative lack of susceptibility to matrix interferences and sensitive mass detection allows most samples to be analyzed without pre-concentration or extraction steps. However,

Table 9-4. Carboxylic acid concentrations of the seepage water of a sewage sludge dump [64].

Analyte	Polyspher OA-HY	IonPac ICE-AS6
Pyruvic acid	<2.2 [f]	b)
Glyoxylic acid	36.3 [a]	d)
Glycerolic acid	63.0	b)
Glycolic acid	239.2	b)
Formic acid	d)	53.0
Lactic acid	c)	55.7
Acetic acid	4368 [a]	4367
Succinic acid	c)	b)
Propionic acid	1545	1622
Glutaric acid	c)	15.7
Adipic acid	<3.7 [f]	d)
i-Butyric acid	b)	196.0
n-Butyric acid	2634	2546
Valeric acid	2290	e)
i-Valeric acid	773.7	e)

a) Quantification interfered (partial peak overlapping)
b) Detectable, but cannot be quantified due to strong peak overlapping
c) Possible or actual co-elution
d) Not detectable (no signal in the characteristic retention range)
e) Not detectable (retention time > run time)
f) Detectable, concentration below minimum determination limit

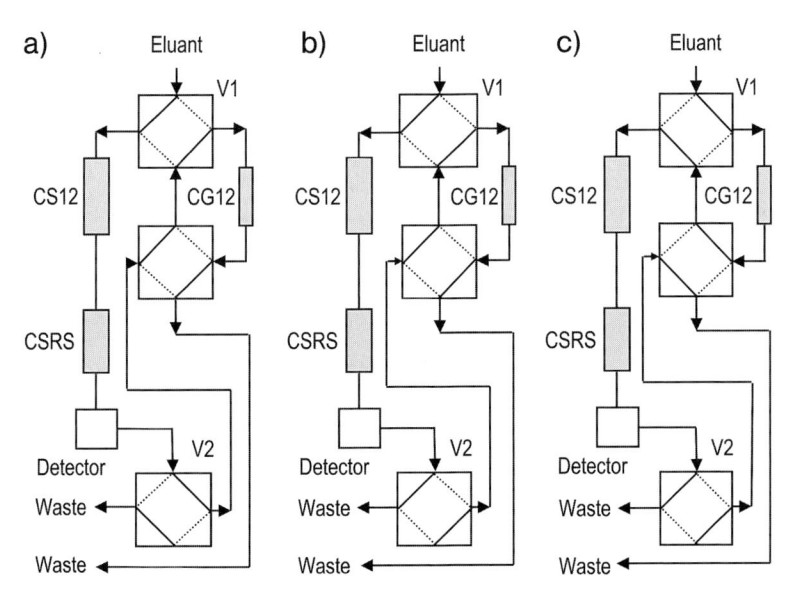

Fig. 9-26. Valve configuration of the column-switching method for determining trace amounts of ammonium in seawater.

Table 9-5. Program for the two-stage procedure.

Time [min]	V$_1$	V$_2$
0.0	On	Off
5.0	On	On
6.1	On	Off
15.0[a]	Off	Off
25.0	Off	Off

[a] Begin data collection

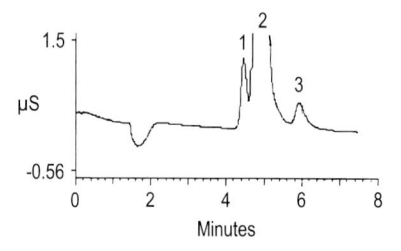

Fig. 9-27. Ammonium analysis in seawater utilizing a column-switching technique. — Separator column: IonPac CS12A; concentrator column: IonPac CG12A; eluent: 25 mmol/L H$_2$SO$_4$; flow rate: 1 mL/min; detection: suppressed conductivity; injection volume: 25 µL; peaks: (1) lithium, (2) sodium, and (3) ammonium; (taken from [66]).

this method does require expensive instrumentation which is not readily available in many analytical laboratories. On the other hand, chromatographic methods for the determination of Be(II) are surprisingly few. Betti and Cavalli [69] developed a cation exchange separation, but it lacked the desired stationary phase selectivity required for the analysis of anything other than relatively simple samples such as drinking water and other low-ionic strength samples. Alternative complexing and chelating stationary phases for use in the ion chromatographic determination of Be(II) have been investigated by Voloschik et al. [70] and Shaw et al. [71]. Voloschik et al. used iminodiacetic acid (IDA) functionalized silica with a nitric acid/dipicolinic acid eluent. Under these conditions, Be(II) was well separated from common transition metals, but eluted immediately after other alkaline-earth metals, so that the method was limited to samples with a relatively low Ca(II)/Mg(II) to Be(II) ratio. Most recently, Shaw et al. developed a method based on an aminomethylphosphonic acid functionalized silica, which exhibited unique selectivity towards Be(II). When used with a strongly acidic eluent, Be(II) could be separated from a large excess of common alkaline-earth and transition metals. Post-column reaction detection based on chromazurol S (CAS) was used for added selectivity. However, the rather broad peak shape obtained for Be(II) resulted in a detection limit of 35 µg/L in stream sediment digest.

Bashir and Paull also used IDA functionalized silica with 8-µm average particle size. In comparison to aminomethylphosphonic acid, IDA exhibits a lesser affinity for Be(II), coordinating with the metal ion through single N and O donor atoms. On such a stationary phase, Be(II) is predominantly retained through surface complexation, while ion-exchange plays a more significant role in the

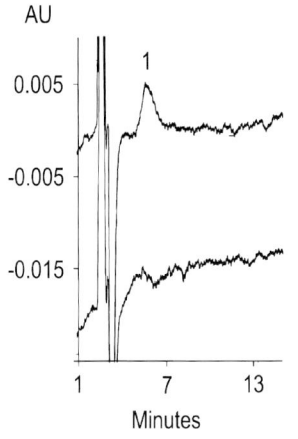

Fig. 9-28. Overlaid chromatograms of simulated seawater and simulated seawater spiked with Be(II). – Separator column: IDA functionalized silica (8-μm); eluant: 0.4 mol/L KNO_3 adjusted to pH 3.0 with HNO_3; flow rate: 1 mL/min; detection: UV at 590 nm after post-column derivatization with chrome azurol S (see text for composition); post-column reagent flow rate: 1.5 mL/min; injection volume: 250 μL; peak: (1) 0.04 mg/L Be(II); (taken from [67]).

retention of common alkaline-earth metals. Its high selectivity for Be(II) makes IDA functionalized silica ideal for the determination of the metal ion in high-ionic strength samples. Therefore, the authors used 0.4 mol/L KNO_3 as an eluant, as this concentration is sufficiently high to suppress any retention due to ion-exchange for Be(II) and other alkaline-earth metals. Bashir et al. optimized the post-column reagent conditions initially developed by Shaw et al. regarding CAS concentration, pH and addition of Triton X-100 as a surfactant. The final post-column reagent solution was 0.26 mmol/L CAS, 2% Triton X-100 and 50 mmol/L MES, adjusted to pH 6.0 with dilute NaOH. Figure 9-28 shows a chromatogram of a simulated seawater sample consisting of 0.52 mol/L NaCl, 1300 mg/L Mg(II), and 400 mg/L Ca(II). The simulated sample was spiked with 0.04 mg/L Be(II). Eluant pH was adjusted to pH 3.0 to increase retention of Be(II) and to resolve the Be(II) peak from a large negative peak close to the system void that results of non-retained matrix ions. Under these chromatographic conditions, the absolute detection limit for Be(II) in simulated seawater is 2.75 ±0.5 ng.

Sample preparation is extremely important in the field of soil analysis. When the sample is a filtrate of a neutral sludge, it may be diluted due to the high electrolyte content. The quality of the resulting chromatogram in Fig. 9-29, therefore, hardly differs from that of a drinking water chromatogram. For cation analysis via non-suppressed conductivity detection [72] (Fig. 9-30) the diluted sample was injected through a cartridge containing an anion exchange resin for neutralization. In addition to the major components, sodium, potassium, magnesium, and calcium, small amounts of lithium, manganese, and strontium could be detected.

In contrast, soil samples, are often extracted with a 10% potassium chloride solution. Due to the chloride matrix, the nitrogen parameters of interest, nitrite and nitrate, cannot be analyzed using a separation system with a carbonate/

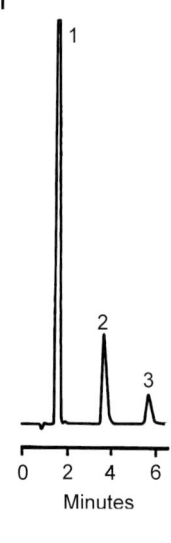

Fig. 9-29. Separation of inorganic anions in the filtrate of a neutral sludge. — Separator column: IonPac AS4A(-SC); eluent: 1.7 mmol/L $NaHCO_3$ + 1.8 mmol/L Na_2CO_3; flow rate: 2 mL/min; detection: suppressed conductivity; injection: 50 µL sample (1:250 diluted); solute concentrations: 17.5 g/L chloride (1), 2.9 g/L nitrate (2), and 0.9 g/L sulfate (3).

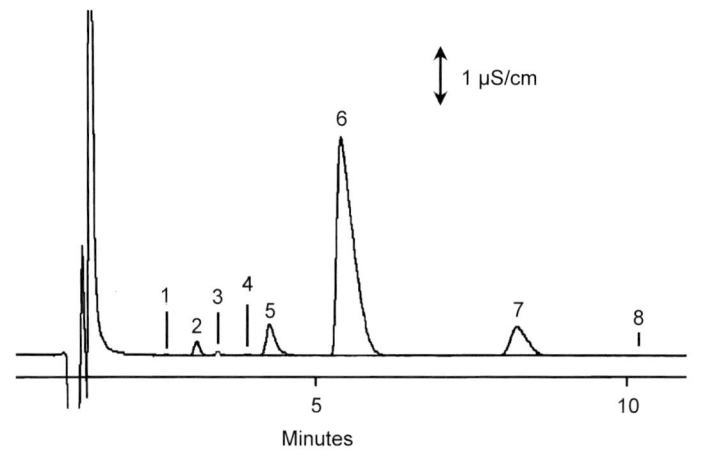

Fig. 9-30. Separation of alkali metals, alkaline-earth metals and ammonium in a sewage sludge after nitric acid digest. — Separator column: Metrosep Cation 1-2; eluent: 4 mmol/L tartaric acid + 1 mmol/L pyridine-2,6-dicarboxylic acid; flow rate: 1 mL/min; detection: non-suppressed conductivity; injection: 10 µL sample (1:25 diluted); solute concentrations: 0.22 mg/L lithium (1), 26 mg/L sodium (2), 4.9 mg/L ammonium (3), 5.8 mg/L manganese (4), 210 mg/L potassium (5), 1320 mg/L calcium (6), 82 mg/L magnesium (7), and 4.5 mg/L strontium (8); (taken from [72]).

bicarbonate eluent. This problem is easily solved, however, by employing potassium chloride as an eluent and determining the analyte ions nitrite and nitrate photometrically. When a stationary phase with both anion and cation exchange capacities (e.g., IonPac CS5A) is used, the ammonium ion is also retained, which may be detected very sensitively by derivatization with o-phthaldialdehyde and subsequent fluorescence detection. By combining both detection methods, all three nitrogen parameters can be determined in one run. Figure 9-31 illustrates this with chromatograms of a real sample, which were obtained by applying simultaneous detection.

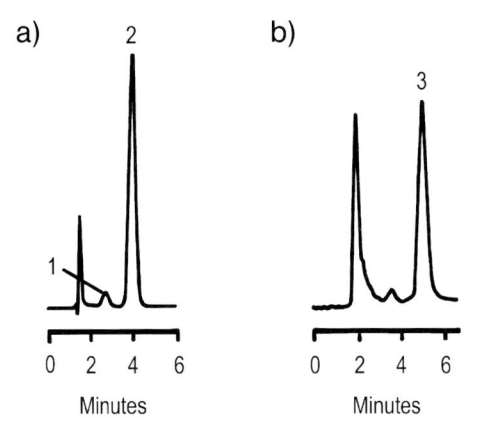

Fig. 9-31. Simultaneous detection of nitrite, nitrate, and ammonium in a KCl soil extract. – Separator column: IonPac CS5A; eluant: 35 mmol/L KCl; flow rate: 1 mL/min; detection: (a) UV (215 nm), (b) fluorescence after reaction with OPA; injection: 50 µL sample of a 1:10 diluted KCl soil extract (10%); solute concentrations: 2 mg/L nitrite (1), 38 mg/L nitrate (2), and 19 mg/L ammonium (3).

In their recent publication, Tucker et al. [73] describe the analysis of iodide in ground water and soil extracts on IonPac AS11 using suppressed conductivity detection and a sodium hydroxide/methanol eluant. The soil samples to be analyzed were extracted with a carbonate/bicarbonate mixture and the resulting extracts diluted with de-ionized water after membrane filtration. As expected, the minimum detection limit for iodide with suppressed conductivity detection is in the mid-µg/L range.

Of topical interest is the investigation of Vermillion and Crenshaw [74] on the analysis of nerve gas degradation products in soil samples. They investigated isopropylmethylphosphonic acid (IMPA) and methylphosphonic acid (MPA) which are formed via hydrolysis of GB (sarin) under environmental conditions (Fig. 9-32). In the vicinity of former production sites and future incineration plants, these two components pose a potential environmental risk, so a simple chromatographic technique was needed to detect nerve gas degradation products in the sub-mg/L level in soil without pre-concentration. Successful separations of IMPA and MPA have already been achieved on special anion exchangers with reversed-phase properties [75], but Vermillion et al. endeavored to develop a

Fig. 9-32. Degradation of GB (sarin) to IMPA and MPA.

chromatographic technique that did not require organic modifiers in the mobile phase. Moreover, they wanted to simplify sample preparation, so that this kind of analysis could be carried out in a field laboratory. Vermillion et al. used two Sarasep AN 300 anion exchangers in series as stationary phases; the desired selectivity was obtained with an eluant mixture of tetraborate and NaOH. The top chromatogram in Fig. 9-33 shows the separation of a standard mixture of IMPA and MPA and other inorganic anions under optimized conditions. Detection was carried out via suppressed conductivity. The bottom chromatogram shows the analysis of a soil sample, which was spiked with 400 ng/g IMPA and MPA for determining the minimum detection limit. Both compounds are separated from the matrix components and can be clearly determined at this concentration level. For extraction Vermillion et al. used a mixture of 5.1 mmol/L sodium carbonate and 5.4 mmol/L sodium bicarbonate, with which only IMPA can be quantitatively extracted; the recovery rate for MPA is only 60%. To keep this analysis simple the authors did not employ a gradient elution technique, with which the total analysis time of more than 40 minutes could certainly be decreased.

Interesting ion chromatographic applications in the area of air hygiene include the analysis of inorganic anions and cations in fly ashes [76] and atmospheric aerosols [77, 78], the analysis of nitrate and nitric acid in the atmosphere using the Denuder technique [79], and the analysis of formaldehyde and acetaldehyde after appropriate sampling [80]. As an example, Fig. 9-34 shows the separation of arsenic(V) in an aqueous extract from fly ash. Under the given chromatographic conditions arsenate elutes shortly after sulfate, which together with chloride and orthophosphate is a major component.

Gases such as cyanic acid [81], sulfur dioxide [82-84], and nitrogen oxides [82, 84-86] can also be determined ion chromatographically. For the determination of sulfur dioxide, for example, Velásquez et al. [83] used a Graseby/Anderson apparatus (Atlanta, GA, USA), with which the air to be analyzed is directed through 50 mL of an absorber solution with a flow rate of 185 mL/min for 24 hours. The absorber solution, which consists of a dilute H_2O_2 solution, is prepared by mixing 20 mL of 30% H_2O_2 solution with 100 µL HCl (0.6 mol/L) in a 1-L volumetric flask, which is then filled up to the mark with de-ionized water. To keep the air flow free of particles, the authors used a 5-µm PTFE filter. The sulfate formed via oxidation of sulfur dioxide can then be determined by means of anion exchange chromatography. The minimum detection limit for sulfate, expressed as threefold standard deviation of the blank value (5.5 µg/m^3), is 0.044 µg/m^3 with this method.

In contrast to sulfur dioxide, nitrogen oxides cannot be trapped in the absorber solution described above due to their lower Ostwald absorption coefficients. For trapping NO_x, the U.S. EPA specifies the use of an evacuated cylinder containing a mixture of dilute sulfuric acid and hydrogen peroxide [87]. Alternatively, nitrogen oxides can also be trapped in an impinger containing an alkaline permanga-

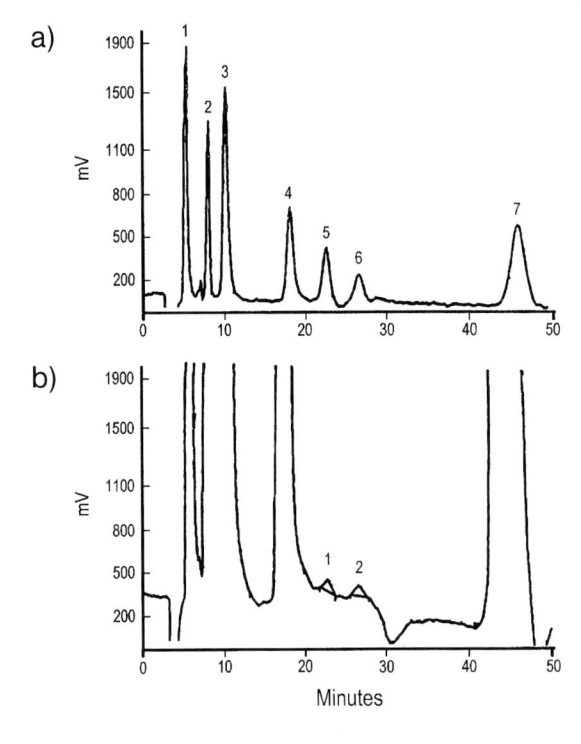

Fig. 9-33. Separation of the GB (sarin) degradation products IMPA und MPA in soil samples. — Separator column: 2 Sarasep AN 300 (100 mm × 7.5 mm i.d.); eluant: 10 mmol/L $Na_2B_4O_7$ + 3.75 mmol/L NaOH; flow rate: 1.5 mL/min; detection: suppressed conductivity; injection volume: 250 µL; (a) standard solution with 100 µg/L each of fluoride (1), chloride (2), carbonate (3), nitrate (4), MPA (5), IMPA (6), and sulfate (7), (b) 400 µg/L MPA (1) and IMPA (2) in a Tilsit soil extract; (taken from [74]).

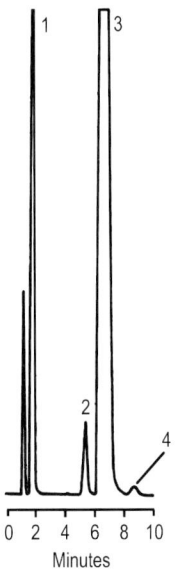

Fig. 9-34. Separation of arsenate in an aqueous extract from fly ash. — Chromatographic conditions: see Fig. 9-30; injection: 50 µL sample (undiluted); peaks: (1) chloride, (2) orthophosphate, (3) sulfate, and (4) arsenate (6.8 mg/L).

nate solution [88]. In both cases, the nitrite and nitrate formed are determined ion chromatographically. Because these methods are very time-consuming, Nonomura et al. [86] developed a technique with which nitrogen oxides are trapped with an arrangement of two absorption cylinders containing 5% triethanolamine. While NO_2 can be directly determined by ion chromatography, the analysis of NO is carried out via oxidation to NO_2. For oxidation, either UV light in the presence of oxygen or a liquid oxidation solution (e.g., a mixture of 5% potassium permanganate solution and 5% orthophosphoric acid) can be used. To determine nitrogen oxides simultaneously, the sample is irradiated with UV light prior to entering the second absorption cylinder. The reaction mechanism between NO_2 and triethanolamine was formulated by Cee et al. [89] as follows:

$$2\ NO_2 \rightarrow N_2O_4 \tag{305}$$

$$N_2O_4 + TEA \rightarrow TEA\text{-}NO^+\ NO_3^- \tag{306}$$

$$TEA\text{-}NO^+\ NO_3^- + H_2O \rightarrow TEA + 2H^+ + NO_2^- + NO_3^- \tag{307}$$

If a gas mixture of NO and NO_2 is directed through the two coupled absorption cylinders, NO_2 will be completely absorbed in the first cylinder to form nitrite and nitrate, while only very small amounts of NO are absorbed as the UV lamp is switched off. However, when applying UV irradiation, NO up to a maximum concentration of 60 ppm is completely oxidized to NO_2 and, consequently, absorbed in the second cylinder. Air and N_2O are not oxidized by UV radiation. Nonomura and Hobo [84] applied this method for the simultaneous determination of sulfur dioxide, nitrogen oxides, and hydrogen chloride in flue gases. Because flue gases contain CO_2, which is also absorbed in triethanolamine, a conventional anion exchanger cannot be used for the subsequent ion chromatographic determination of the formed nitrite-, nitrate-, sulfite-, and sulfate anions, because the nitrite determination is interfered with by carbonate under standard conditions. For this reason, the authors used an IonPac AS7 separator column with an eluant mixture of NaOH and p-cyanophenol. As can be seen from the chromatogram in Fig. 9-35, the nitrogen and sulfur species of interest can be perfectly separated in the presence of a high carbonate concentration under these conditions.

Ion chromatography is also employed as a detection method in the sum detection of organically bonded halogen and sulfur compounds (AOX and AOS) [90-92]. In these methods the water sample is treated with nitrate solution and nitric acid to displace interfering chloride and sulfate ions. An activated carbon low in chlorine, bromine, and sulfur and almost ash-free is added, at which the organically bonded halogen and sulfur compounds are adsorbed. After this concentration step the loaded activated carbon is filtered off through a polycarbonate filter and incinerated together with the filter at 1000 °C in an oxygen stream. Until now, the resulting reaction gases were introduced into the titration cell of a micro-coulometer, where chloride, bromide, and iodide are precipitated with silver ions present in the electrolyte. The silver ion concentration is determined potentiometrically. The silver ion concentration is re-adjusted to its initial value

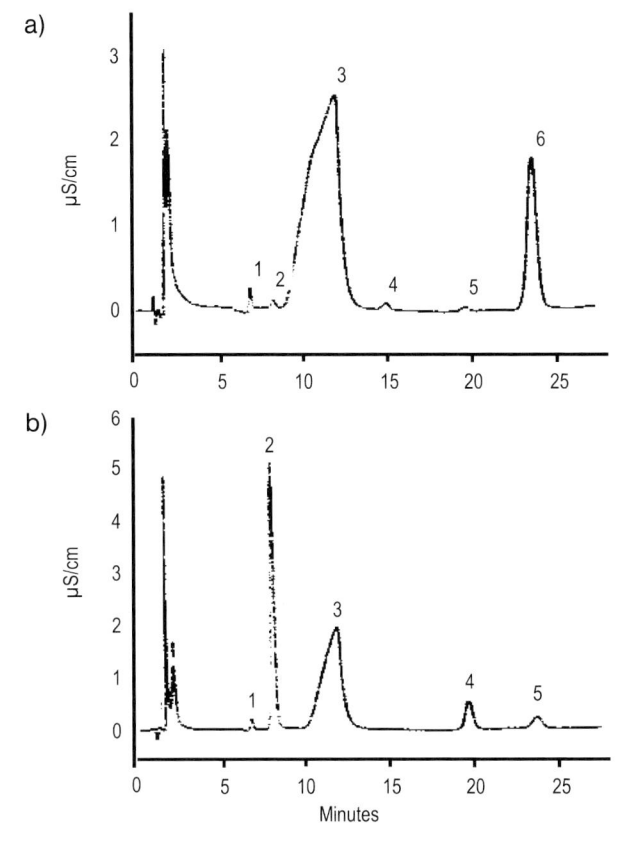

Fig. 9-35. Analysis of NO, NO_2, and SO_2 in a flue gas sample. – Separator column: IonPac AS7; eluant: 0.02 mol/L NaOH + 0.01 mol/L p-cyanophenol; flow rate: 1 mL/min; detection: suppressed conductivity; absorber solution: 5% triethanolamine (10 mL each); (a) 1st absorption cylinder, (b) 2nd absorption cylinder; peaks: chloride (1), nitrite (2), carbonate (3), sulfite (4), nitrate (5), and sulfate (6); concentrations in the flue gas: 0.5 ppmv NO_2, 9.3 ppmv NO, and 3.5 ppmv SO_2; (taken from [84]).

via anodic oxidation of a silver electrode. The amount of charge measured during the electrolysis is equivalent to the amount of chloride, bromide, and iodide which entered the cell. In this type of analysis, chlorine, bromine, and iodine are detected as molar sum; it is impossible to differentiate the species. For the AOS determination, the reaction gases are passed into an aqueous solution and are analyzed using an appropriate detection method.

However, if ion chromatography is used as a detection method, the AOX and AOS parameters may be determined in one working step. Moreover, it is possible to differentiate between organically bonded chlorine, bromine, and iodine. A basic buffer solution made of sodium bicarbonate and sodium carbonate used for ion chromatographic analysis is applied as absorption solution for the acidic reaction gases. A small amount of hydrogen peroxide is added to achieve complete oxidation of sulfite to sulfate. A typical chromatogram obtained in the simultaneous determination of AOX and AOS is shown in Fig. 9-36. In addition to chloride and sulfate, signals for nitrite and nitrate are detected. The latter are due to the application of nitrate, which was added to prevent chloride and sulfate adsorption.

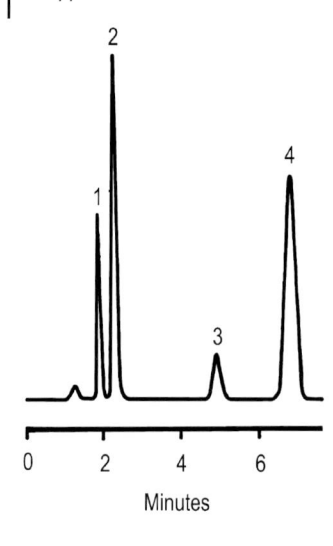

Fig. 9-36. Typical chromatogram of a simultaneous AOX/AOS determination. — Separator column: IonPac AS4; eluant: 2.8 mmol/L $NaHCO_3$ + 2.2 mmol/L Na_2CO_3; flow rate: 2 mL/min; detection: suppressed conductivity; injection: 100 µL absorption solution; peaks: (1) chloride, (2) nitrite, (3) nitrate, and (4) sulfate; (taken from [92]).

9.2
Ion Chromatography in Power Plant Chemistry

The evaluation of the water-, steam-, and condensate quality is one of the most important applications of ion chromatography in power plant technology. In the past, water quality was monitored via registration of the electrical conductance. Routine plant control by this method allows leaks in condensers to be identified and the functioning of water purification devices to be checked. However, conductivity detection is unspecific; an increase in signal does not provide information about the type of contaminant. In addition, sodium chloride or sodium sulfate impurities resulting from inflowing cooling water or insufficiently regenerated condensate purification filters cannot be reliably detected in the sub-µg/L range.

By means of ion chromatography, high-purity waters can be assayed for their chloride- and sulfate- and sodium content. To do this, the instrument should be directly connected to the sampling line in order to analyze the high-purity water free of contamination. If the direct connection of the instrument to the sampling line is not possible, samples must be transported in FEP-Teflon (Nalgene) containers, as only this material has proven to exhibit no wall adsorption effects. Today, sensitivities in the lowest µg/L range are obtained by direct injections of large sample volumes (typically 750 µL). Thus, sample pre-concentration on concentrator columns with the necessary additional equipment such as a sampling pump and a switching valve can be eliminated and, consequently, so can the required time for the pre-concentration step [93]. However, large volume injection requires special separator columns. For anion analysis, hydroxide-selective anion exchangers have proved to be suitable. Anions are focused at the column head at low eluant ion concentration and subsequently eluted by increasing the ionic strength in the mobile phase. Typically, microbore columns with 2-mm internal diameter are used to obtain maximum sensitivity under the given chromatographic conditions.

Such an analysis usually starts with the injection of a high-purity water sample in order to determine the blank values for the ions to be analyzed. The sample is taken from the water purification system and serves as a blank. The achievable minimum detection limits directly depend on the purity of this water. As an example, Fig. 9-37 shows the anion chromatogram of such a blank water sample, which was obtained by injecting a volume of 750 µL and by applying a conventional NaOH concentration gradient. With the exception of fluoride, acetate, and formate no corrosive anions such as chloride and sulfate could be detected in the concentration range below 200 ng/L. The presence of these two anions is the major cause for corrosion in steam generators, reactor vessels, heat exchangers, turbines, and pipes. In conventional power stations, therefore, the concentrations of these anions in the feed water of steam generators are monitored in the concentration range below 1 µg/L; in some nuclear power stations, concentrations below 0.2 µg/L are monitored. A representative chromatogram with the separation of anions in the 100-2000 ng/L range is illustrated in Fig. 9-38. This chromatogram shows that inorganic and organic anions can be quantitated in the sub-µg/L range.

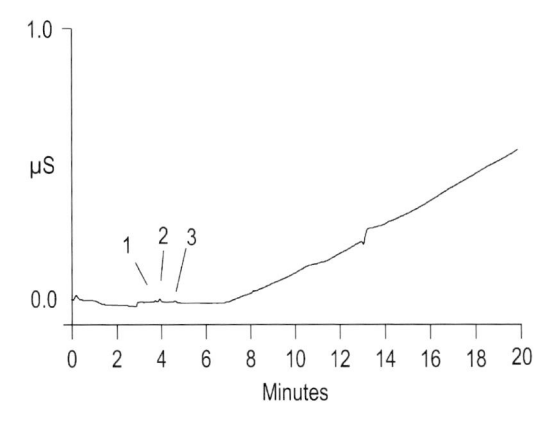

Fig. 9-37. Separation of anions in a high-purity water blank by direct injection. — Separator column: IonPac AS11 (2-mm); eluant: NaOH; gradient: 0.5 mmol/L for 2.5 min isocratic, then linearly to 5 mmol/L in 3.5 min, then to 26 mmol/L in 14 min; flow rate: 0.5 mL/min; detection: suppressed conductivity; injection volume: 750 µL; solute concentrations: <200 ng/L fluoride (1), <200 ng/L acetate (2), and <200 ng/L formate (3).

With the introduction of an Eluant Generator™ (see Section 3.9) the technique of large volume injection could be significantly improved, because electrolytic hydroxide generation prevents carbonate contamination which, in turn, minimizes background conductivity and baseline drift even under gradient conditions [94]. Figure 9-39 illustrates the application of this technology at trace level, with the gradient elution of anions on IonPac AS15. As can be seen from this chromatogram, this novel hydroxide-selective anion exchanger exhibits higher selectivity for early eluting organic acids such as glycolate, acetate, and

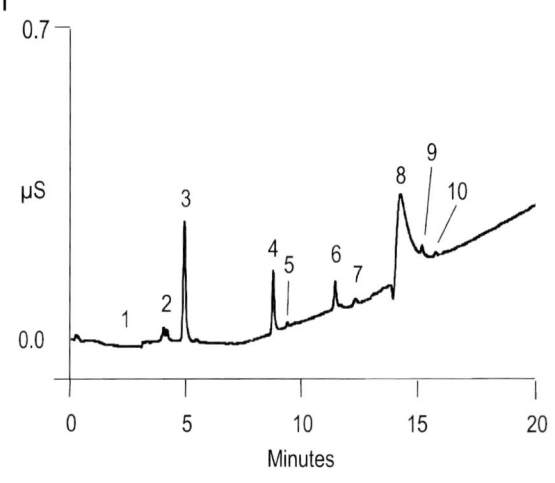

Fig. 9-38. Separation of anions in a nuclear power plant polisher effluent by direct injection. — Chromatographic conditions: see Fig. 9-37; solute concentrations: 0.09 µg/L fluoride (1), 0.05 µg/L acetate (2), 0.46 µg/L formate (3), 0.34 µg/L chloride (4), <0.02 µg/L nitrite (5), 0.40 µg/L nitrate (6), unknown component (7), carbonate (8), 0.10 µg/L sulfate (9), and 0.07 µg/L oxalate (10).

formate than does the AS11 column. The detection of low-molecular weight organic acids such as acetate and formate at trace level is of great interest, because these acids affect the operation of a power station in a negative way, increasing the conductivity of the steam and lowering the pH value [95]. Their presence in the feed water for steam generators is attributed to organic impurities and the thermal degradation of chemicals that are added to adjust the pH value.

Large injection volumes can also be applied in cation analysis to obtain detection limits in the lowest µg/L range by direct injection. Figure 9-40 exemplifies this with the analysis of alkali- and alkaline-earth metals using an IonPac CS12A weak acid cation exchanger. Because the background conductivity of this separation is close to zero when using a suppressor system, analyte signals are not interfered with despite the large injection volume of 500 µL.

For analyses at ultra-trace levels a pre-concentration technique is usually employed. A defined volume of the sample to be analyzed is pumped with a sampling pump through the respective concentrator column, which is connected to the injection valve in place of the sample loop (Fig. 9-41). By switching the valve, all ions that have accumulated on the concentrator column are flushed onto the analytical separator column where they are separated. A tandem switching with two concentrator columns allows a simultaneous analysis and loading.

As examples, Figures 9-42 and 9-43 shows anion and cation chromatograms of a high-purity water sample with concentrations in the sub-µg/L range that were obtained by pre-concentrating 100 mL and 40 mL high-purity water. Anion analysis was carried out with an online ion chromatography system in close proximity to the sampling location using a gradient elution technique with electrolytic hydroxide eluant generation. With this method, the limits of determination for anions and cations of around 10 µg/L by direct injection of 50 µL sample can be lowered to 5 ng/L after pre-concentration [96, 97].

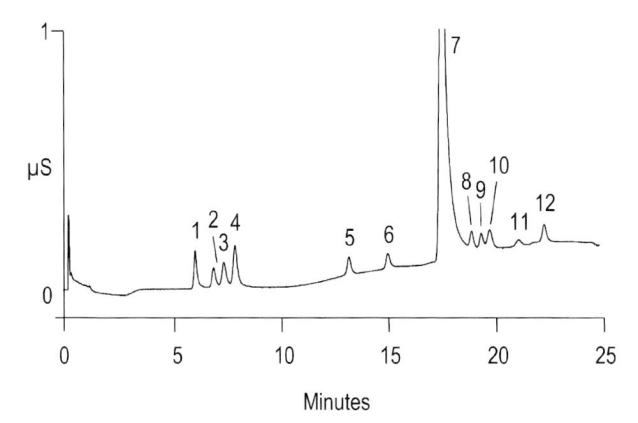

Fig. 9-39. Trace anion analysis with IonPac AS15 via direct injection utilizing an Eluant Generator. — Separator column: IonPac AS15 (2-mm); eluant: KOH; gradient: 8 mmol/L for 6 min isocratic, then linearly to 60 mmol/L in 10 min; flow rate: 0.5 mL/min; detection: suppressed conductivity; injection volume: 1000 µL; solute concentrations: 1.08 µg/L fluoride (1), 3.35 µg/L glycolate (2) 3.86 µg/L acetate (3), 3.63 µg/L formate (4), 1.03 µg/L chloride (5), 1.17 µg/L nitrite (6), carbonate (7), 0.91 µg/L sulfate (8), 0.97 µg/L oxalate (9), 2.87 µg/L bromide (10), 0.89 µg/L nitrate (11), and 3.07 µg/L orthophosphate (12).

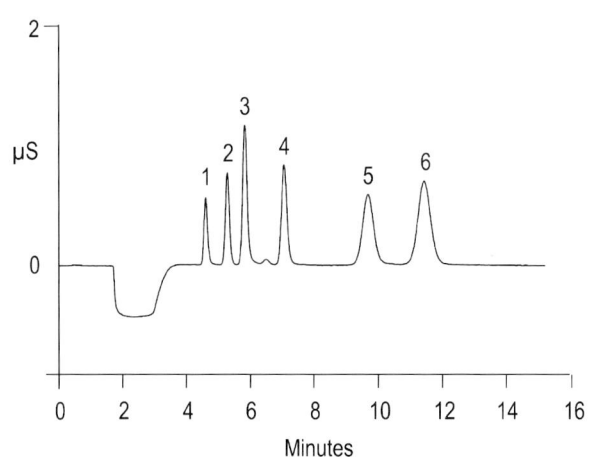

Fig. 9-40. Trace analysis of alkali- and alkaline-earth metals and ammonium by direct injection. — Separator column: IonPac CS12A (2-mm); eluant: 11 mmol/L H_2SO_4; flow rate: 0.25 mL/min; detection: suppressed conductivity; injection volume: 500 µL; solute concentrations: 0.25 µg/L lithium (1), 1 µg/L sodium (2), 1.2 µg/L ammonium (3), 2.5 µg/L potassium (4), 1.2 µg/L magnesium (5), and 2.5 µg/L calcium (6).

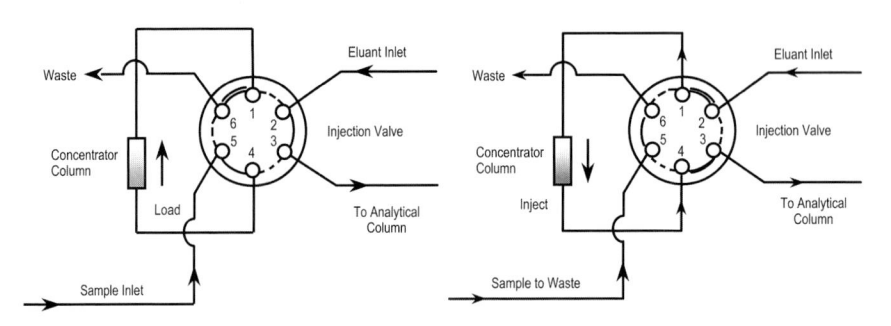

Fig. 9-41. Valve switching for applications utilizing a pre-concentration technique.

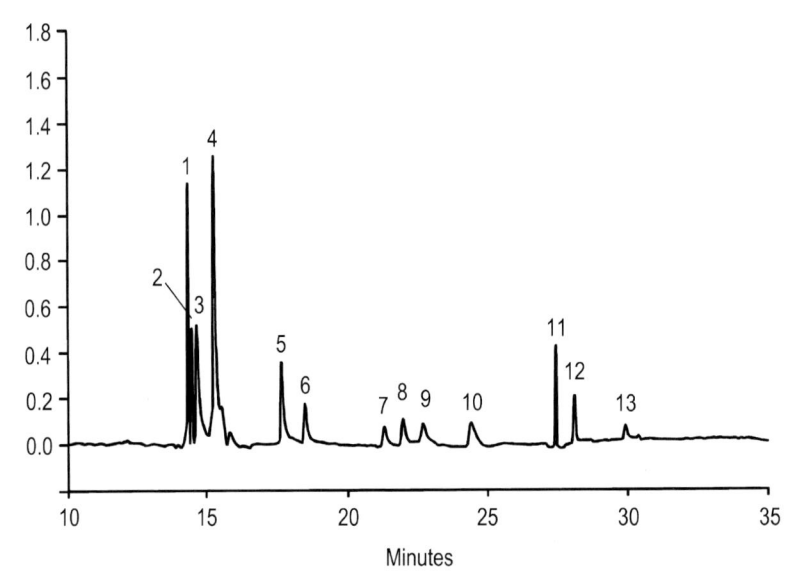

Fig. 9-42. Baseline-corrected anion chromatogram of a high-purity water sample after preconcentration. − Separator column: IonPac AS17 (2-mm); eluant: NaOH; gradient: 0.3 mmol/L for 6 min isocratic, then linearly to 15 mmol/L in 13 min, then to 60 mmol/L in 10 min; flow rate: 0.25 mL/min; detection: suppressed conductivity; pre-concentrated volume: 100 mL; solute concentrations: 10 ng/L each of fluoride (1), acetate (2), propionate (3), formate (4), chloride (5), nitrite (6), bromide (7), nitrate (8), chlorate (9), carbonate (10), sulfate (11), oxalate (12), and orthophosphate (13); (taken from [97]).

Ion chromatography can also be applied for ultra-trace analysis of transition metals. In comparison to ICP-MS it offers the advantageous ability to specify the oxidation state of metals and, moreover, to carry out multi-element determinations in the lowest ng/L range after pre-concentration. Transition metal separation is performed with ion exchangers that have defined anion and cation exchange capacities; detection is carried out photometrically after derivatization of the column effluent with PAR (see Section 4.5.3). Figure 9-44 shows a chromatogram of transition metals in the sub-µg/L range after pre-concentrating 40 mL of high-purity water. As can be seen from this chromatogram, dissolved manganese, which is regarded as a very effective corrosion indicator [98, 99], can also be detected under the given chromatographic conditions. Manganese concentrations in feed waters conditioned with ammonia or morpholine are clearly below 100 ng/L, which does not pose any problem for the ion chromatographic determination. On the other hand, manganese can also be analyzed together with alkali- and alkaline-earth metals using weak acid cation exchangers such as IonPac CS12A with suppressed conductivity detection.

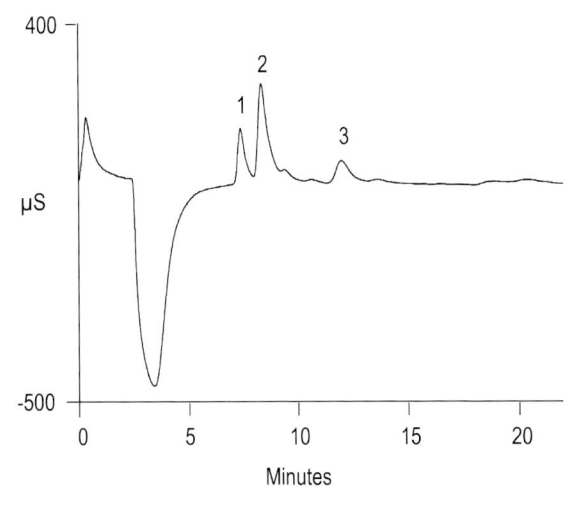

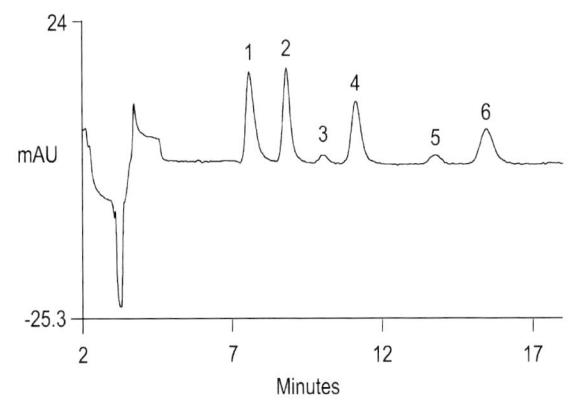

Fig. 9-43. Cation analysis of a high-purity water sample after pre-concentration. — Separator column: IonPac CS12A (2-mm); eluant: 10 mmol/L H_2SO_4; flow rate: 0.25 mL/min; detection: suppressed conductivity; pre-concentrated volume: 40 mL; solute concentrations: <10 ng/L each of sodium (1), ammonium (2), and trimethylamine (3).

Fig. 9-44. Trace analysis of transition metals in a high-purity water sample after pre-concentration. — Separator column: IonPac CS5A (2-mm); eluant: PDCA; flow rate: 0.25 mL/min; detection: photometry at 520 nm after derivatization with PAR; concentrated volume: 40 mL; solute concentrations: <75 ng/L each of the metals iron (1), copper (2), nickel (3), zinc (4), cadmium (5), and manganese (6).

A typical procedure for nuclear power stations is the monitoring of dissolved transition metals in the cooling circle (310 °C) of a light-water reactor. Cobalt is of special interest, because naturally occurring cobalt-59 is the source for activated cobalt-60 which, in turn, represents the largest contribution to the radiation outside the reactor core in water-cooled reactors. Reactor coolants are a perfect transport medium for corrosion products as well as for activated substances in dissolved and particulate form. They can be investigated for their transition metal content by means of ion chromatography [100]. The analysis of dissolved elemental cobalt down to the level of 1 ng/L is of great importance. To achieve this sensitivity spectrophotometrically, 1 L coolant has to be pre-concentrated. For the analysis of non-dissolved components, 1500 L coolant has to be directed through a membrane filter. Prior to ion chromatographic analysis, those particles are solubilized via a potassium bisulfate digest. Looking for a much more sensitive detection method for cobalt that would reduce the time needed for pre-concentration chemiluminescence methods were investigated for their suitability in cobalt analysis. Based on a method developed by Boyle et al. [101] of post-column derivatization of cobalt with Luminol (5-amino-2,3-dihydro-1,4-phthalazindione) and hydrogen peroxide, where the chemiluminescence yield is proportional to the metal concentration over several orders of magnitude [102], Jones et al. [103] developed a separation technique compatible with this kind of post-column derivatization. Instead of pyridine-2,6-dicarboxylic acid, whose cobalt complexes are too stable to allow the release of cobalt ions for catalyzing the Luminol/peroxide reaction, Jones et al. used an eluant based on lactic acid. The derivatization reagent is prepared as follows:

0.1 g Luminol and 3 g boric acid are dissolved in 500 mL high-purity de-ionized water containing 5 mL hydrogen peroxide. The pH of this solution is adjusted to 12 with KOH.

Figure 9-45 shows a typical chromatogram of cobalt in a pressurized water reactor coolant by means of chemiluminescence analysis. The specificity of this method is demonstrated by the absence of other component signals. The detection limit of this method is around 0.3 pg and, thus, three orders of magnitude more sensitive than the spectrophotometric method via derivatization with PAR.

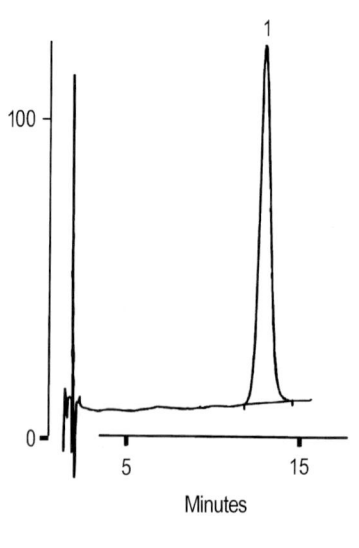

Fig. 9-45. Chemiluminescence analysis of cobalt at ultra-trace level in a pressurized water reactor coolant. — Separator column: IonPac CG2; eluant: 0.1 mol/L lactic acid, pH 3.8 with ammonium hydroxide; flow rate: 1.2 mL/min; detection: chemiluminescence detector with a photomultiplier tube (Thorn EMI Type 6097, spectral range: 330-680 nm) after derivatization with Luminol; flow cell volume: 200 µL; solute concentration: 15 pg cobalt (1); (taken from [100]).

High-purity water can also be analyzed for silicate by means of ion chromatography. Because silicate detection also involves a post-column derivatization technique (see Sections 5.5 and 7.2.1), a conventional anion exchanger can be used for the separation. However, carbonate eluants cannot be employed, because silicate will not be retained on a concentrator column in the carbonate form. For this reason, a mixture of boric acid and sodium hydroxide is used for the elution of silicate. Figure 9-46 shows a typical chromatogram of silicate at trace level after pre-concentrating 20 mL of high-purity water. Theoretically, lower detection limits should be obtained when pre-concentrating a larger volume. However, when plotting the peak area as a function of the pre-concentrated volume (Fig. 9-47), it is clear that silicate is no longer completely retained at pre-concentration volumes above 25 mL.

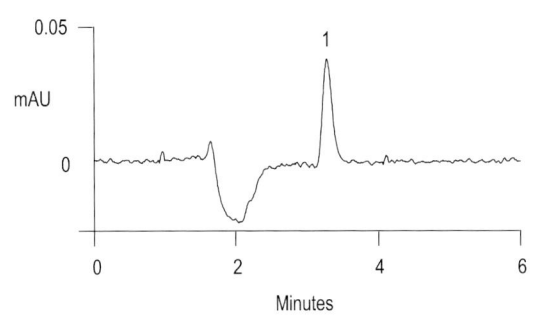

Fig. 9-46. Trace analysis of silicate in a high-purity water sample after pre-concentration. – Separator column: IonPac AS4A-SC (2-mm); eluant: 10 mmol/L boric acid + 10 mmol/L NaOH; flow rate: 0.25 mL/min; detection: photometry at 410 nm after derivatization with molybdate; reaction coil volume: 375 µL; pre-concentrated volume: 20 mL; solute concentration: 50 ng/L silicate (1).

For trace analysis of borate in high-purity water a special pre-concentrator column, TBC-1 (Trace Borate Concentrator), was developed. The resin material contains *cis*-diol groups, on which borate is retained via complexation. When pre-concentrating large volumes, detection limits around 100 ng/L are achieved. A typical chromatogram of a 500-ng/L borate standard, obtained after pneumatic pre-concentration of 160 mL of the respective standard solution, was already shown in Fig. 5-7 (see Section 5.5). An IonPac ICE-Borate column, which is an IonPac ICE-AS1 conditioned for borate analysis, was used as a separator.

Another application of ion-exclusion chromatography is the trace analysis of carbonic acid in the water-steam cycle of conventional and nuclear power plants, which has so far been a failure because of poor sensitivity and contamination risk by atmospheric CO_2. Gilbert et al. [104] overcame the contamination problem by preparing eluants and standards in a glove box modified for this purpose, which was connected to a two-stage air purification system. In such a glove box eluants and standards can be kept in special containers (Nalgene PETG bottles) under inert gas. Eluants and standards are directed into the chromatograph via stainless-steel tubing, as tubing made of polymers is not completely gas-tight. Samples have to be taken as quickly as possible to minimize the contamination risk. Thus, Tygon tubing is used for connecting the sampling locations to sample containers made of glass, which are carefully rinsed with the sample. Carbonate separation is carried out on a totally sulfonated cation exchanger with high-purity de-ionized water as a mobile phase, so that using a suppressor system as a preparation step for conductivity detection is obsolete. To determine carbonate in the lower µg/L range 25-40 mL water must be pre-concentrated on an anion exchanger. After pre-concentration, the anions are transferred onto the analytical separator column with 6 mmol/L sodium octanesulfonate. Under the given chromatographic conditions, short-chain fatty acids such as formic acid and acetic acid are not completely separated. However, the carbonate peak can be quantitated free of any interferences. According to Gilbert et al. the carbonate detection limit based on a pre-concentrated volume of 40 mL is 0.9 µg/L.

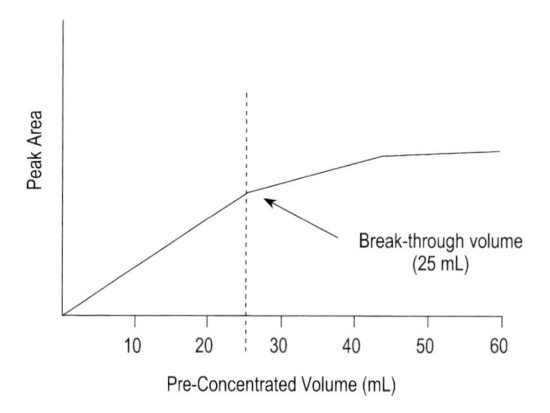

Fig. 9-47. Graphic determination of the break-through volume in silicate pre-concentration.

9.2.1
Analysis of Conditioned Waters

In addition to high-purity water analysis, ion chromatography is the primary method used for analyzing conditioned waters, which is a lot more difficult, because the concentrations of the chemicals used for conditioning exceed the analyte concentrations by many orders of magnitude.

Typical conditioning agents include amines such as ammonium, ethanolamines, morpholine, and others with concentrations in the lower mg/L range. In some cases, mixtures of various amines are used for corrosion inhibition, which renders the analysis of corrosive ions more difficult. Two different strategies can be followed to increase resolution between matrix components and analytes with their disparate concentration levels. On one hand, special stationary phases such as the crown ether modified IonPac CS15 [105] or the high-capacity cation exchanger CS16 can be used, with which a significantly better resolution between sodium and ammonium is obtained. A similar effect can be achieved with a column switching technique, in which the order of the two separator columns with different selectivities can be changed during the run [106] (Fig. 9-48). Due to this change of order, magnesium and calcium elute between sodium and ammonium, thus allowing concentration differences up to 1:50,000 between analyte- and matrix ions to be detected. The change in elution order of mono- and di-valent cations that results from this column switching technique is illustrated in Fig. 9-49.

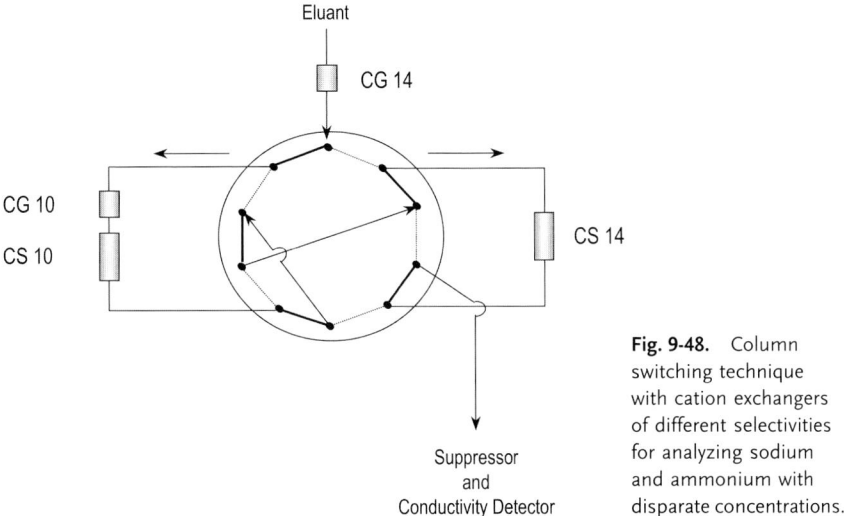

Fig. 9-48. Column switching technique with cation exchangers of different selectivities for analyzing sodium and ammonium with disparate concentrations.

A characteristic example for trace analysis of sodium in water containing ammonia on IonPac CS15 is illustrated in Fig. 9-50. Sodium concentrations around 1 µg/L can only be detected when injecting a large volume of 1000 µL. The high resolution between sodium and ammonium on this stationary phase allows sodium quantification even in the presence of a 10,000-fold excess of ammonium.

The same stationary phase can also be used for the trace analysis of sodium in the presence of monoethanolamine (Fig. 9-51), although the latter elutes a little bit earlier than ammonium. If monoethanolamine and ammonium are present in comparable concentrations, both compounds can be separated to baseline. In the given example, however, the monoethanolamine concentration is so high, that ammonium (as an impurity) merely elutes as a shoulder peak on the slope of the monoethanolamine peak. With the column switching technique described above, this problem can be solved by using a column combination of an IonPac CS12A weak acid cation exchanger and an IonPac CS10 totally sulfonated cation exchanger (Fig. 9-52). Elution is carried out with dilute methanesulfonic acid. While divalent cations only pass the CS12A column and, consequently, reach the conductivity cell first, monovalent cations have to pass both columns due to the change in the column order. This leads to a significant

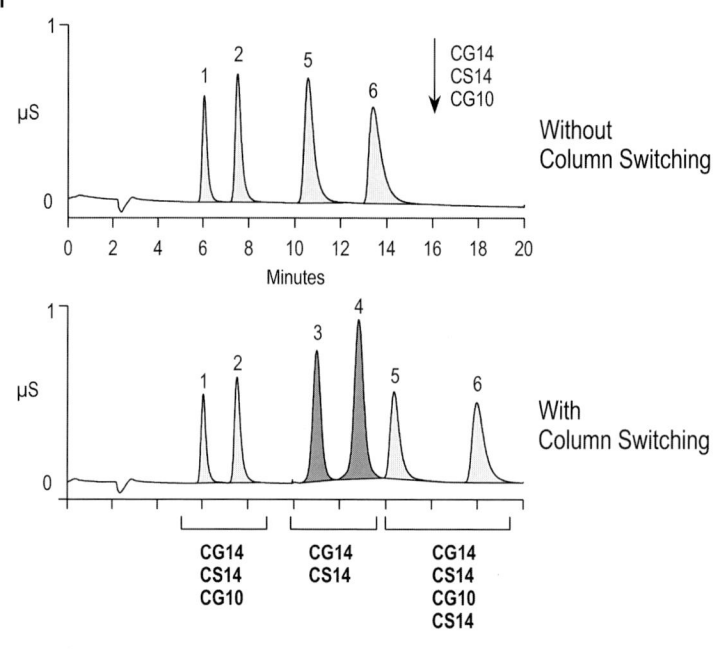

Fig. 9-49. Change in elution order of mono- and di-valent cations when applying a column switching technique; peaks: (1) lithium, (2) sodium, (3) magnesium, (4) calcium, (5) ammonium, and (6) potassium.

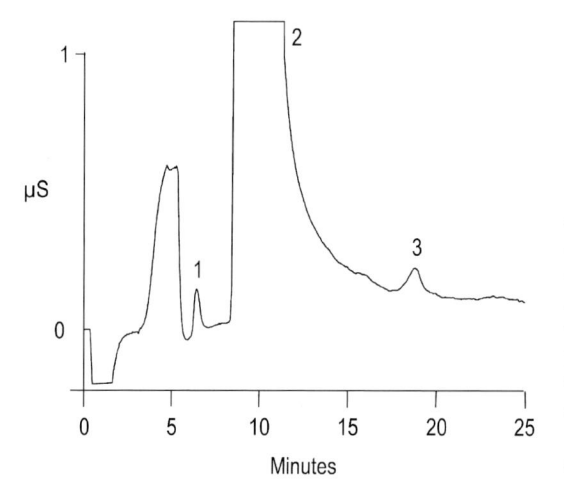

Fig. 9-50. Trace analysis of sodium in excess of ammonium on a crown ether modified stationary phase. — Separator column: IonPac CS15 (2-mm); column temperature: 40 °C; eluant: 5 mmol/L H_2SO_4 — acetonitrile (91:9 v/v); flow rate: 0.3 mL/min; detection: suppressed conductivity; injection volume: 1000 µL; solute concentrations: 1 µg/L sodium (1), 10 mg/L ammonium (2) and calcium (3, not quantitated).

increase in resolution between ammonium and monoethanolamine. Thus, small amounts of ammonium can be determined in presence of a large excess of monoethanolamine. (This application can be transferred to the 2-mm format without any problem if an IonPac CS11 is used instead of an IonPac CS10.)

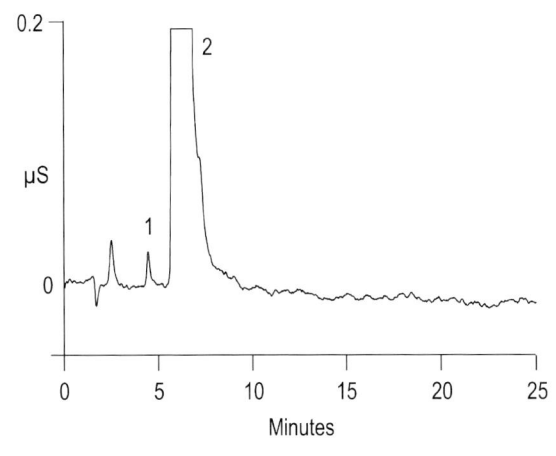

Fig. 9-51. Trace analysis of sodium in presence of a high excess of monoethanolamine on a crown ether modified stationary phase. — Chromatographic conditions: see Fig. 9-50; injection volume: 25 µL; solute concentrations: 10 µg/L sodium (1) and 200 mg/L monoethanolamine (2).

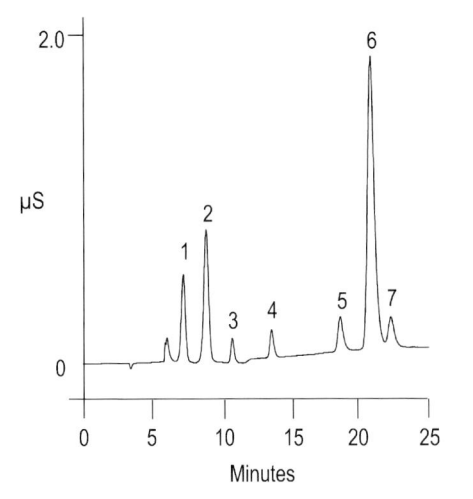

Fig. 9-52. Trace analysis of sodium and ammonium in presence of a large excess of monoethanolamine using a column switching technique. — Separator columns: IonPac CG12A/CS12A and CS10; column switching valve is switched after 6 min; eluant: 24 mmol/L methanesulfonic acid; flow rate: 1 mL/min; detection: suppressed conductivity; injection volume: 25 µL; solute concentrations: 250 µg/L magnesium (1), 500 µg/L calcium (2), 50 µg/L lithium (3), 200 µg/L sodium (4), 250 µg/L ammonium (5), 10 mg/L monoethanolamine (6), and 500 µg/L potassium (7).

For the conditioning of feed water, ammonia is often used in combination with morpholine; morpholine acts as corrosion inhibitor and ammonia is used for adjusting pH (Fig. 9-53). Both compounds can be separated from common inorganic cations on an IonPac CS14 weak acid cation exchanger. The tailing of the morpholine peak is reduced by adding 5% (v/v) acetonitrile to the mobile phase.

Trace analysis of anions in amine-containing waters is relatively simple as long as the desired sensitivity can be obtained by direct injection of a large sample volume. Depending on the type of anion to be analyzed and the existing instrument hardware, isocratic techniques, step gradients, or gradient elution techniques are used. The application of gradient elution is ideal; nowadays it is only a matter of instrument programming if an Eluant Generator™ is available, because early eluting short-chain organic acids can be analyzed together with mineral acids at very low concentrations. Figure 9-54 illustrates this, taking anion analysis in the presence of morpholine as an example. An IonPac AS11 hydroxide-selective anion exchanger in the 2-mm format was used as a stationary phase. The relatively strong baseline drift is attributed to the conventional eluant preparation and can be diminished by using an Eluant Generator™. Another advantage of the hydroxide eluant is the significantly higher sensitivity in comparison to a carbonate/bicarbonate eluant mixture.

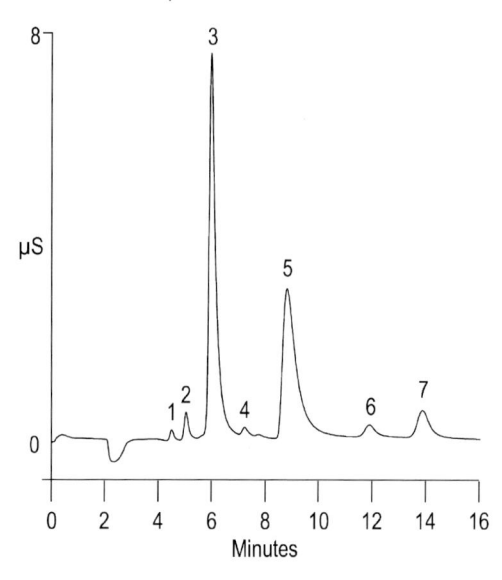

Fig. 9-53. Trace analysis of cations in presence of a large excess of ammonium and morpholine. – Separator column: IonPac CS14 (2-mm); eluant: 8 mmol/L methanesulfonic acid – acetonitrile (95:5 v/v); flow rate: 0.25 mL/min; detection: suppressed conductivity; injection volume: 1 mL pre-concentrated on IonPac CG14 (2-mm); solute concentrations: 0.5 µg/L lithium (1), 2 µg/L sodium (2), 150 µg/L ammonium (3), 2 µg/L potassium (4), 2000 µg/L morpholine (5), 2 µg/L magnesium (6), and 10 µg/L calcium (7).

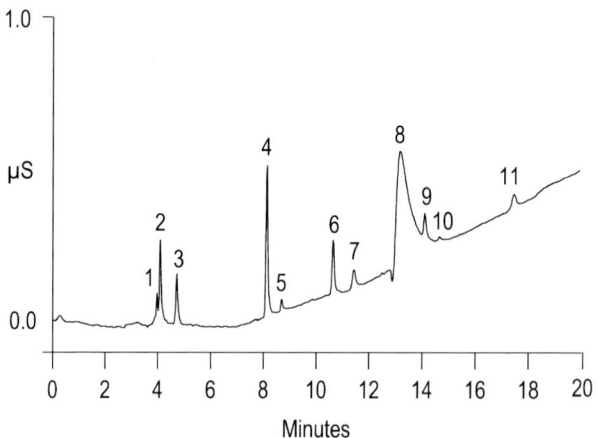

Fig. 9-54. Trace analysis of anions in presence of morpholine utilizing direct injection. – Separator column: IonPac AS11 (2-mm); eluant: NaOH; gradient: 0.5 mmol/L for 2.5 min isocratic, then linearly to 5 mmol/L in 3.5 min, then to 26 mmol/L in 20 min; flow rate: 0.5 mL/min; detection: suppressed conductivity; injection volume: 750 µL; solute concentrations: 0.51 µg/L fluoride (1), 3.2 µg/L acetate (2), 1.7 µg/L formate (3), 3.3 µg/L chloride (4), 0.26 µg/L nitrite (5), 2.1 µg/L nitrate (6), unknown component (7), carbonate (8), 1.1 µg/L sulfate (9), 0.47 µg/L oxalate (10), and unknown component (11).

In contrast, poor recoveries are obtained when pre-concentrating anions in the presence of amines, especially for early eluting anions such as fluoride, acetate, formate, and chloride. This effect is demonstrated with an anion standard (Chromatogram A in Fig. 9-55), which was prepared with pure de-ionized water and chromatographed on IonPac AS10 (2-mm format) with a hydroxide eluant under isocratic conditions after pre-concentration of a 47 mL sample. As can be seen from this chromatogram, fluoride and chloride can be detected under these conditions at the sub-µg/L level. If the same experiment is carried out with an anion standard that additionally contains 8 mg/L monoethanolamine (see chromatogram B in Fig. 9-55), fluoride and chloride are not retained in the concentrator column because of the matrix and, consequently, are not observed in the resulting chromatogram. The sulfate and nitrate signals are also much smaller as compared with Chromatogram A. This means that reliable data for the pre-concentration of amine-containing waters can only be obtained after matrix elimination.

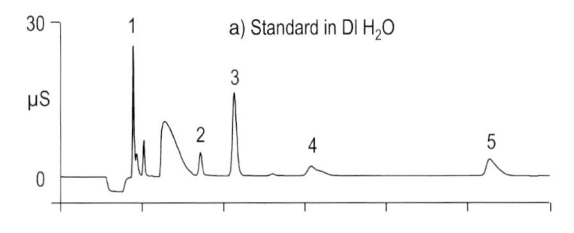

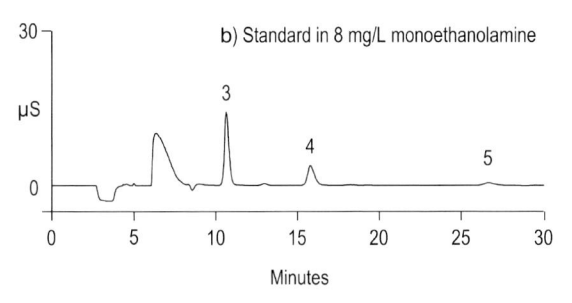

Fig. 9-55. Trace analysis of anions in amine-containing waters without preceding matrix elimination. – Separator column: IonPac AS10 (2-mm); eluant: 97 mmol/L NaOH; flow rate: 0.25 mL/min; detection: suppressed conductivity; concentrator column: IonPac AC10; pre-concentrated volume: 43 mL; solute concentrations: (A) 0.27 µg/L fluoride (1), 0.37 µg/L chloride (2), 1.26 µg/L sulfate (3), 1.93 µg/L orthophosphate (4), and 1.65 µg/L nitrate (5), (B) 1.18 µg/L sulfate (3), 2.3 µg/L orthophosphate (4), and 0.22 µg/L nitrate (5).

AutoNeutralization [107] is a suitable sample preparation technique for samples of this kind. The term AutoNeutralization describes an automated neutralization of acidic or basic samples with a special membrane suppressor to analyze traces of anions in strongly basic samples and traces of cations in strongly acidic samples, respectively. In the present case, the cations of the sample are replaced by hydronium ions, the sample being neutralized in that way. The schematic setup of an AutoNeutralization unit – coupled to an ion chromatograph – is depicted in Fig 9-245 in Section 9-11. The positive effects of such a sample preparation are revealed in Fig. 9-56, which compares the chromatograms of the two anion standards with their different matrices when applying AutoNeutralization. The recovery rates for the five anions in a matrix of 8 mg/L monoethanolamine are summarized in Table 9-6; some are above 100%.

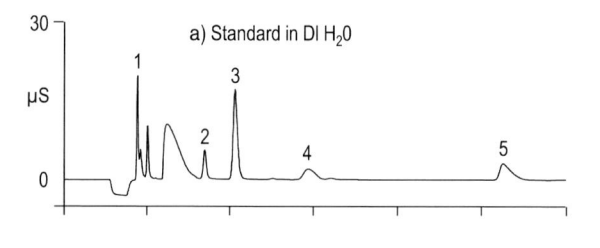

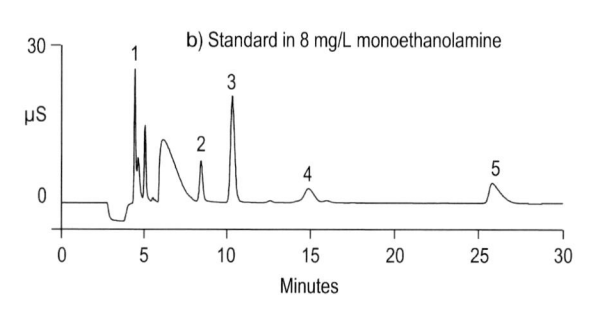

Fig. 9-56. Trace analysis of anions in amine-containing waters with preceding matrix elimination. – Chromatographic conditions: see Fig. 9-55; solute concentrations: (A) 0.62 µg/L fluoride (1), 0.75 µg/L chloride (2), 3.90 µg/L sulfate (3), 7.39 µg/L orthophosphate (4), and 2.46 µg/L nitrate (5), (B) 0.85 µg/L fluoride (1), 1.07 µg/L chloride (2), 4.54 µg/L sulfate (3), 7.64 µg/L orthophosphate (4), and 2.86 µg/L nitrate (5).

Table 9-6. Recovery rates of anions in presence of 8 mg/L monoethanolamine after matrix elimination via AutoNeutralization.

Analyte	Spiked amount [µg/L]	Measured amount* [µg/L]	Recovery rate [%]
Fluoride	0.62	0.72 ± 0.07	116
Chloride	0.68	0.94 ± 0.13	137
Sulfate	3.60	3.98 ± 0.45	110
Orthophosphate	6.92	6.70 ± 0.70	97
Nitrate	2.22	2.38 ± 0.28	107

*) for $n = 6$

9.2.2
Cooling Water Analysis

Figure 9-57 shows a chromatogram of a cooling water sample; suppressed conductivity detection was used. A Metrosep Anion Dual 2 anion exchanger was selected as the stationary phase, with a carbonate/bicarbonate mixture as an eluant. The sample to be analyzed was diluted 1:10 with de-ionized water.

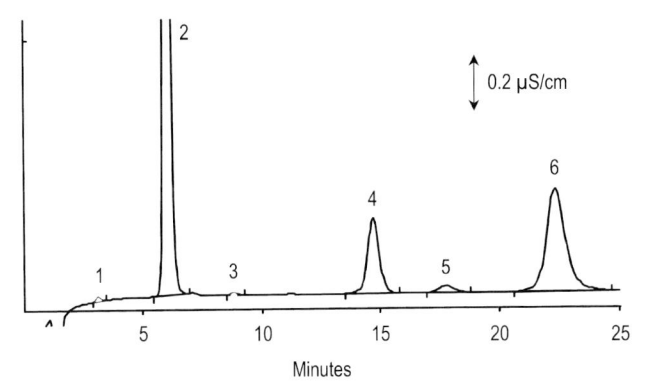

Fig. 9-57. Anion analysis of cooling water. — Separator column: Metrosep Anion Dual 2; eluant: 1.3 mmol/L Na_2CO_3 + 2 mmol/L $NaHCO_3$; flow rate: 0.8 mL/min; detection: suppressed conductivity; injection: 20 µL sample (1:10 diluted); solute concentrations: 0.08 mg/L fluoride (1), 75.3 mg/L chloride (2), 0.26 mg/L nitrite (3), 18.3 mg/L nitrate (4), 4.3 mg/L orthophosphate (5), and 31.7 mg/L sulfate (6); (taken from [108]).

Trace analysis of anions in the primary cooling cycle of nuclear power plants is more difficult, because this water contains boric acid in the g/L range. Boric acid acts as a soluble neutron absorber for long-term reactivity control of pressurized water reactors. Thus, matrix elimination is equally important in this case. After the pre-concentration step the concentrator column is rinsed with high-purity de-ionized water to remove boric acid from the interstitial volume of the concentrator column. Only then can the concentrator column be put inline with the analytical separator column to elute the pre-concentrated anions. The column switching for this application is schematically depicted in Fig. 9-58. It is extremely important to use de-ionized water of the highest possible quality for rinsing the concentrator column. For further purification, the water is directed through an Anion Trap Column (ATC) to exclude contaminations as much as possible. Despite matrix elimination, the concentration of boric acid that gets onto the analytical separator column is still very high, so a high-capacity anion exchanger is used as a stationary phase. Optimum selectivity for this type of analysis is provided by IonPac AS10, on which borate and chloride can be separated to baseline. The chromatogram of a high-purity water sample containing around 2% boric acid is shown in Fig. 9-59. Fluoride, chloride, and sulfate were detected at the lowest µg/L level. Without matrix elimination, the boric acid peak preceding chloride would make it impossible to quantitate the chloride signal.

Another interesting application is the trace analysis of sequestering agents in cooling water. Typically, these products are mixtures of polycarboxylic acids and polyphosphonic acids, which are added to cooling waters as corrosion inhibitors.

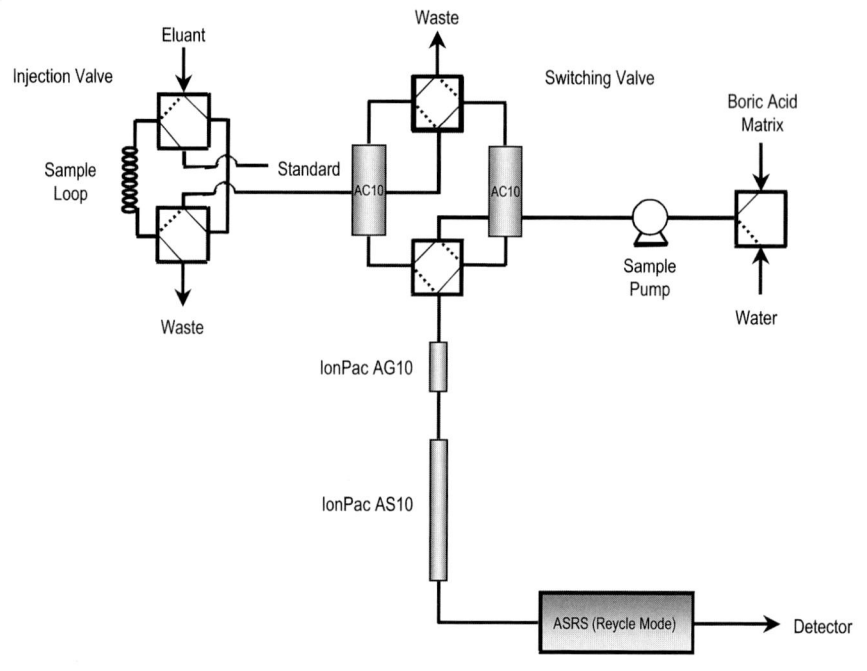

Fig. 9-58. Column switching for trace analysis of anions in presence of high boric acid concentrations.

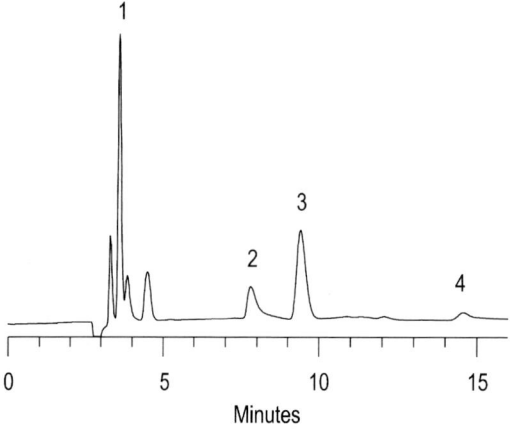

Fig. 9-59. Trace analysis of anions in the primary cooling circuit of a nuclear power plant after matrix elimination. – Separator column: IonPac AS10; eluant: 85 mmol/L NaOH; flow rate: 1 mL/min; detection: suppressed conductivity; concentrator column: IonPac AC10; pre-concentrated volume: 12 mL; solute concentrations: fluoride (1), boric acid (2), 5 µg/L chloride (3), and sulfate (4).

When using these conditioning agents, calcium bicarbonate is stabilized at thermally stressed positions in the system, and scaling on metal surfaces is prevented.

Polyphosphonic acids are separated on a special latexed anion exchanger and detected photometrically after complexation with iron(III) nitrate [109] (see Section 3.8.2). With this method qualitative analysis of inorganic and organic phosphates in conditioning agents is possible. As an example, Fig. 9-60 shows the chromatogram of a commercial product. As can be seen from this chromatogram, this product is a multi-component mixture with two major constituents and several impurities. Because the concentration of these products in the cooling water circuit is in the upper µg/L range, the sensitivity of this method is insufficient for direct injection of a 50-µL sample. Therefore, a pre-concentration technique has to be employed for determining low concentrations of these compounds. The chromatogram of cooling tower circulation water in Fig. 9-61 is obtained when pre-concentrating a 2-mL sample. The polyphosphonic acid therein is clearly detected. The other two signals visible in the chromatogram are chloride and sulfate, which are present at much higher concentration and are also enriched. Because both anions are complexed by iron(III), they appear as large signals in the chromatogram. If the polyphosphonic acid being used co-elutes with one of the inorganic anions, phosphorus-specific detection [110] will eliminate this interference. Hence, phosphorus-specific detection also facilitates polyphosphonate quantification, as it hydrolyzes to orthophosphate and the concentration of the polyphosphonate component does not have to be determined via external calibration with appropriately pure reference substances.

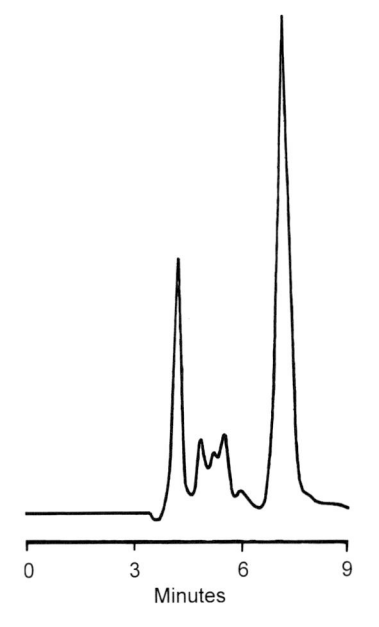

Fig. 9-60. Analysis of a cooling water conditioning agent. – Separator column: IonPac AS7; eluant: 0.03 mol/L HNO_3; flow rate: 0.5 mL/min; detection: photometry at 330 nm after reaction with iron(III) nitrate; injection: 50 µL sample (1:1000 diluted).

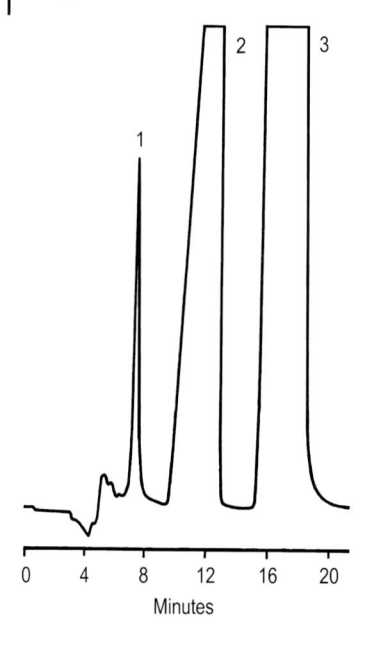

Fig. 9-61. Analysis of a cooling water conditioning agent in a cooling tower circulation water. – Chromatographic conditions: see Fig. 9-60; pre-concentrated volume: 2 mL; peaks: (1) polyphosphonate, (2) chloride, and (3) sulfate.

9.2.3
Flue Gas Scrubber Solutions

Today, a predominant number of fossil fuel power plants in Germany are equipped with a multi-stage flue gas scrubber system consisting of denitrification, de-dusting and desulfurization. Thus, power plant chemistry has gained a new and wide field of activity and has to make a contribution to the satisfactory operation of such systems [111]. In Germany, more than 90% of the desulfurization plants operate according to the lime- or limestone washing technique. The final product is gypsum, which in many cases is directed to a material utilization. Routine monitoring of a flue gas desulfurization is more or less limited to the control of some chemical process parameters such as pH value, concentration of solids, and sulfite content. As in the monitoring of the water-steam cycle, the control of a few key parameters is sufficient to monitor flue gas desulfurization. In addition, product monitoring (e.g., the gypsum quality) and chemical monitoring of the wastewater processing systems that exist in many power plants have been introduced. Today, the chemical process of wet flue gas desulfurization with lime or limestone is considerably matured. The washing liquids contain a number of ionic components, which can be analyzed by means of ion chromatography. In flue gas desulfurization based on gypsum, sulfite is oxidized to sulfate, which can be determined together with mineral acids in the same run, as shown in Fig. 9-62. Experience has shown that sulfate cannot be analyzed reproducibly with the method originally specified by the authorities (precipitation with barium sulfate under acidic conditions). The chemical composition of the flue gas scrubber solution differs greatly depending on the type of power plant and is characterized, above all, by the various types of furnaces and the operation mode of the DENOX system. This is substantiated by the fact that sulfur-nitrogen compounds formed from nitrous oxides and sulfur dioxide in the absorber cycle also appear in dissolved form in scrubber solutions. The formation of hydroxylamine-disulfonic acid is mentioned as an example:

$$2NO_2 + H_2O \rightarrow HNO_2 + HNO_3 \tag{308}$$

$$HNO_2 + 2SO_2 + H_2O \rightarrow HO\text{-}N(SO_3H)_2 \tag{309}$$

In subsequent reactions other SN compounds are formed. Via hydrolysis, those compounds can release sulfate, so a reliable sulfate determination can only be performed by ion chromatography.

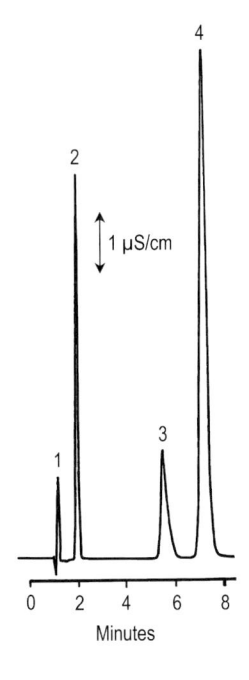

Fig. 9-62. Analysis of inorganic anions in a flue gas scrubber solution. — Separator column: IonPac AS4; eluant: 2.2 mmol/L Na_2CO_3 + 2.8 mmol/L $NaHCO_3$; flow rate: 2 mL/min; injection: 50 μL sample (1:100 diluted); solute concentrations: 110 mg/L fluoride (1), 650 mg/L chloride (2), 1300 mg/L nitrate (3), and 4000 mg/L sulfate (4).

Anion exchange chromatography can also be used for the analysis of the toxic chromium(VI) in such scrubber solutions, but their high sulfate loading requires the application of UV detection at 364 nm. Alternatively, chromate can be derivatized with 1,5-diphenylcarbazide; the formed complex is detected photometrically at 520 nm.

Today, the determination of sodium, potassium, and magnesium in flue gas scrubber solutions is carried out with a weak acid cation exchanger utilizing suppressed conductivity detection (Fig. 9-63). However, ammonium present at low concentration cannot be determined in the same run under these conditions. To detect ammonium selectively, it is appropriate to apply the derivatization technique using o-phthaldialdehyde with subsequent fluorescence detection, which has been described repeatedly. The two detection methods can, of course, be combined.

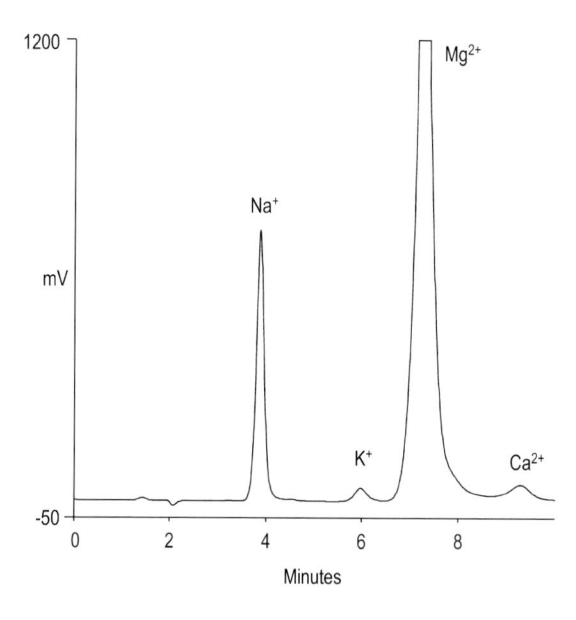

Fig. 9-63. Cation analysis in a flue gas scrubber solution. — Separator column: IonPac CS12; eluant: 20 mmol/L methanesulfonic acid; flow rate: 1 mL/min; detection: suppressed conductivity; injection: 25 µL flue gas scrubber solution.

As already mentioned above, sulfur compounds other than inorganic sulfate may be detected in scrubber solutions from flue gas desulfurization. For example, dithionate, $S_2O_6^{2-}$, is formed via recombination of two bisulfite radicals and can also be separated by ion chromatography [112]. Dithionate exhibits a strong affinity towards the stationary phase of an anion exchanger; hence, either special stationary phases with extremely hydrophilic ion-exchange groups or high electrolyte concentrations must be used for its elution. In both cases, sensitive detection by measuring electrical conductivity is extremely difficult. Because dithionate is UV-transparent, there is no alternative to this detection method. Ion-pair chromatography is commonly employed for this application, because this method allows the retention behavior of the analyte species to be influenced by a variety of experimental retention-determining parameters. As dithionate is a surface-inactive substance, tetrabutylammonium hydroxide is used as an ion-pair reagent. Organic solvents such as acetonitrile or methanol serve to adjust the polarity of the mobile phase. Figure 9-64 shows an example chromatogram for the determination of dithionate in a flue gas scrubber sample. For the analysis of polythionates, $S_nO_6^{2-}$ ($n > 2$), only the acetonitrile content in the mobile phase must be enhanced.

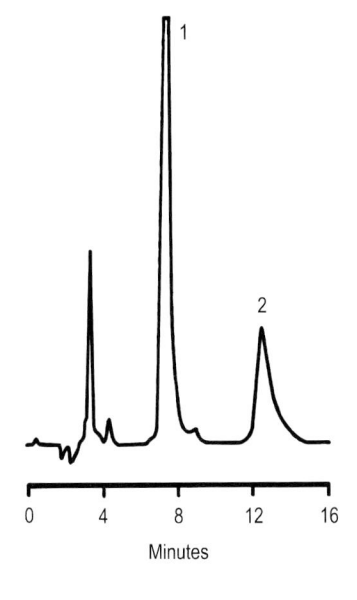

Fig. 9-64. Analysis of dithionate in a scrubber solution from flue gas desulfurization. — Separator column: IonPac NS1 (10-µm); eluant: 2 mmol/L TBAOH + 1 mmol/L Na_2CO_3 / acetonitrile (85:15 v/v); flow rate: 1 mL/min; detection: suppressed conductivity; injection: 50 µL sample (1:100 diluted); peaks: (1) sulfate and (2) dithionate.

Under similar chromatographic conditions, it is possible to detect the class of sulfur-nitrogen compounds [113], which includes compounds such as amidosulfonic acid, iminodisulfonic acid, hydroxylaminedisulfonic acid, hydroxylaminetrisulfonic acid, and nitrilotrisulfonic acid, all of which were clearly identified in the scrubber solution from flue gas desulfurization. While amidosulfonic acid elutes close to chloride, nitrilotrisulfonic acid as a trivalent species exhibits a much higher retention than dithionate. The appearance of these compounds is not only of interest for the analytical chemist but also of practical relevance for the desulfurization process. Some of these compounds inhibit the oxidation from sulfite to sulfate, so that non-oxidized sulfite remains in the absorber cycle; in turn, this changes the degree of desulfurization for the worse. With the commissioning of a DENOX system in series with a desulfurization system, NO_x intrusion was diminished and the desulfurization performance increased. The reason for this is the decreased formation of oxidation inhibitors, which can be attributed to nitrous oxides. Ion chromatography may also contribute analytically to the monitoring of the flue gas denitrification process. The most important parameter — the concentration ratio between nitrite and nitrate — can be established within five minutes, as revealed in Fig. 9-65.

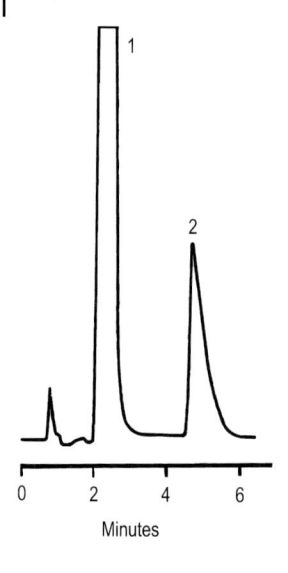

Fig. 9-65. Analysis of nitrite in a scrubber solution from flue gas denitrification. — Separator column: IonPac AS3; eluent: 2.8 mmol/L $NaHCO_3$ + 2.2 mmol/L Na_2CO_3; flow rate: 2.3 mL/min; detection: suppressed conductivity; injection: 50 µL sample; solute concentrations: 197 mg/L nitrite (1) and 14.9 mg/L nitrate (2).

9.2.4
Analysis of Chemicals

As outlined in the previous section, a number of amines are used for conditioning and pH adjustment of cooling waters, which can also contain corrosive ions. Because these amines are added at the mg/L level, their purity also has to be investigated. The technique of Autoneutralization has proved to be suitable for identifying anions; thus, the correlative amine is neutralized prior to its analysis. As an example, Fig. 9-66 shows the analysis of anions in 40% monoethanolamine by direct injection of a 100-µL sample after two neutralization cycles [114]. An IonPac AS12A anion exchanger was used as a stationary phase and a carbonate/bicarbonate mixture as an eluant. If a pre-concentration technique is used instead of direct injection, those anions can also be detected at a sub-µg/L level in such matrix.

Ion chromatography is also suitable for determining the purity of boric acid used in the primary cooling circuit of nuclear power plants. The analysis of alkali- and alkaline-earth metals is usually carried out by direct injection, with a weak acid cation exchanger. Due to the low dissociation of boric acid, samples are allowed to contain relatively large boric acid concentrations. Figure 9-67 illustrates this with a chromatogram of a 10% boric acid solution, in which mono- and di-valent cations were detected at the lowest µg/L level. In contrast, to reach this sensitivity, a pre-concentration technique has to be employed for anion analysis on IonPac AS14 (Fig. 9-68). To minimize interferences by the borate matrix, elution is not carried out with a carbonate/bicarbonate mixture but with sodium tetraborate in this case.

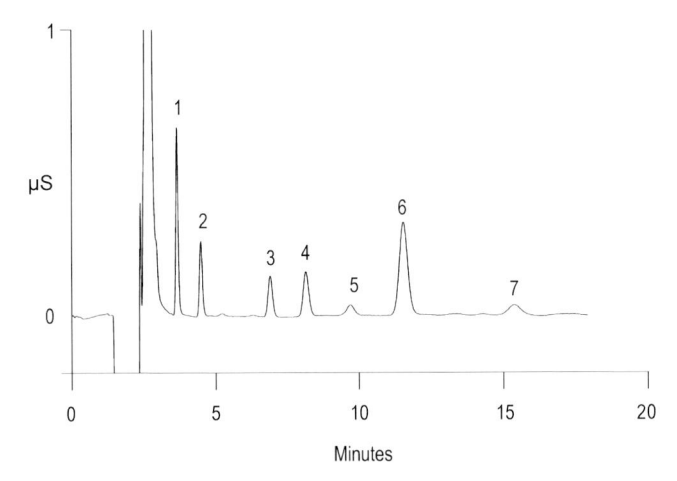

Fig. 9-66. Trace analysis of anions in 40% monoethanolamine. – Separator column: IonPac AS12A; eluant: 2.7 mmol/L Na_2CO_3 + 0.3 mmol/L $NaHCO_3$; flow rate: 1.5 mL/min; detection: suppressed conductivity; injection: 100 µL sample after two neutralization cycles; solute concentrations: 0.14 mg/L chloride (1), 0.12 mg/L nitrite (2), 0.13 mg/L bromide (3), 0.18 mg/L chlorate (4), 0.14 mg/L orthophosphate (5), 0.34 mg/L sulfate (6), and 0.09 mg/L oxalate (7).

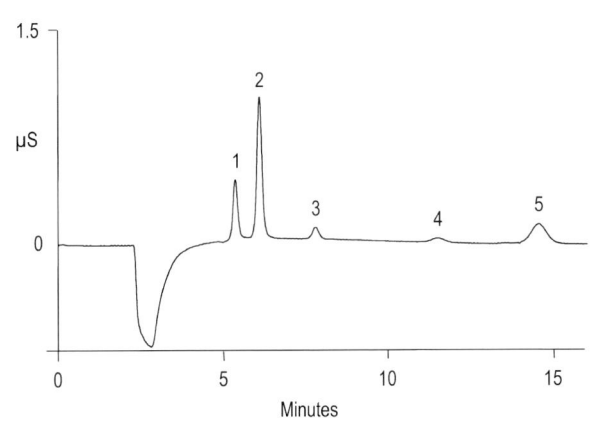

Fig. 9-67. Trace analysis of cations in high-purity boric acid. – Separator column: IonPac CS12A; eluant: 20 mmol/L methanesulfonic acid; flow rate: 1 mL/min; detection: suppressed conductivity; injection: 500 µL boric acid solution (10 g/100 mL DI water); solute concentrations: 9.5 µg/L sodium (1), 33.8 µg/L ammonium (2), 3.4 µg/L potassium (3), 1.4 µg/L magnesium (4), and 12 µg/L calcium (5).

Traces of aluminum and arsenate at the lowest µg/L level were also detected by ion chromatography in high-purity boric acid. Aluminum is separated by cation exchange chromatography on IonPac CS5A and detected photometrically after derivatization with Tiron (see Section 4.5.3). Based on an injection volume of 500 µL of a 4% boric acid solution, 5 µg/L aluminum can be detected in this

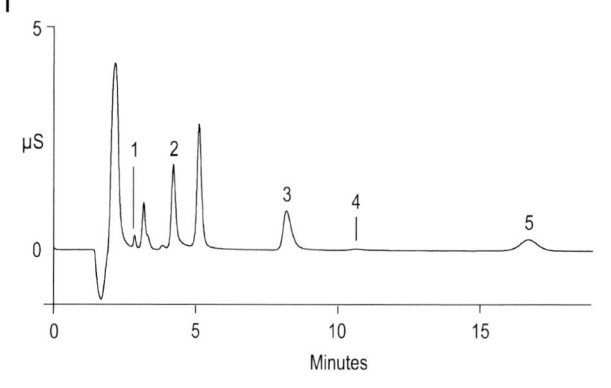

Fig. 9-68. Trace analysis of anions in high-purity boric acid. — Separator column: IonPac AS14; eluant: 9 mmol/L sodium tetraborate; flow rate: 1 mL/min; detection: suppressed conductivity; injection: 500 µL boric acid solution (10 g/100 mL DI water); solute concentrations: 17 µg/L fluoride (1), 101 µg/L chloride (2), 842 µg/L nitrate (3), 62 µg/L orthophosphate (4), and 393 µg/L sulfate (5).

matrix without any problem. When injecting 500 µL of the same sample onto an IonPac AS16 anion exchanger, which was especially developed for the analysis of polarizable anions (see Section 3.4.2), arsenate can be detected down to the 1-µg/L level via suppressed conductivity; a shallow hydroxide gradient (preferably using an Eluant Generator™ to avoid a strong baseline drift) is employed.

Finally, it should be mentioned that many of the examples described above may be monitored online with an appropriate process monitoring instrument [115]. Especially in nuclear power plants, where breakdowns in the cooling and steam circuit may have fatal consequences and where downtimes for related repairs are much more costly than in fossil fuel power plants, online ion chromatography is the only method with which corrosive ions can be measured in real-time down to the µg/L range. Moreover, different oxidation states of ions such as nitrite/nitrate, sulfate/thiosulfate, etc. can be differentiated. This is the reason that today more than 50% of all U.S. nuclear power plants are equipped with online ion chromatographs. It was quickly recognized that online IC is an extremely effective method to monitor condensate clean-up and to localize possible malfunctions. Sulfate analysis plays the most important role, followed by short-chain fatty acids such as acetate, glycolate, and formate. Prior to the introduction of online IC, all organic acids in the steam circuit were thought to be degraded to carbon dioxide. With the commissioning of online ion chromatographs it was discovered that this is not true. Acetate, for example, was first detected in the steam circuit in the lowest µg/L range with an online ion chromatograph. Today, it is known that acetate in the steam circuit of pressurized water reactors is more ubiquitous than sulfate and extremely stable even at steam temperatures up to 320 °C. Another characteristic example of the utility of online IC is the early warning of possible leaks between the primary and the secondary circuit of pressurized water reactors. In this case, it is the extreme sensitivity of cation exchange chromatography for lithium (down to the lowest ng/L level) that enables the identification of a possible leak, long before it is detected by a radioactivity monitor. (^{7}Li ions are the natural product of the $^{10}B(n,\alpha)$ reaction; their concentration in the primary circuit is in the mg/L range.)

In comparison, cooling waters in boiling water reactors are less complex than those of pressurized water reactors, because ionic additives are not used. In the past, the quality of these cooling waters was monitored by online measurement of conductivity and pH. Online IC, on the other hand, was successfully used to find the reasons for conductivity increases, which could not be attributed to the sum of all known impurities such as sodium, chloride, nitrate, and sulfate. As a result of online- and offline investigations by IC, chromate was identified. Chromate is released from the solid chromium(III) oxide layer of stainless-steel pipe surfaces via corrosion when sudden changes in the electrochemical potential occur (caused, for example, by dissolved oxygen and hydrogen). Only by means of online measurements was it also found that such a chromate spike can time-wise be very well correlated to the concentration of dissolved hydrogen.

If one extrapolates the number of applications developed for online ion chromatography systems in nuclear power plant analysis over the last 15 years, the field of applications for this technique is expected to grow even further. Online IC applications that will receive the most attention include: monitoring organic acids in cooling circuits; monitoring wastewater treatment for recycling purposes as well as for identifying trace elements (e.g., manganese) that could serve as corrosion indicators for other metals and alloys.

9.3
Ion Chromatography in the Semiconductor Industry

A key problem in the semiconductor industry is the lifetime of manufactured semiconductor components. Their durability is decisively affected by ionic contaminations caused by
- Insufficient quality of the water being used
- Incomplete rinsing processes
- Impurities in the chemicals being used
- Contaminations in the passivation layers
- Quality and type of polymers being used

The complexity, speed of development, and diversification of this industry demand the use of the highest quality water, chemicals, and solvents.

Ion chromatography offers an efficient method for analyzing ionic species for the production of semiconductor components and printed circuit boards. Conventional methods based on the SEMI (**S**emiconductor **E**quipment and **Ma**terials Institute, Inc.) regulations for identifying and quantifying ionic species responsible for corrosion are labor-intensive and time-consuming, whereas it is possible to obtain complete anion and cation profiles within ten minutes using ion chromatography. The various application areas and typical analytical examples are summarized in Table 9-7. In the following sections, the many possible applications for ion chromatography in the field of microelectronics are illustrated by characteristic examples.

Table 9-7. Application areas and typical analytical examples in the semiconductor industry.

Application area	Analytical example
Water analysis	Determination of mineral acids in high-purity water, process liquors, rinsing- and wastewater
Etching solutions	Determination of main components and impurities in HF/HNO_3- and $HF/HAc/HNO_3$ mixtures
Solvents	Determination of ionic impurities (directly or after matrix elimination)
Acids	Determination of anionic impurities such as chloride, nitrate, and sulfate in orthophosphoric acid (electronic grade)
Hydroxides	Determination of sodium in 46% KOH
Polymers	Determination of anionic and cationic impurities in encapsulation plastics

9.3.1
High-Purity Water Analysis

The quality of the ultra-pure water required in high quantities in the manufacture of printed circuit boards is of great importance [116]. As in the field of power plant chemistry, ion chromatographic methods for determining mineral acids [117-120], silicate, borate [121], alkali metals and transition metals [122, 123] have been used in the semiconductor industry for some years to ensure this quality. Today, these methods are fully automated. The necessary instrument hardware may be configured for both laboratory and online operation. Online operation allows continuous monitoring of the ultra-pure water quality at several sampling locations that are connected to the chromatograph via appropriately dimensioned tubing. A personal computer with the appropriate software serves as the control unit of such an online chromatograph. With the software, individual sample streams can be selected in any sequence and frequency, and subsequently analyzed. The calibration of the system can also be fully automated. In this case, the required standards are prepared by the programmable dilution of concentrated standard solutions in a specially designed module and then delivered to the chromatograph as a separate stream. The analysis report is usually obtained not in form of actual chromatograms, but rather as numeric printouts of the determined solute concentrations or graphic depictions of trending diagrams.

With an increase in integration density of semiconductor components on the printed circuit boards, the demands on the purity of water and other chemicals also increase rapidly. While at present the specifications for ionic contaminations in high-purity water are in the lowest ng/L range [124], the pg/L level is under consideration as the quality profile in the coming years. Large volume injection with detection limits in the mid ng/L range as described in Section 9.2 cannot be applied in the semiconductor industry for lack of sensitivity. Therefore, sample pre-concentration techniques using concentrator columns are invaluable in this type of industry. Corresponding examples were presented in Section 9-2. Given this connection, the online procedure developed by Johnson et al. [125] for the simultaneous analysis of mineral acids and silicate is very interesting. As depicted in Fig. 9-69, the sample is passed through two different concentrator columns. The first one, TAC-1, retains mineral acids, which are separated on a conventional anion exchanger by using a carbonate/bicarbonate eluant mixture after switching the respective injection valve; detection is carried out by suppressed conductivity. Silicate is not retained by this concentrator column because the capacity of a TAC-1 column in the carbonate/bicarbonate form is too low. Therefore, after passing the TAC-1 column, the sample to be analyzed is passed through an IonPac AG5 concentrator column, which is suited for concentrating silicate due to its high capacity. Elution is then accomplished with a mixture of boric acid and sodium hydroxide; sensitive detection in the lowest µg/L range is performed by measuring the light absorption at 410 nm after reaction with sodium molybdate. Some glycerol is added to the reagent to prevent the precipitation of sodium molybdate in the pump head of the delivery pump. Any precipitates that might arise can be rendered soluble with 0.1 mol/L sodium hydroxide solution. Figure 9-70 shows the two chromatograms, obtained with this method, of a standard containing fluoride, chloride, bromide, nitrate, orthophosphate, sulfate, and silicate with mass concentrations in the lowest µg/L range. If the silicate standard is prepared from ultra-pure de-ionized water that has been additionally passed through an IonPac AG5 column to remove traces of silicate, a detection limit in the sub-µg/L range may also be achieved for this analyte ion.

Upon application of a pre-concentration technique, transition metals may also be determined in the sub-µg/L range [122]. Based on the photometric detection of metal ions after derivatization with PAR, detection limits of about 10 ng/L are obtained for metals such as iron, copper, nickel, zinc, cobalt, and manganese when pre-concentration is applied. Hence, the sensitivity of this method is comparable to ICP-MS, which is today the traditional method for trace analysis of metals in this concentration range. Ion chromatography has the additional advantage of being able to differentiate the oxidation state of some metals. The chromatogram shown in Fig. 9-71, with metal concentrations between 2 ng/L and 38 ng/L, was obtained after concentrating 310 mL of a de-ionized water sample. In this sample, the concentration of zinc clearly exceeds the 4M DRAM SEMI guidelines.

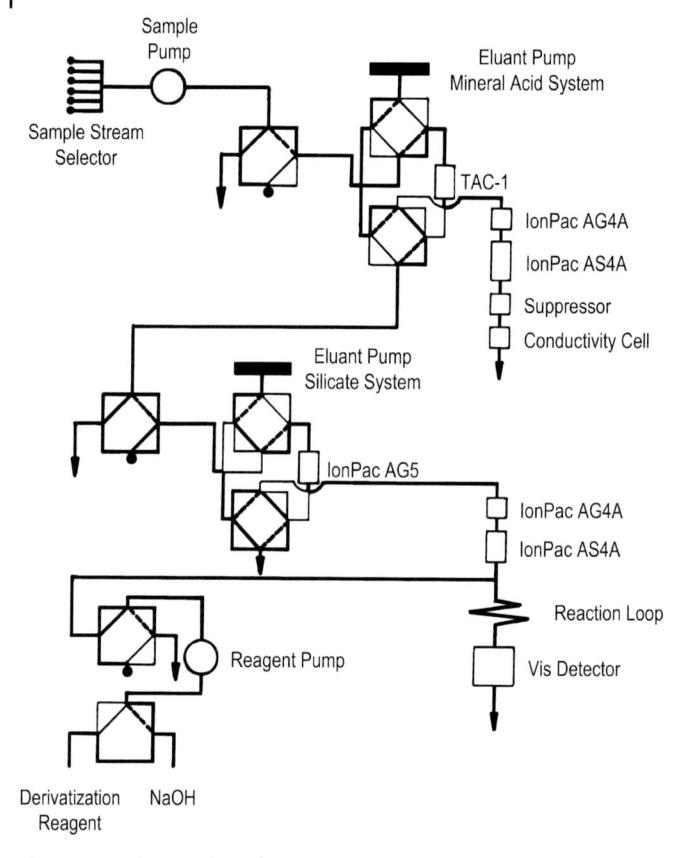

Fig. 9-69. Valve switching for the simultaneous analysis of mineral acids and silicate at trace level.

9.3.2
Surface Contaminations

It is well known that ionic contaminations on the surface of electronic components impair their functioning [126, 127]. Corrosion of aluminum after etching, for example, is regarded to be one of the major problems in the dry, chlorine gas-based plasma etching of aluminum.

$$Al_2O_3 + 3H_2O \rightleftharpoons 2Al(OH)_3 \tag{310}$$

$$Al(OH)_3 + Cl^- \rightleftharpoons Al(OH)_2Cl + OH^- \tag{311}$$

$$Al + 4Cl^- \rightleftharpoons [AlCl_4]^- + 3e^- \tag{312}$$

$$2[AlCl_4]^- + 6H_2O \rightleftharpoons 2Al(OH)_3 + 6H^+ + 8Cl^- \tag{313}$$

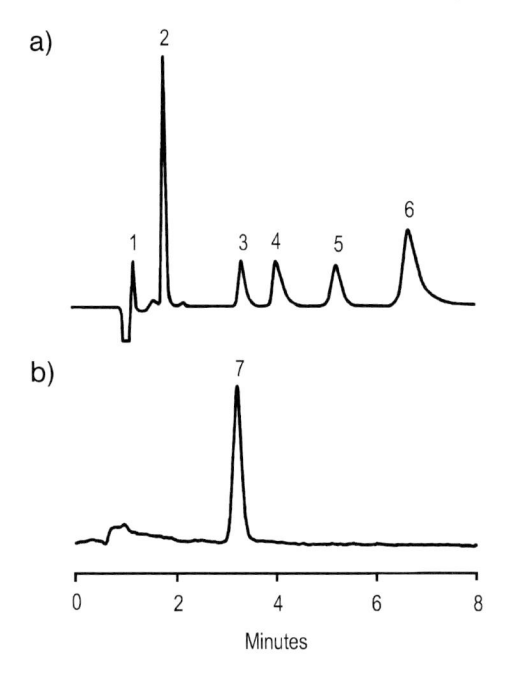

Fig. 9-70. Online determination of mineral acids and silicate. — Separator column: 2 IonPac AS4A; concentrator columns: (a) TAC-1, (b) IonPac AG5; eluant: (a) 1.7 mmol/L $NaHCO_3$ + 1.8 mmol/L Na_2CO_3, (b) 15 mmol/L H_3BO_3 + 15 mmol/L NaOH; flow rates: (a) 2 mL/min, (b) 1 mL/min; detection: (a) suppressed conductivity, (b) photometry at 410 nm after reaction with sodium molybdate; pre-concentrated volume: 23 mL; solute concentrations: 10 µg/L fluoride (1), 10 µg/L chloride (2), 10 µg/L bromide (3), 10 µg/L nitrate (4), 20 µg/L orthophosphate (5), 20 µg/L sulfate (6), and 23 µg/L silicate (7); (taken from [125]).

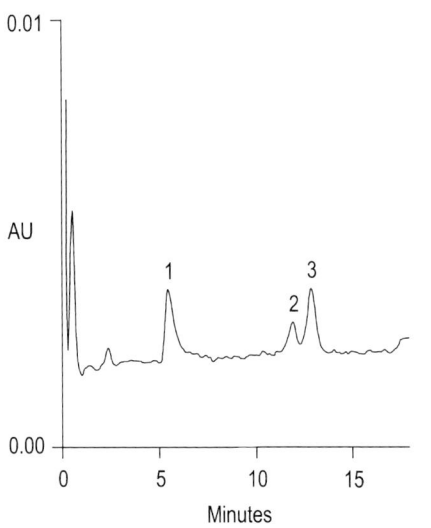

Fig. 9-71. Trace analysis of metals in a high-purity water sample — Separator column: IonPac CS5; concentrator column: IonPac CG2; eluant: 6 mmol/L pyridine-2,6-dicarboxylic acid + 86 mmol/L LiOH; flow rate: 1 mL/min; detection: photometry at 520 nm after reaction with PAR; pre-concentrated volume: 310 mL; solute concentrations: 8 ng/L iron(III) (1), 2 ng/L nickel (2), and 38 ng/L zinc (3).

Traces of chlorine that remain at the surface of the silicon wafers after the etching process, when exposed to ambient air, are the major origin of corrosion phenomena. This typically leads to bubble-type outgrowths on aluminum conductors (Fig. 9-72). With an increased copper content in aluminum, this problem is aggravated because this system then acts like a galvanic cell, which is characterized by HCl-catalyzed chemical reactions. Two different strategies are feasible to prevent these corrosion phenomena: the silicon wafers are not to be exposed to ambient air, or the chlorine residues have to be completely removed from their surface. Because contact with ambient air cannot be avoided in practice, the second option is the only one possible. Current procedures for decreasing corrosion phenomena include intensive rinsing after the etching processes and heating of the silicon wafers at high temperatures. Ion chromatography is a suitable tool for identifying and quantifying ionic contaminations [128, 129]. In the past, however, extractions with relatively large volumes (25 mL) limited its applicability. Thus, contaminations on the silicon wafer surface could only be investigated on relatively large sections of the wafer. If small extraction volumes are applied, on the other hand, corrosion phenomena can be investigated on single integrated circuits on the silicon wafer [118]. To do this, 100 µL of high-purity water, for example, is pipetted with an Eppendorf pipette onto the silicon wafer surface. Then the area of the drop is measured for subsequent calculations. After two minutes, the volume is picked up and transferred into a thoroughly rinsed polyethylene vessel. With the same pipette tip, 100 µL high-purity water is charged twice and added to the sample, which is then diluted with high-purity water to a final volume of 1300 µL. This volume can then be directly injected into the ion chromatograph using the large volume injection technique described by Kaiser et al. [120] for trace analysis without pre-concentration. In this way, anionic contaminants down to the 20-ng/L range can be localized on the silicon wafer surface and thus, single process steps can be optimized in terms of corrosion. As an example, Fig. 9-73 shows a chromatogram of such a wafer extract, which was obtained with an IonPac AS14 anion exchanger in the microbore format and a borate step gradient.

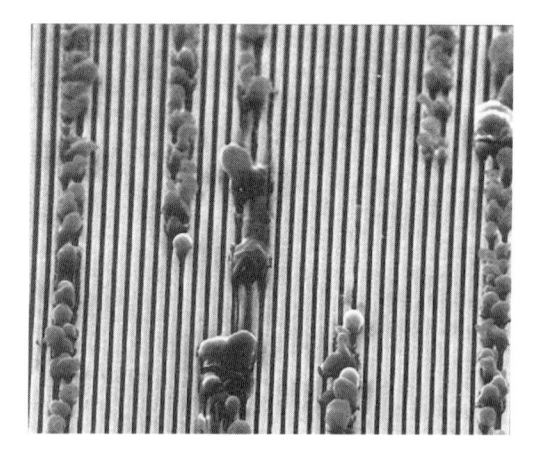

Fig. 9-72. SEM of corroded aluminum conductors.

As an alternative to the extraction procedure described above, small components can be sealed into a double-layered plastic bag made of polypropylene/polyethylene together with 100 mL of ultra-pure water (SEMI standard method G52-90) and heated in a water bath to 95 °C for 30 minutes. After cooling it

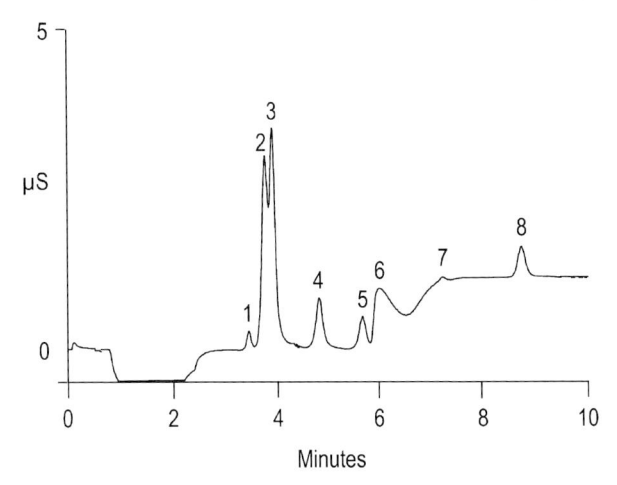

Fig. 9-73. Trace analysis of anions in a wafer extract. — Separator column: IonPac AS14 (2-mm); eluant: 6.75 mmol/L NaOH + 9 mmol/L H_3BO_3 after 3.5 min to 30 mmol/L NaOH + 40 mmol/L H_3BO_3; flow rate: 0.75 mL/min; detection: suppressed conductivity; injection volume: 1 mL; solute concentrations: 1.5 µg/L each of fluoride (1), acetate (2), and formate (3), 11 µg/L chloride (4), unknown component (5), carbonate (6), 2.3 µg/L nitrate (7), and 10 µg/L sulfate (8).

down to room temperature, an aliquot of the extract can be directly injected into the ion chromatograph. In co-operation with the Read-Rite Company Ltd. in Thailand, Heberling and Siriraks [131] developed a micro-extraction technique that is especially suited for corrosion studies on hard disk reading heads. According to this technique, the reading head to be investigated, which has an area of 0.097 cm² (nano slider) or 0.038 cm² (pico slider), is placed into the micro-extraction cell (see Fig. 9-74) comprising a cell body and a screw made of PEEK. Using a pair of tweezers, the reading head is positioned in an O-ring, which tightens the cell at the same time. After closing the cell, high-purity water with a flow rate of 0.8 mL/min is directed through the cell for three minutes. The extractable ions are trapped on a concentrator column and subsequently separated on an analytical column.

Due to their small dimensions, 20 reading heads per sample were needed when applying conventional extraction techniques in order to obtain the required specifications for ionic impurities of 10-100 ng/cm². These low thresholds are indispensable today, because the development of magnetic reading heads tends towards smaller and smaller dimensions and the distances between reading head and storage medium also become constantly smaller. For instance, if a nano slider is contaminated with ions responsible for corrosion in the range of 10 ng/cm², a 1-mL extract only contains 1 µg/L of these ions, which is close to the detection limit of ion chromatography using 4-mm columns. Moreover, the surface of the reading head with thin layers of NiFe or cobalt alloys is extremely

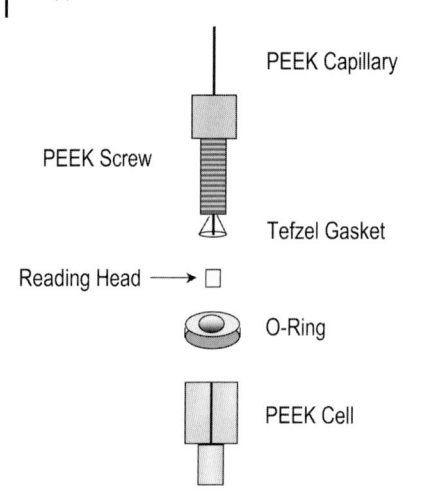

Fig. 9-74. Schematic representation of a cell for micro-extraction of magnetic reading heads.

susceptible to corrosion in the presence of anions such as chloride, nitrate, and sulfate as well as cations such as sodium, potassium, and ammonium. The micro-extraction technique presented above allows corrosion studies on single components, through which the manufacturing process can be optimized. Figure 9-75 illustrates the valve switching that is necessary in the coupling of micro-extraction and ion chromatography. The procedure is divided into three sub-steps: loading, extraction, and chromatography.

The parallel anion chromatogram of a micro-extract is depicted in Fig. 9-76. Usually, fluoride, chloride, orthophosphate, nitrate, and sulfate are found as anionic contaminants on a reading head. IonPac AS12A in the microbore format has proved to be a suitable stationary phase for separating these anions, which are eluted with a carbonate/bicarbonate eluant mixture. The minimum detection limits for a single reading head obtained with this technique are summarized in Table 9-8.

9.3.3
Solvents

In the semiconductor industry, a number of organic solvents such as isopropanol, acetone, and N-methylpyrrolidone (NMP) are used for cleaning purposes. Because these solvents are coming into contact with micro-electronic circuits, they also have to be investigated for ions that can cause corrosion. Until recently, a large sample volume was evaporated and the anions were separately determined by colorimetry and turbidimetry [132]. This is not only time-consuming and labor-intensive but also very insensitive. Specifications in the µg/L range that are required today cannot be achieved with this method. As early as 1994, Kaiser et al. [133] demonstrated the applicability of ion chromatography for the determination of anions in isopropanol. Although modern anion exchangers are

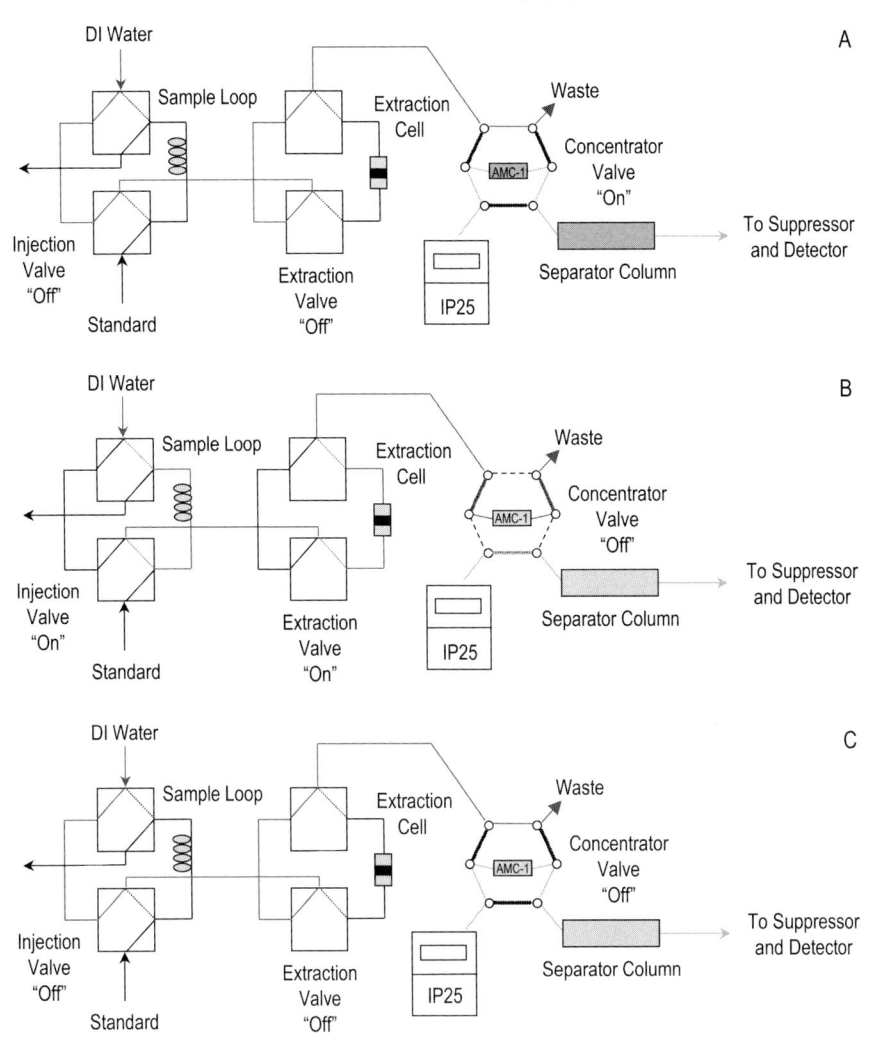

Fig. 9-75. Valve switching for coupling micro-extraction and ion chromatography. (A) Loading of sample or standard, (B) extraction of the component or injection of the standard, (C) chromatography.

100% solvent-compatible, direct determination by injecting a large volume of solvent is not recommended for lack of reproducibility. Also, calibration via standard addition is very cumbersome. For this reason, Kaiser et al. [134] utilized the technique of matrix elimination comprising four working steps:
1. Filling of the injection loop with sample
2. Loading of the concentrator column
3. Elimination of the solvent matrix
4. Separation of the pre-concentrated ions

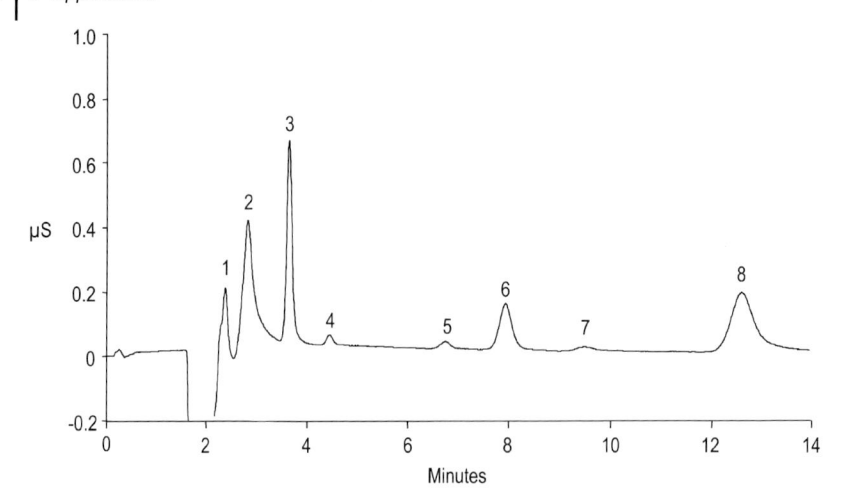

Fig. 9-76. Anion chromatogram of a micro-extract of a single reading head. — Separator column: IonPac AS12A (2-mm); eluant: 2.7 mmol/L Na_2CO_3 + 0.3 mmol/L $NaHCO_3$; flow rate: 1 mL/min; concentrator column: 15 mm × 2 mm i.d. AMC-1; pre-concentrated volume: 2.4 mL; detection: suppressed conductivity; peaks: (1) fluoride, (2) carbonate, (3) chloride, (4) unknown component, (5) unknown component, (6) nitrate, (7) orthophosphate, and (8) sulfate.

Table 9-8. Minimum detection limits for anionic contaminants on a single magnetic reading head after micro-extraction coupled to microbore ion chromatography.

Anion	Detection limit [ng]	Detection limit [ng/cm^2]
Fluoride	0.49	5.6
Chloride	0.23	2.3
Nitrate	0.72	7.4
Orthophosphate	2.02	20.8
Sulfate	0.37	3.7

Figure 9-77 illustrates the required valve switching for this technique. In the first step (Fig. 9-77A), the injection loop (5 mL) is filled with the solvent sample. Sample delivery is carried out with a simple isocratic pump at a flow rate of 1.5 mL/min for seven minutes out of a reservoir with a small head pressure, so that the loop is rinsed several times with sample. After loading the loop, the sample is transferred with ultra-high purity water to the concentrator column, in which the anions or cations are retained (Fig. 9-77B). The remaining solvent is rinsed off with ultra-high purity water (flow rate: 1.7 mL/min; rinse time: 10 minutes). Afterwards, the concentrator column is put inline with the analytical separator column. The pre-concentrated ions are then eluted from the concentrator column in opposite directions and separated on the analytical separator column (Fig. 9-77C). The ultra-pure water used for the sample transfer is directed through a trap column prior to entering the switching valve. In this trap column, anionic or cationic impurities are retained. This is necessary to avoid contamination of the concentrator column during the rinse step.

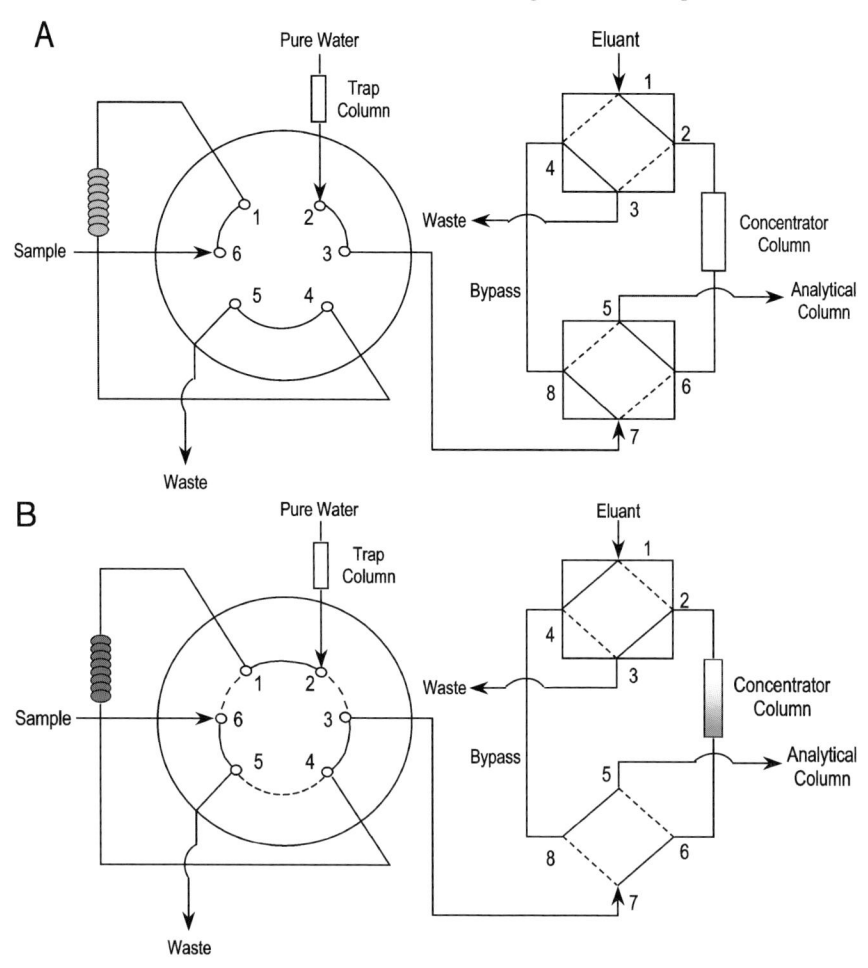

Fig. 9-77. Configuration of an IC system for matrix elimination. (A) Filling of the injection loop, (B) loading of the concentrator column and elimination of the matrix, and (C) separation of the retained ions.

The great advantage of matrix elimination is the ability to prepare standards in de-ionized water and to calibrate externally. Kaiser et al. obtained detection limits between 0.2 and 1 µg/L for anions such as chloride, sulfate, orthophosphate, and nitrate. A representative chromatogram of a sample spiked with 10 µg/L each of the anions of interest is shown in Fig. 9-78. The IonPac AS10 high-capacity anion exchanger used as a stationary phase was operated with a NaOH eluant.

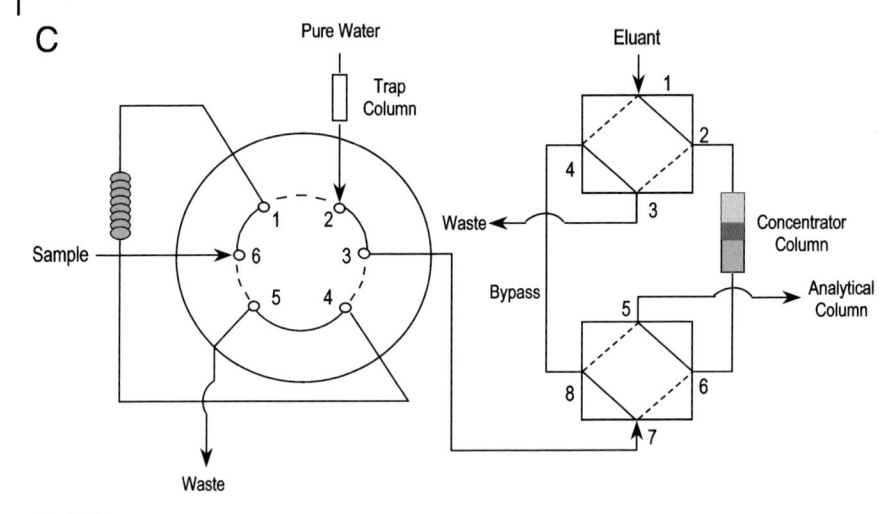

Fig. 9-77c.

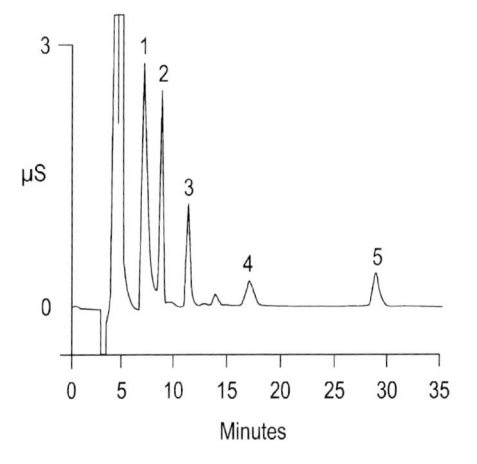

Fig. 9-78. Separation of anions in a spiked isopropanol sample after matrix elimination. — Separator column: IonPac AS10 with guard column; eluant: 0.1 mol/L NaOH; flow rate: 1 mL/min; detection: suppressed conductivity; pre-concentrated volume: 5 mL; concentrator column: IonPac AC10; sample: isopropanol (electronic grade); solute concentrations: carbonate (1) and 10 µg/L each of chloride (2), sulfate (3), orthophosphate (4), and nitrate (5).

Recently, the matrix elimination technique has been applied by Kaiser et al. [134] and Viehweger et al. [135] to a number of other organic solvents, utilizing different kinds of stationary phases. For their investigations on acetone and N-methylpyrrolidone Kaiser et al. [134] used an IonPac AS9-HC carbonate-selective, high-capacity anion exchanger [136], on which the anions of interest can be separated under isocratic conditions with a carbonate-based eluant in less than 30 minutes. The pH of the standard eluant (9 mmol/L Na_2CO_3) was slightly raised with NaOH to achieve a baseline-resolved separation between chloride and carbonate. To obtain maximum sensitivity a microbore column was used, which has the additional advantage of lower eluant consumption [137]. In comparison to IonPac AS10, the AS9-HC column exhibits better resolution between chloride and short-chain organic acids such as glycolate, acetate, and formate.

The chromatogram of an acetone sample obtained under these chromatographic conditions is shown in Fig. 9-79. Even after pre-concentration of 5 mL of sample, the chloride concentration is below the detection limit. In comparison, Fig. 9-80 illustrates anion analysis in dimethylformamide, which is significantly more contaminated. In this case, a Metrosep Anion SUPP 1 column was used as a stationary phase, with a sodium carbonate eluant.

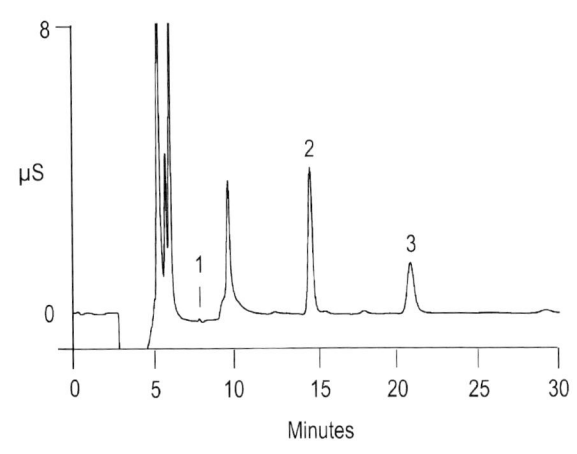

Fig. 9-79. Separation of anions in acetone after matrix elimination. – Separator column: IonPac AS9-HC (2-mm) with guard column; eluant: 8 mmol/L Na_2CO_3 + 1.5 mmol/L NaOH; flow rate: 0.25 mL/min; detection: suppressed conductivity; pre-concentrated volume: 5 mL; concentrator column: IonPac AG9-HC (4-mm); sample: acetone (electronic grade); solute concentrations: chloride (1), 20 µg/L nitrate (2), and 5.5 µg/L sulfate (3).

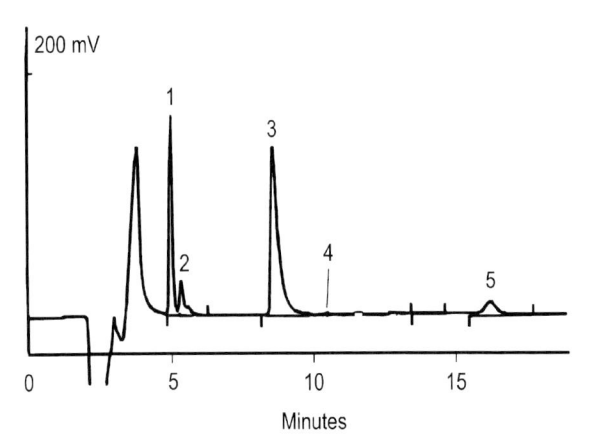

Fig. 9-80. Separation of anions in dimethylformamide after matrix elimination. – Separator column: Metrosep Anion SUPP 1; eluant: 4 mmol/L Na_2CO_3; flow rate: 1.2 mL/min; detection: suppressed conductivity; pre-concentrated volume: 250 µL; concentrator column: IC concentrator Metrosep Anion (spheric hydroxyethylmethacrylate); sample: dimethylformamide (electronic grade); solute concentrations: 82 µg/L chloride (1), 24 µg/L nitrite (2), 342 µg/L nitrate (3), 4 µg/L orthophosphate (4), and 23 µg/L sulfate (5); (taken from [135]).

Cation analysis in organic solvents such as methyl-*n*-amylketone (MAK), *n*-butylacetate (NBA), and propyleneglycol-monomethylether acetate (PMA) was described by Sanders [138]. Because this solvent is not water-miscible, the cations have to be extracted prior to IC analysis. To simplify sample preparation as much as possible, Sanders carried out the extraction in an autosampler vial. He used an 8-mL Nalgene vial (HDPE), in which he pipetted 3.5 mL sample and 3.5 mL internal standard solution. The vial is then sealed with a silicon/PTFE septum, shaken for 10 minutes, and placed in the autosampler. The autosampler needle is adjusted in the way that it charges the aqueous extract short of the bottom of the vial. The investigation of the extraction yields revealed a dependence of the extraction on the organic sample matrix and on the pH of the extraction agent. When comparing the extraction yield of a neutral aqueous standard with those of the investigated organic solvents, the extraction of cations from an organic matrix seems to be more complete. Organic acids present in organic solvents at trace level are responsible for this; they cause a slight acidification of the aqueous extraction medium. If the aqueous extraction medium is acidified with high-purity acetic acid to a final concentration of 0.6 mmol/L, the extraction yield of the aqueous standards and the organic solvents is comparable. Possible volume changes are taken into consideration by using rubidium as an internal standard. For the separation of mono- and di-valent cations Sanders used an IonPac CS12A cation exchanger with a dilute sulfuric acid eluant. An elevated temperature of 50 °C was applied to improve the peak efficiencies of magnesium and calcium, which did not have any negative effect on the column lifetime but shortened total analysis time to less than eight minutes. Figure 9-81 shows the chromatogram of a 5-µg/L cation standard under the given chromatographic conditions. Sanders verified the results obtained with this method using ICP-MS, taking the sodium analysis in MAK as an example.

9.3.4
Acids, Bases, and Etching Agents

Concentrated acids also play an important role in the manufacturing of semiconductor devices. Hydrofluoric acid, for example, is used to remove oxide layers from wafer surfaces [139]. Glycolic acid is a major component of non-aqueous soldering agents [140] and concentrated orthophosphoric acid is used for etching silicon nitride on wafer surfaces [141].

The extremely large excess of matrix ions renders the analysis of anionic impurities very difficult. Concentration differences between matrix and analyte ions of $10^6:1$ are not unusual. (Example: chloride contamination in the range of 100 µg/L in 24.5% (v/v) HF) Although the matrix problem can be solved by diluting the acids, the sample then falls below the detection limits for the analyte ions. Siriraks et al. [142], for example, obtained detection limits between 0.04 and 0.08 mg/L for chloride, bromide, orthophosphate, and sulfate in 0.25% HF by direct injection. Utilizing an alternative matrix elimination technique – methanol/water rinse of the concentrator column after pre-concentration of HF – the tolerable HF concentration could be increased to 5%, while maintaining the detection limit. However,

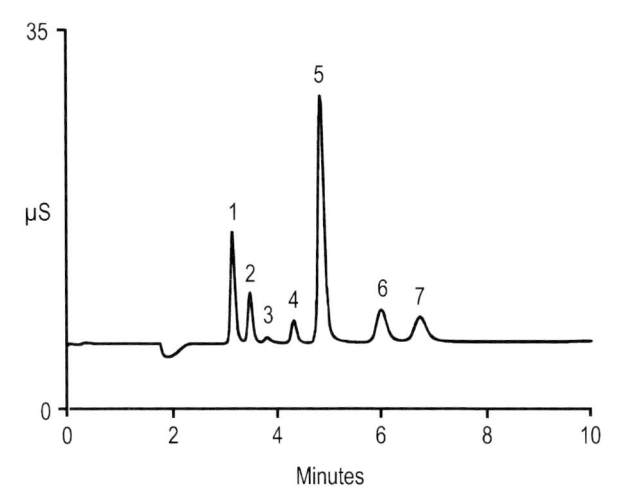

Fig. 9-81. Trace analysis of a cation standard after matrix elimination. — Separator column: IonPac CS12A; eluant: 11 mmol/L H_2SO_4; flow rate: 1 mL/min; detection: suppressed conductivity; pre-concentrated volume: 2 mL; concentrator column: IonPac CG12A; solute concentrations: 5 µg/L each of lithium (1), sodium (2), ammonium (3), potassium (4), magnesium (6), and calcium (7), 140 µg/L rubidium (5) as an internal standard; (taken from [138]).

this sensitivity is by far insufficient to meet today's demands. On the other hand, 5 % HF is the maximum concentration that can be injected directly onto conventional ion exchangers. To improve this, Watanabe et al. [143] developed a two-dimensional technique with a pre-separation between weak and strong acids on an ion-exclusion column. In a second step, the fraction with the anionic contaminants of interest will be separated on an anion exchanger. With this technique, detection limits of 30 µg/L in 25 % HF were achieved for the first time. Chen and Wu [144] optimized the experimental conditions of this technique and applied it to the determination of anionic contaminants in orthophosphoric acid [145]. Finally, Kaiser et al. [146] modified this technique by using an anion exchanger in the microbore format, which increased sensitivity even further. The basic principle of this technique was already described in Section 5.7. Here, the required valve switching is outlined; see Fig. 9-82. The starting configuration (Fig. 9-82A) serves the loading of the injection loop. Thus, the sample valve is in the "load" position and the injection valve in the "inject" position. The concentrated acid is pneumatically delivered through the sample loop from a pressurized (helium) reservoir. In this way it is ensured that a representative sample is taken and that it reaches the sample loop free of any contaminants. Configuration B depicts the valve switching for the pre-separation between the acid and the anionic contaminants on an ion-exclusion column. For this, the sample valve is switched to the "inject"-position. The first part of

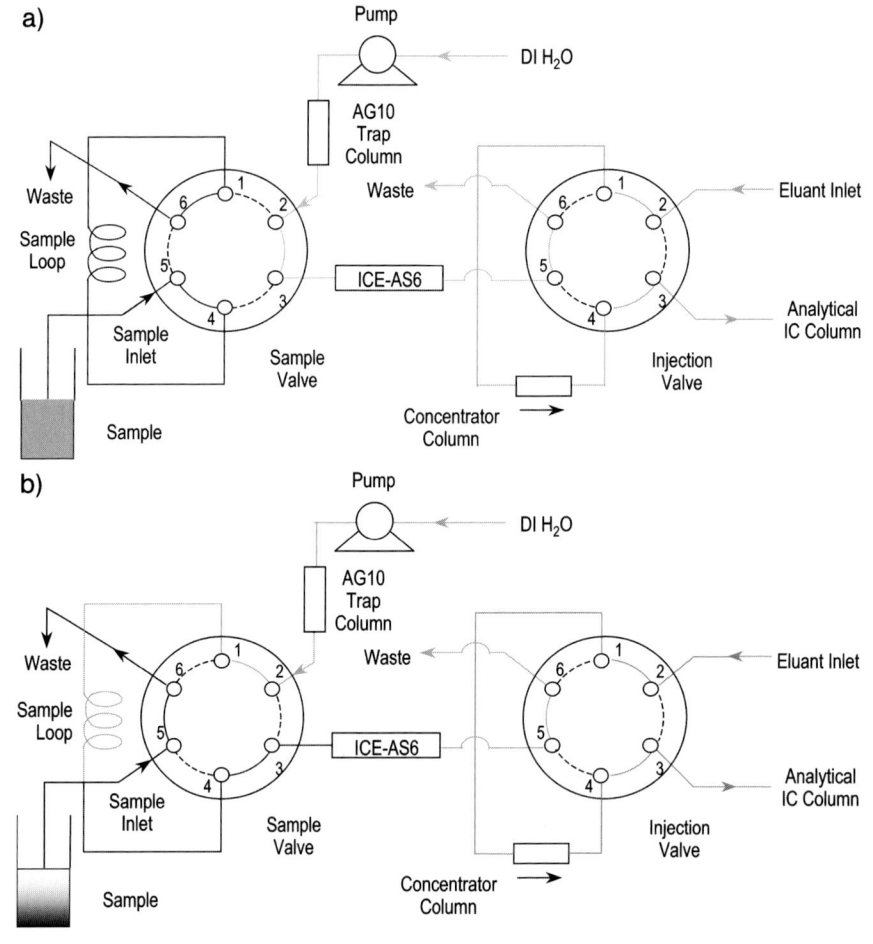

Fig. 9-82. Valve switching for the trace analysis of anions in concentrated acids of low acid strength. – (aa) Loading the sample loop, (b) pre-separation on an ion-exclusion column, (c) pre-concentration of the hard-cut fraction, and (d) separation of the pre-concentrated ions.

the column effluent is discarded. After switching the injection valve to the "load"-position, the concentrator column is inline with the ion-exclusion column (configuration C), so that the fraction with the anionic contaminants can be pre-concentrated. In the last step, the concentrator column is put inline with the anion exchanger and the pre-concentrated anions are eluted.

The fractionation scheme for the separation of strong and weak acids on an ion-exclusion column was already shown in Fig. 5-18 (Section 5.7), taking chloride and fluoride as an example. The chromatograms illustrating the anion analysis in concentrated hydrofluoric acid (24.5% v/v) and orthophosphoric acid (85% w/w) are found in the same section (Figures 5-10 and 5-20). In Fig. 9-83,

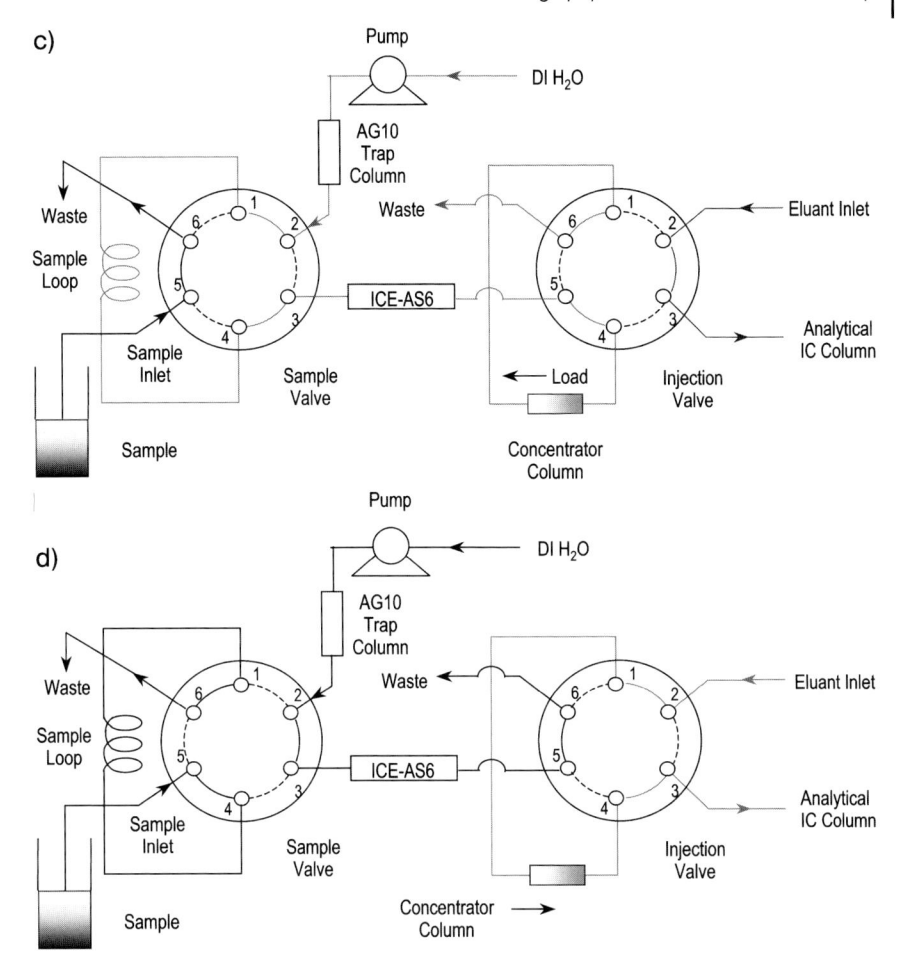

Fig. 9-82c, d.

these two chromatograms are complimented by the anion chromatogram of glycolic acid (70% w/w), which is much more strongly contaminated and, therefore, has been diluted 1:100 with ultra-high purity water. In analogy to the HF analysis, an IonPac AS9-HC high-capacity anion exchanger in the microbore format was also used for the glycolic acid analysis, because the anions of interest can be separated in less than 30 minutes on this stationary phase with a carbonate/hydroxide eluant mixture.

For the determination of cationic impurities in concentrated acids, the AutoNeutralization technique must be used; this technique was outlined above in Section 9.2. The direct injection of the acids is not possible for lack of sensitivity. Moreover, if a weak acid cation exchanger was used for the simultaneous analysis of mono- and di-valent cations, the high acid concentration would lead to a complete protonation of the ion-exchange groups which, in turn, would result in a loss of separation power. Figure 9-84 shows the cation analysis in 12% hydrofluoric acid, which is representative for all the other acids mentioned above. Separation was carried out on an IonPac CS12 cation exchanger with methanesulfonic acid as an eluant. In addition to alkali- and alkaline-earth metals, ammonium and methylamine could be detected. The latter were not quantified, because they are partly released from the suppressor membrane used for AutoNeutralization; this is attributed to the high acid strength in the original sample.

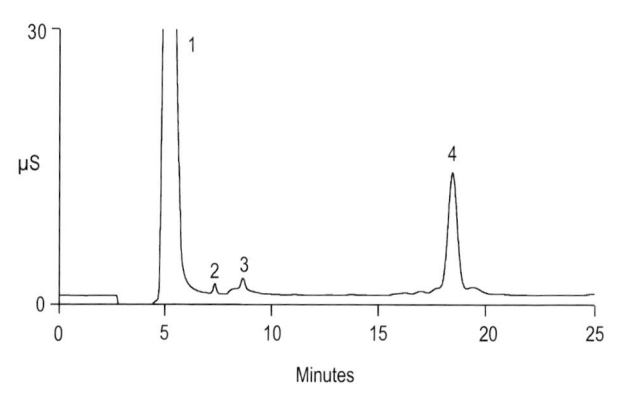

Fig. 9-83. Separation of chloride and sulfate in 0.7% glycolic acid after pre-separation on IonPac ICE-AS6. – Eluant for the pre-separation: de-ionized water; flow rate: 0.55 mL/min; concentrator column: IonPac AG9-HC (4-mm); separator column: IonPac AS9-HC (2-mm); eluant: 8 mmol/L Na_2CO_3 + 1.5 mmol/L NaOH; flow rate: 0.25 mL/min; detection: suppressed conductivity; injection volume: 750 µL; solute concentrations: glycolate (1), 11 µg/L chloride (2), carbonate (3), and 336 µg/L sulfate (4).

AutoNeutralization also has to be applied for trace analysis of anions in bases such as sodium hydroxide, ammonium hydroxide, tetramethylammonium hydroxide, and tetrabutylammonium hydroxide, because the high concentration of hydroxide in the matrix would in practice act as an eluant and thus, render impossible the pre-concentration of anions in a concentrator column [147]. In the semiconductor industry quaternary ammonium bases serve as developers for light-sensitive paints and, therefore, also have to be investigated for anions responsible for corrosion. As shown in Fig. 9-85, detection limits in the mid-µg/L range are obtained without difficulty when applying AutoNeutralization.

One of the most important chemicals in the semiconductor industry is hydrogen peroxide, which is used in a number of etching and cleaning processes. Due to the high demands for the purity of this chemical, it also has to be investigated for ionic impurities. While hydrogen peroxide solutions with content below 3% can be analyzed by ion chromatography without any sample preparation, a significant decrease in the lifetime of conventional ion exchangers is observed at higher H_2O_2 concentrations [148]. In the case of anion analysis, the high eluant

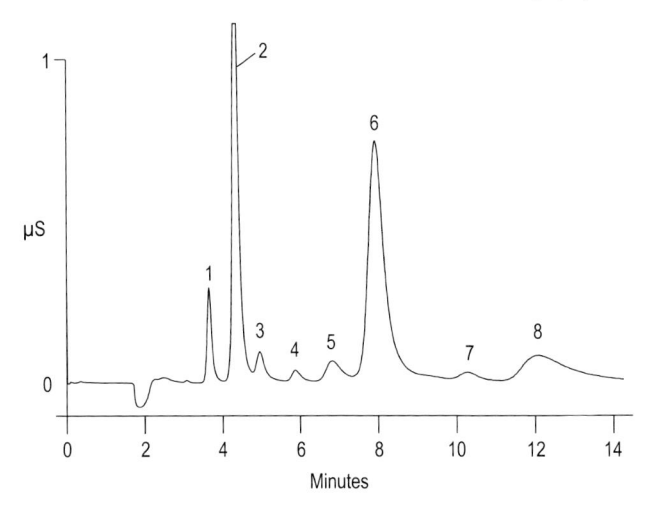

Fig. 9-84. Separation of cations in 12% hydrofluoric acid after Auto-Neutralization. — Separator column: IonPac CS12; eluant: 19 mmol/L methanesulfonic acid; flow rate: 1 mL/min; detection: suppressed conductivity; injection volume: 100 µL; solute concentrations: 8 µg/L sodium (1), ammonium (2), monomethylamine (3), 5 µg/L potassium (4), dimethylamine (5), 69 µg/L magnesium (6), 4.5 µg/L calcium (7), and trimethylamine (8).

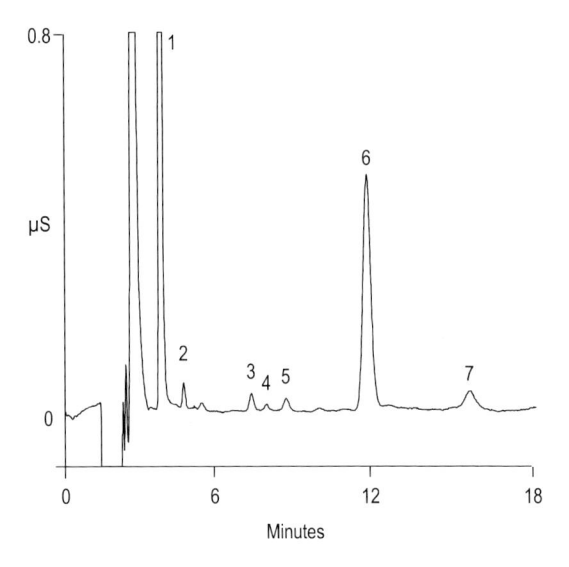

Fig. 9-85. Separation of anions in 25% tetramethylammonium hydroxide after Auto-Neutralization. — Separator column: IonPac AS12A; eluant: 2.7 mmol/L Na_2CO_3 + 0.3 mmol/L $NaHCO_3$; flow rate: 1.5 mL/min; detection: suppressed conductivity; injection volume: 100 µL with 2 neutralization cycles; solute concentrations: 2.3 mg/L chloride (1), 0.02 mg/L nitrite (2), 0.04 mg/L bromide (3), 0.01 mg/L chlorate (4), 0.02 mg/L nitrate (5), 0.55 mg/L sulfate (6), and 0.07 mg/L oxalate (7).

pH also supports the formation of peroxohydroxide anions, HO_2^-, and thus, causes oxidation of the resin material. Because dilution is not an option due to the required detection limits, the matrix has to be eliminated in some way. In the past, the sample was transferred into a platinum crucible, in which hydrogen peroxide is broken down to water and oxygen under the catalytic effect of platinum. However, this kind of sample preparation is very time-consuming and prone to contamination. If the hydrogen peroxide to be analyzed contains a stabilizer, there is a high risk that this stabilizer will be degraded by the heat generated during the platinum treatment.

Today, online matrix elimination is the sample preparation method of choice, as described above for anion analysis in organic solvents. Due to the manufacturing process of hydrogen peroxide, inorganic anions and traces of organic acids are to be expected as anionic contaminants. Therefore, Kerth and Jensen [149] applied a gradient elution technique on an IonPac AS11 anion exchanger with a hydroxide eluant. Because the hydrogen peroxide used in the semiconductor industry does not contain stabilizers such as pyrophosphate, analysis time is on the order of 15 minutes. Figure 9-86 shows the chromatogram of a 35% hydrogen peroxide solution (electronic grade), which was obtained after pre-concentration of a 750 µL sample. Besides the very low concentrations of chloride, nitrate, and sulfate in the lowest µg/L range, the sample shows a relatively complex pattern of organic acids in the retention range of sulfate. Under the given chromatographic conditions, sulfate is separated from these components and can be quantitated without difficulty. Kerth and Jensen identified the main components close to the system void by means of capillary electrophoresis utilizing indirect UV detection. According to their investigations these peaks are acetate and formate with estimated concentrations of 33 mg/L and 4 mg/L, respectively.

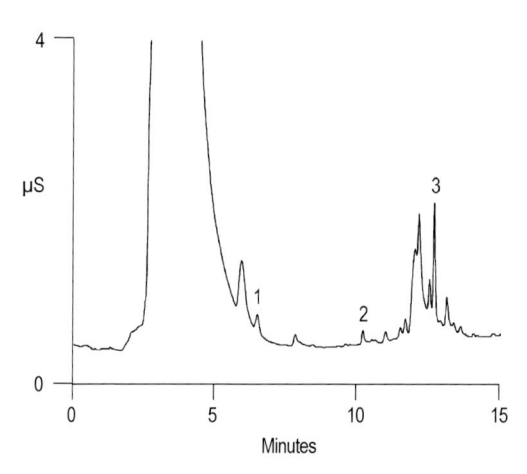

Fig. 9-86. Trace analysis of anions in 35% hydrogen peroxide solution after matrix elimination. − Separator column: IonPac AS11; eluant: (A) 50 mmol/L NaOH, (B) water; gradient: 3% A for 3 min isocratic, then linearly to 80% A in 13 min; flow rate: 1 mL/min; detection: suppressed conductivity; injection volume: 750 µL; concentrator column: IonPac AG11; sample: 35% hydrogen peroxide (electronic grade); solute concentrations: 2.5 µg/L chloride (1), 1.4 µg/L nitrate (2), and 2.9 µg/L sulfate (3); (taken from [149]).

Another application area for ion chromatography is the analysis of etching solutions [150] (i.e., mixtures of different acids), with which metal oxides and other impurities can be removed from metal surfaces. The choice of acids depends on the type of materials to be etched. While a mixture of hydrofluoric acid, nitric acid, and acetic acid etches silicon dioxide without affecting elemental silicon itself, mixtures of orthophosphoric acid, acetic acid, and nitric acid are used for etching aluminum. The efficiency of the etching process depends on the temperature of the etching solution and on the concentration of the acids. Because the depth of penetration in the etching process has to be thoroughly controlled, the concentration of the individual acids has to be determined with an absolute tolerance between 0.5% and 1%. In the past, titration and spectrophotometric techniques were applied, which were very time-consuming due to extensive sample preparation. Figure 9-87 shows the chromatogram of a HF/HNO$_3$/HOAc etching mixture diluted 1:500 with de-ionized water, which was obtained by ion-exclusion chromatography with suppressed conductivity detection. It has to be noted that with ion chromatography, only the total fluoride concentration and not the concentration of the free hydrofluoric acid is detected, because metal-fluoride complexes (for example FeF^{2+}) are formed during the etching process. At alkaline pH, these complexes decompose to form metal hydroxides. Indirect analytical methods such as the one developed by Dulski [151] show good agreement with conventional wet-chemical methods [152].

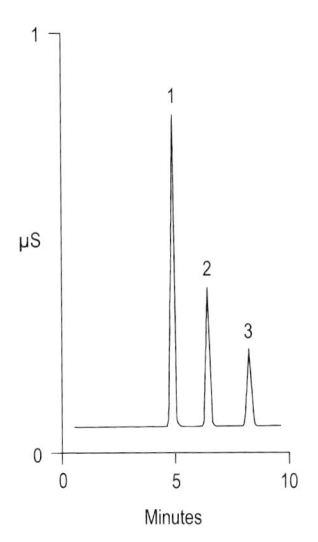

Fig. 9-87. Analysis of a HF/HNO$_3$/HOAc etching mixture. — Separator column: IonPac ICE-AS1; eluant: 1 mmol/L octanesulfonic acid — 2-propanol (95:5 v/v); flow rate: 1 mL/min; detection: suppressed conductivity; injection: 10 µL sample (1:500 diluted); solute concentrations: 200 mg/L HNO$_3$ (1), 200 mg/L HF (2), and 800 mg/L HOAc (3).

Anion exchange chromatography contributed to the identification of a fourth component in HF/H$_3$PO$_4$/HNO$_3$ etching mixtures that was previously unknown [153]. The component elutes behind orthophosphate when using a carbonate/bicarbonate eluant. Because the appearance of this compound goes along with a decrease of peak area for fluoride and orthophosphate, it seemed likely that the unknown component might be monofluorophosphate. Using a corresponding reference component, this supposition was confirmed. The reaction between HF and H$_3$PO$_4$ is described in literature by Jacobson [154]:

$$HF + H_3PO_4 \rightleftharpoons H_2PO_3F + H_2O \tag{314}$$

However, investigations of concentrated acids or etching mixtures containing nitric acid have not yet been published. The existence of an additional component is of great importance for the user of such etching mixtures because, as outlined above, the concentration of every single component has to be precisely determined. All four components can be separated in the same run within 10 minutes by anion exchange chromatography. Vanetta et al. chose IonPac AS14 as a stationary phase, with which fluoride is well separated from the system void. Optimum separations were obtained with an eluant mixture comprising 2.5 mmol/L sodium carbonate and 3.4 mmol/L sodium bicarbonate.

Finally, attention is drawn to the analysis of BPSG (borophosphosilicate glass)-films [155], which are intermediately used for protecting certain structures [156]. The flow properties of such films are determined by their boron and phosphorus content. Until recently, these films have been analyzed with element-specific methods such as ICP-AES, ICP-MS, XRFA, etc., with which only the total content of the element can be determined. Only with ion chromatographic techniques it was discovered that the flow properties depend only on the content of phosphorus(V) and boron [157]. In reality, BPSG-films additionally contain larger amounts of phosphorus(III)-compounds, oxygen-containing phosphorus(V)-polymers, boron oxide, and boric acid, none of which influence product properties; however, they do affect the result of elemental analysis.

The individual modifications and oxidation states cannot be differentiated without sample preparation. However, when dissolving BPSG-films in dilute hydrofluoric acid, orthophosphate, monofluorophosphate, polyphosphates, and tetrafluoroborate are formed according to Eqs. (315) to (318):

$$=P-O-P= \;+\; H-F \;\longrightarrow\; =P-F \;+\; =P-O-H \quad (315)$$

$$=P-F \;+\; H-O-H \;\longrightarrow\; =P-O-H \;+\; H-F \quad (316)$$

$$B-O-Si- \;+\; 2\,H-F \;\xrightarrow{min}\; B-F \;+\; -Si-F \;+\; H-O-H \quad (317)$$

$$B-O-H \;+\; H-F \;\xrightarrow{h}\; B-F \;+\; H-O-H \quad (318)$$

Typical anion chromatograms of two different BPSG-films (A, B) and a standard (C) are illustrated in Fig. 9-88. Separation was carried out on IonPac AS4A with a carbonate/bicarbonate eluant mixture. Chromatogram A was obtained after dissolving an APCVD-film (APCVD: Atmospheric Pressure Chemical Vapor Deposition) that exhibited good flow properties in 0.2% HF solution. Chromatogram B results from a PECVD-film (PECVD: Plasma Enhanced Chemical Vapor

Deposition) with significantly worse flow properties, and Chromatogram C represents a corresponding reference standard with 2 mg/L orthophosphite, 4 mg/L orthophosphate, and 4 mg/L tetrafluoroborate in 0.2% HF solution.

If BPSG-films contain more than 5% phosphorus, oxygen-containing phosphorus, depending on the deposition parameters, can be formed as well, yielding monofluorophosphate and polyphosphates after treatment with HF. Today, polyphosphates can be analyzed without difficulty on IonPac AS16 with a hydroxide gradient. Moreover, one can easily distinguish among borosilicates, boron oxide or boric acid, because their conversion to tetrafluoroborate according to Equations (317) and (318) is not equally fast.

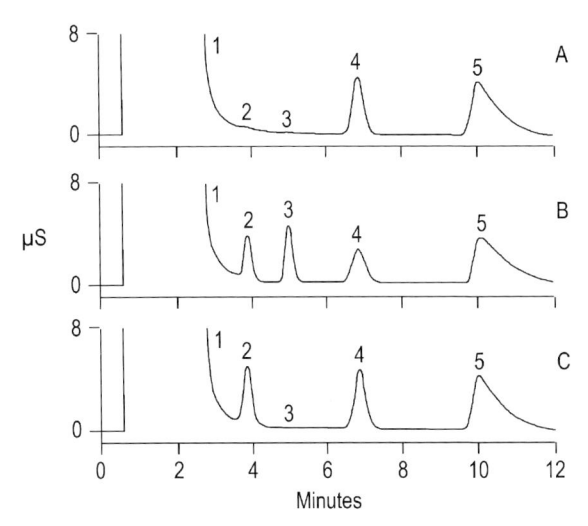

Fig. 9-88. Anion chromatograms of various BPSG-films. — Separator column: IonPac AS4A; eluant: carbonate/bicarbonate mixture; flow rate: 2 mL/min; detection: suppressed conductivity; samples: (A) APCVD-film with good flow properties, (B) PECVD-film with bad flow properties, (C) reference standard with 2 mg/L orthophosphite, 4 mg/L orthophosphate, and 4 mg/L tetrafluoroborate; peaks: fluoride (1), orthophosphite (2), monofluorophosphate (3), orthophosphate (4), and tetrafluoroborate (5); (taken from [156]).

9.3.5
Other Applications

Ion chromatography is not only used for monitoring the purity of water and chemicals, but also for analyzing various plating baths such as acid and electroless copper baths, tin/lead baths, electrolytic nickel baths, and gold baths. The analysis of the key components in such baths will be described in more detail in Section 9.4.

Even after the manufacturing of printed circuit boards and semiconductor devices, ion chromatography can be fruitfully utilized. Malfunctions are often caused by anions responsible for corrosion at the printed circuit board surface or inside the enclosure in which the microchip is embedded. Identification and quantification of those compounds by ion chromatographic techniques is often a significant aid in localizing contamination sources. One of the most important contamination sources is the welding process, because the welding agents used may contain corrosive anions. After the welding process, such anions cause corrosion inside the microchip housing. Figure 9-89 shows this effect with the analysis of mineral acids in semiconductor housings. For determining anions, the housing of malfunctioning semiconductor components is forced open and extracted with 2 mL of de-ionized water. The chromatogram in Fig. 9-89 is obtained after direct injection of such an extract. Besides relatively high concentrations of chloride and nitrate, a number of other mineral acids as well as oxalic acid could be detected in small concentrations. In this particular case, ion chromatography helped to optimize the welding process by informing the selection of suitable welding agents.

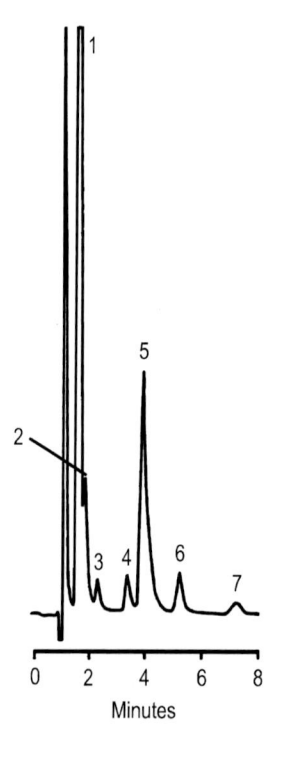

Fig. 9-89. Analysis of mineral acids in a semiconductor housing. — Separator column: IonPac AS4; eluant: 2.8 mmol/L $NaHCO_3$ + 2.2 mmol/L Na_2CO_3; flow rate: 2 mL/min; detection: suppressed conductivity; injection: 50 µL sample (undiluted); solute concentrations: 2.2 mg/L chloride (1), nitrite (2), 0.1 mg/L orthophosphate (3), 0.4 mg/L bromide (4), 2.8 mg/L nitrate (5), 0.4 mg/L sulfate (6), and oxalate (7).

In the manufacturing of integrated circuit boards with geometries in the sub-µm range, one can only counteract the contamination risk by completely monitoring all processes in wafer manufacturing. This not only includes the quality of water and chemicals but also the quality of air in clean rooms (class 1). In the past, air contamination did not attract much interest, as indicated by the lack of a SEMI directive on the subject. In the course of characterizing health risks by contaminations in air, a vast number of analytical methods were suggested. For the analysis of inorganic acids, for example, titrimetric, colorimetric, gravimetric, potentiometric, and ion chromatographic techniques have been suggested, of which only ion chromatography has the potential for analyzing multi-component mixtures at trace level. Based on the work of Cassinelli [158], Lue et al. [159] developed a method for measuring acidic contaminations in clean rooms. For collecting the sample they used a glass pipe (11 cm × 7 mm o.d.) that was filled in the upper part with 400 mg silica. The lower part, separated from the upper one by a plug of polyurethane foam, contains another 200 mg silica. The inlet of the glass pipe is covered with a fiber glass filter for retaining particles. For collecting samples the glass pipe is connected to a pump and the air sample to be analyzed is sucked through the glass pipe at a flow rate of 0.216 L/min. After 24 hours, the two silica segments are transferred into clean vessels. 10 mL of eluant is added, and the vessel is heated in a boiling water bath for 10 minutes to desorb the analytes. After the solution is cooled down, it is filtrated and directly injected into the ion chromatograph. Alternatively, Lue et al. tested an arrangement with two wash bottles in series, each filled with 7 mL eluant as an absorber solution, maintaining flow rate and sampling time. The authors used an IonPac AS4A conventional anion exchanger as a stationary phase for anion separation with a standard mixture of 1.8 mmol/L sodium carbonate and 1.7 mmol/L sodium bicarbonate as an eluant. The injection volume was 500 µL. Depending on the sampling location, almost all common mineral acids can be detected in clean room air. Concentrations vary between 100 and 900 nmol/m^3. The main contaminations are hydrochloric acid and sulfuric acid, both of which can be detected with either of the sampling techniques described above. In contrast, hydrofluoric acid can only be detected with the silica absorber. Presumably, fluoride is retained in the wash bottle due to interactions with the glass material. Separate analysis of the two silica segments allows a statement of the efficiency of the gas absorption by silica, which is 100% for HF and 91-100% for HCl.

An automated system for analyzing air was introduced by Roehl and Doyle [160] as a result of a research project. The system is schematically depicted in Fig. 9-90 and comprises a sampling part (framed segment) and an ion chromatograph. The chromatography setup consists of an analytical pump, a conductivity detector, an injection valve with a concentrator column, a separator column and a suppressor. The sampling part consists of a vacuum pump, a syringe pump, a flow meter, and a gas wash bottle. As in the arrangement by Lue et al., clean room air is sucked through the gas wash bottle, which contains an absorber solution with a vacuum pump. The flow rate is adjusted to 1 L/min with the flow meter. The great advantage of such an arrangement is the ability to calibrate with liquid standards. A defined volume of this standard is pumped from an

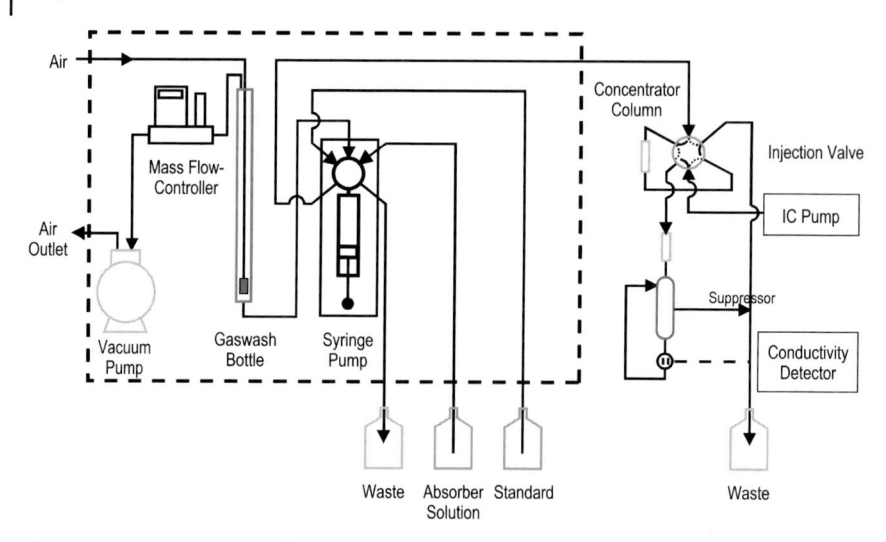

Fig. 9-90. Schematic setup of an automated system for air analysis.

appropriate reservoir through the concentrator column with the syringe pump. Prior to taking the sample, the gas wash bottle is rinsed several times with absorber solution and finally filled with 20 mL of this solution. After a defined volume of air has been sucked through a bottle, an aliquot of the absorber solution is transferred onto the concentrator column with the syringe pump. Vacuum pump and syringe pump are controlled by the ion chromatograph via TTL relays. While the vacuum pump is only switched on and off, a number of operations are carried out with the syringe pump that are controlled with special software.

Roehl and Doyle used a dilute hydrogen peroxide solution as an absorber solution, in which various gases can be absorbed with an efficiency of more than 90%. The only exception is NO_2, which exhibits an absorption efficiency in hydrogen peroxide of 20% at best. This efficiency can be increased to more than 90% with a mixture of triethanolamine and 2-propanol, but the higher pH of this mixture also leads to the absorption of large amounts of CO_2, and thus interferes with the determination of chloride. Therefore, the mixture of triethanolamine and 2-propanol can only be used for analyzing NO_2.

The separation of cations is usually carried out on IonPac CS12A with sulfuric acid or methanesulfonic acid as an eluant. For anion analysis, carbonate-selective anion exchangers such as IonPac AS12A or AS14 have proved to be suitable. Alternatively, gradient elution on IonPac AS11 can be used. As a matter of principle, 2-mm columns are utilized in order to reach the required sensitivity. With a gas flow of 1 L/min, sampling times are between 10 and 40 minutes; the preconcentrated volume ranges from 1 mL to 10 mL. Metal cations are rarely detected in clean room air because they are usually bound to particles. Because clean room air contains very few particles, only ammonia is detected in measurable amounts. Figure 9-91 shows such a chromatogram with 11 nL/L ammonia, which was obtained after pre-concentration of a 5 mL absorber solution. The detection limits for cations that can be obtained with this technique are summarized in Table 9-9.

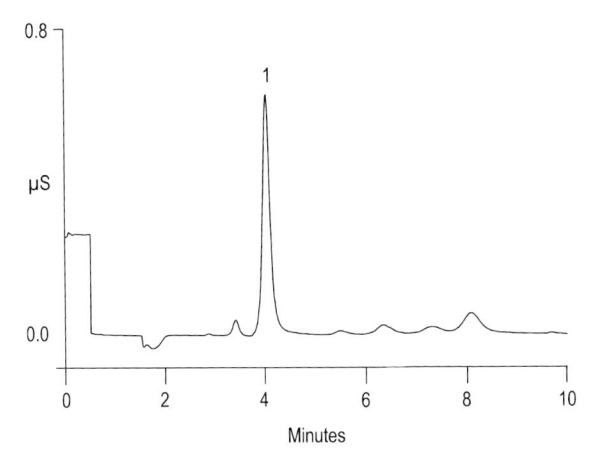

Fig. 9-91. Trace analysis of ammonia in clean room air.
— Separator column: IonPac CS12; eluant: 10 mmol/L methanesulfonic acid; flow rate: 1 mL/min; detection: suppressed conductivity; pre-concentrated volume: 5 mL; concentrator column: TCC-LP1; peaks: 11 nL/L ammonia (1) as ammonium.

Table 9-9. Detection limits for cations in clean room air with an automated system for analyzing air (based on a gas volume of 10 L and a pre-concentrated volume of 5 mL).

Analyte	Detection limit [ng/L air]
Lithium	0.02
Sodium	0.02
Ammonium	0.06
Potassium	0.03
Magnesium	1.0
Calcium	1.0

Finally, the typical anion chromatogram in Fig. 9-92 exemplifies the trace analysis of sulfur dioxide; it was obtained under standard conditions on IonPac AS4A after pre-concentration of 5 mL absorber solution. The sulfate peak represents 10 nL/L sulfur dioxide. The detection limits for various acidic gases are summarized in Table 9-10.

Table 9-10. Detection limits for acidic gases with an automated system for analyzing air (based on a gas volume of 10 L and a pre-concentrated volume of 5 mL).

Analyte	Detection limit [ng/L air]
HCl	0.2
HF	2.5
NO_2	2.0
Acetic acid	8.0
Formic acid	4.0
SO_2	3.0

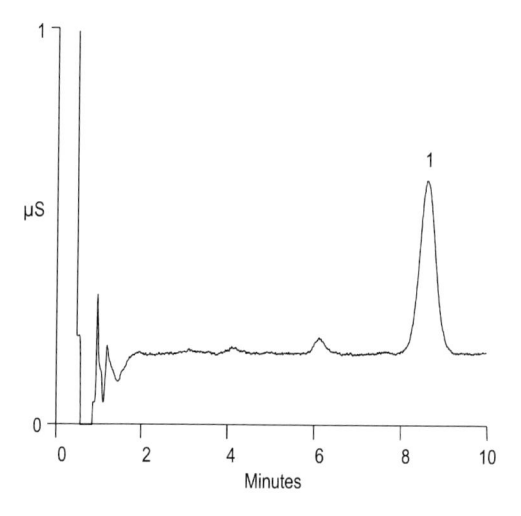

Fig. 9-92. Trace analysis of sulfur dioxide. — Separator column: IonPac AS4A; eluant: 1.8 mmol/L Na_2CO_3 + 1.7 mmol/L $NaHCO_3$; flow rate: 1 mL/min; detection: suppressed conductivity; pre-concentrated volume: 5 mL; concentrator column: TAC-LP1; peaks: 10 nL/L sulfur dioxide (1) as sulfate.

9.4
Ion Chromatography in the Electroplating Industry

Within a short time, the electroplating industry has become an important field of application for ion chromatography [161, 162]. Attention is being focused on the routine monitoring of plating baths and the improvement of the plating quality in dependence on the chemical composition. As production costs rise for high quality products, it has become necessary to use novel analytical methods that contribute to a better understanding of the complex chemical processes that occur in plating baths.

Ion chromatographic analysis methods ensure speed and high precision in the analysis of main components as well as of reaction and decomposition products in electroplating baths. The advantage of ion chromatography as compared to the somewhat unspecific wet-chemical methods utilized in the past lies in the selectivity of the stationary phases and the detection systems used. In most cases, sample preparation is limited to a simple dilution with de-ionized water and subsequent filtration. A variety of applications concerning electrodeposition and electroless plating are summarized in Table 9-11.

Table 9-11. Application areas and examples of analyses in the electroplating industry.

Type of plating bath	Examples of Analysis
Copper (electroless)	Determination of formic acid, tartaric acid, triethanolamine, EDTA, and $Cu(EDTA)^{2-}$
Copper sulfate (electrolytic)	Determination of the Cl^-/SO_4^{2-} ratio, metallic impurities such as Ni^{2+} and Zn^{2+}, organic additives
Copper pyrophosphate (electrolytic)	Determination of ammonium, orthophosphate, and nitrate
Copper cyanide (electrolytic)	Determination of Cu^{2+} after preparation with strong acids, hexacyano ferrates
Nickel sulfamate (electrolytic)	Determination of sulfamate, chloride, sulfate, and ammonium
Nickel/Iron (electrolytic)	Determination of the Ni^{2+}/Fe^{2+} ratio, Na^+, Cu^{2+}, boric acid, saccharin, and lauryl sulfate
Nickel Nickel/Cobalt (electroless)	Determination of hypophosphite, orthophosphite, citric acid, succinic acid, Ni^{2+}/Co^{2+} ratio
Gold cyanide (electrolytic)	Determination of $Au(CN)_2^-$, $Au(CN)_4^-$, $Co(CN)_6^{3-}$, cyanide, chloride, orthophosphate, and hexacyano ferrates

9.4.1
Analysis of Inorganic Anions

There are numerous applications for the determination of mineral acids in the field of plating bath analyses. In copper sulfate plating baths based on electrodeposition, it is possible to simultaneously analyze the small amount of chloride contained in these baths along with the main component, sulfate (Fig. 9-93).

An important constituent in copper pyrophosphate baths is nitrate, which enhances the maximum permissible current density [163]. The chromatogram in Fig. 9-94 shows the separation of nitrate and orthophosphate. The latter is the hydrolysis product of pyrophosphate and is formed during the plating process. If gradient elution with a hydroxide gradient is applied instead of an isocratic technique, the main component, pyrophosphate, may also be separated on a suitable anion exchanger and detected via suppressed conductivity.

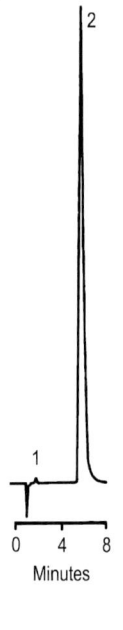

Fig. 9-93. Trace analysis of chloride in an acid copper sulfate plating bath. — Separator column: IonPac AS4; eluant: 2.8 mmol/L NaHCO$_3$ + 2.2 mmol/L Na$_2$CO$_3$; flow rate: 2 mL/min; detection: suppressed conductivity; injection: 10 µL sample (1:1000 diluted); peaks: (1) chloride and (2) sulfate.

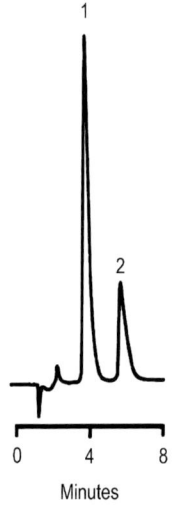

Fig. 9-94. Analysis of mineral acids in a copper pyrophosphate bath. — Separator column: IonPac AS3; eluant: 3 mmol/L NaHCO$_3$ + 2.8 mmol/L Na$_2$CO$_3$; flow rate: 2.3 mL/min; detection: suppressed conductivity; injection: 10 µL sample (1:1000 diluted); peaks: (1) orthophosphate and (2) nitrate.

Nickel plating baths based on electroless deposition contain a reducing agent and a catalyst. The reducing agent acts as an electron donor for the reduction of metal ions to metal. Hypophosphite, formaldehyde, hydrazine, and boron hydride are suited for this. A colloidal Pd/Sn^{2+} solution usually serves as a catalyst. This solution is adsorbed on the material surface and catalyzes the deposition of a monolayer of the corresponding metal. Further plating occurs autocatalytically to form hydride ions.

The following reaction scheme has been proposed for the electroless deposition of nickel with hypophosphite as reducing agent:

$$H_2PO_2^- + H_2O \xrightarrow{Pd/Sn^{2+}} HPO_3^{2-} + 2H^+ + H^-$$

$$2H^- + Ni^{2+} \longrightarrow Ni + H_2 \uparrow \qquad (319)$$

$$\overline{2H_2PO_2^- + 2H_2O + Ni^{2+} \longrightarrow Ni + H_2 \uparrow + 4H^+ + 2HPO_3^{2-}}$$

It is obvious from this reaction scheme that the deposition rate depends on the concentration of the reducing agent. Because both nickel and hypophosphite are exhausted in the plating process, both compounds must be added occasionally to the bath as corresponding salts. The electroless plating process is accompanied by an increase in the concentration of phosphite as a hypophosphite oxidation product, which limits the effectiveness of such a plating bath. Both phosphorus species can be determined simultaneously with the aid of ion chromatography, thus allowing the plating bath control to be optimized. Figure 9-95 shows the difference in the amounts of both compounds between a new and an exhausted plating bath after one and five metal turnovers, respectively.

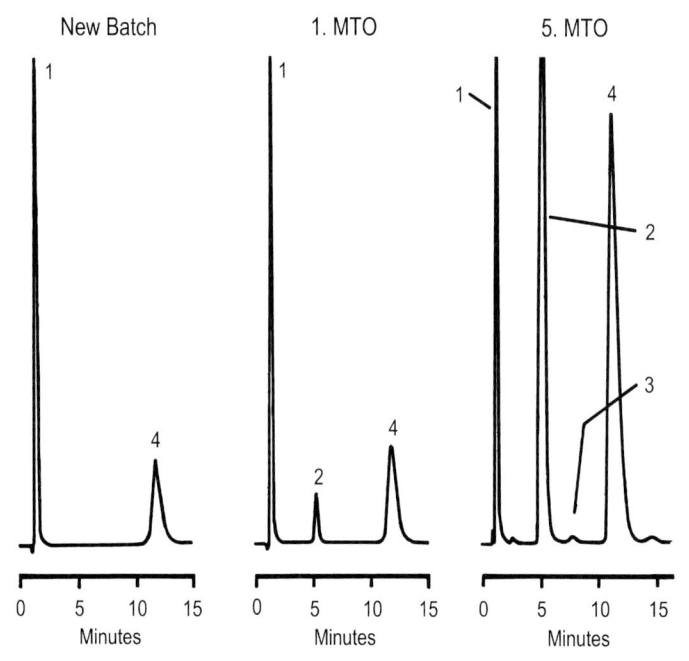

Fig. 9-95. Analysis of hypophosphite and orthophosphite in an electroless nickel bath. — Separator column: IonPac AS3; eluant: 3 mmol/L Na_2CO_3; flow rate: 2.3 mL/min; detection: suppressed conductivity; injection: 50 µL sample (1:200 diluted); peaks: (1) hypophosphite, (2) orthophosphite, (3) orthophosphate, and (4) sulfate.

If the nickel bath also contains organic acids such as lactic acid, it is necessary to apply a gradient technique, because lactic acid and hypophosphite have the same retention time under the isocratic conditions chosen in Fig. 9-95.

In nickel sulfamate baths the main component, sulfamate, can be determined together in the same run with sulfate as decomposition product and other bath constituents such as chloride and bromide. An example chromatogram is shown in Fig. 9-96. To achieve sufficient separation between sulfamate and chloride, both of which elute close to the system void, two identical anion exchangers were used in series. The total analysis time increased to about 20 minutes.

Fig. 9-96. Analysis of chloride, bromide, and sulfate in a nickel sulfamate bath. – Separator columns: 2 IonPac AS4A; eluant: 1.7 mmol/L NaHCO$_3$ + 1.8 mmol/L Na$_2$CO$_3$; flow rate: 1.5 mL/min; detection: suppressed conductivity; injection: 50 µL sample (1:1000 diluted); solute concentrations: sulfamate (1), 0.4 g/L chloride (2), 8.2 g/L bromide (3), and 1.1 g/L sulfate (4).

Large concentration differences between the individual anions are observed in nickel/zinc plating baths, in which both metals are introduced as sulfates with a total concentration of 130 g/L. In order to determine traces of other anions such as fluoride, chloride, and nitrate in such a matrix, the bath has to be strongly diluted and the excess of metal removed using a cation exchange cartridge. This is illustrated by the chromatogram in Fig. 9-97, which employed a Metrosep Anion Dual 2 anion exchanger.

In chromium baths, ion chromatography allows the determination of the major and minor anionic components [165, 166]. Major components include chromic acid itself and sulfuric acid acting as an anionic catalyst; minor components include fluoride or hexafluorosilicate, chloride, and organic acids such as methanesulfonic acid. Methanesulfonic acid is added to the bath to improve the deposition on nickel. Analytical monitoring of chromic acid plating baths is of

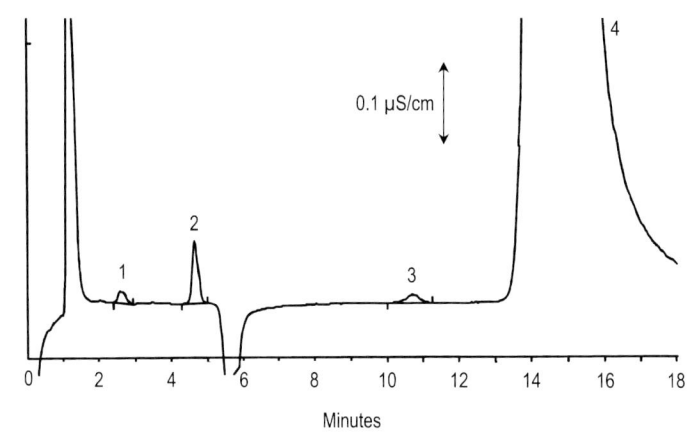

Fig. 9-97. Separation of fluoride, chloride, and nitrate in a nickel/zinc plating bath. — Separator column: Metrosep Anion Dual 2; eluant: 1.5 mmol/L Na_2CO_3 + 2 mmol/L $NaHCO_3$; flow rate: 0.8 mL/min; detection: suppressed conductivity; injection: 20 µL sample (1:400 diluted); sample preparation: treatment with a cation exchanger to remove metals; solute concentrations: 9.6 mg/L fluoride (1), 38.1 mg/L chloride (2), and 21.6 mg/L nitrate (3) in a matrix of 50 g/L nickel and 80 g/L zinc as sulfate (4); (taken from [164]).

great importance, because the mass ratio between chromic acid and the anionic catalyst determines the quality of deposition and, thus, is an important bath criterion. All major and minor components can be analyzed in the same run by ion chromatography. Due to the strongly different retention behavior of fluoride and chromate, an analysis under isocratic conditions would be very time-consuming. If the bath also contains disulfonic acids, the required resolution between disulfonic acids and chromate leads to a total analysis time of 40 minutes (see Fig. 9-98). However, total analysis time can be decreased significantly by applying gradient elution. A simple hydroxide gradient and a hydroxide-selective stationary phase are sufficient. The chromatogram of a chromic acid bath of a slightly different composition and a total analysis time of about 15 minutes (including re-equilibration to the initial conditions after the gradient run) shown in Fig. 9-99 clearly demonstrates the advantage of gradient elution. Depending on the bath composition, the chromatographic conditions can still be optimized for shorter analysis times. In the analysis of chromic acid plating baths, attention should be paid to adjusting the pH of the diluted sample to be in the alkaline range. Because chromium(VI) is present as dichromate in the acidic pH range, a repeated injection of this strongly oxidizing substance would damage the separator column irreversibly.

Besides various anions, the Cr(III)/Cr(VI) ratio also affects the deposition quality. If the Cr(III) content is too high, for example, the deposition is incomplete. In ion chromatography, two different methods exist for the simultaneous analysis of Cr(III) and Cr(VI). The first one is based on pre-column derivatization

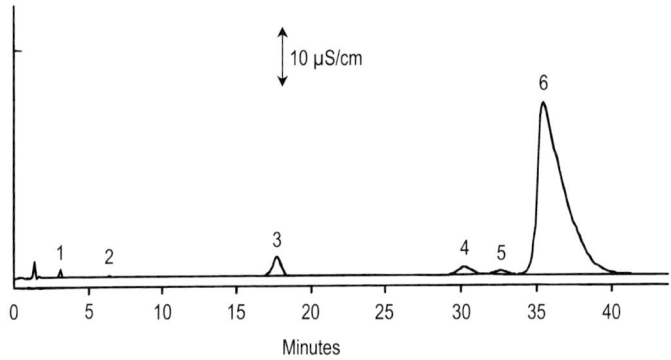

Fig. 9-98. Isocratic elution of anions in a chromic acid plating bath. – Separator column: Metrosep Anion Dual 2; eluant: 1.2 mmol/L Na_2CO_3 + 2 mmol/L $NaHCO_3$ – acetone (82:18 v/v); flow rate: 0.85 mL/min; detection: suppressed conductivity; injection: 5 µL sample (1:500 diluted); solute concentrations: 47 g/L methanesulfonic acid (1), 5 g/L chloride (2), 16 g/L sulfate (3), 28 g/L methanedisulfonic acid (4), 16 g/L ethanedisulfonic acid (5), and 1100 g/L chromate (6); (taken from [167]).

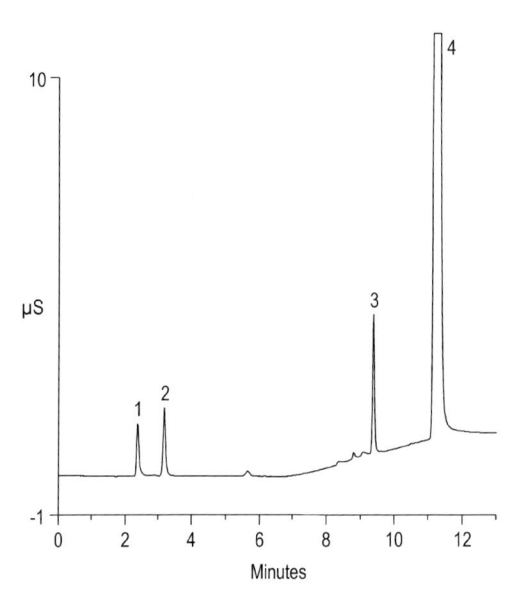

Fig. 9-99. Gradient elution of anions in a chromic acid plating bath. – Separator column: IonPac AS11; eluant: NaOH gradient; flow rate: 1 mL/min; detection: suppressed conductivity; injection: 25 µL sample (diluted 1:50 with 0.1 mol/L NaOH); solute concentrations: 11.4 mg/L fluoride (1), 81.5 mg/L methanesulfonic acid (2), 53.3 mg/L sulfate (3), and 6.64 g/L chromate (4).

with pyridine-2,6-dicarboxylic acid to form a stable monovalent anion, which can be separated together with chromate on a suitable anion exchanger. However, the derivatization reaction only occurs at elevated temperature and the pH value of the sample has to be adjusted very precisely. The more modern second method is based on the separation of both species on an ion exchanger that has anion- and cation exchange capacities, so that Cr(III) can be separated as a cation and Cr(VI) as an anion (see Section 4.5.3). Detection is carried out photometrically after post-column oxidation of Cr(III) to Cr(VI) with peroxodisulfate and a silver catalyst. The example chromatogram is illustrated in Fig. 9-100.

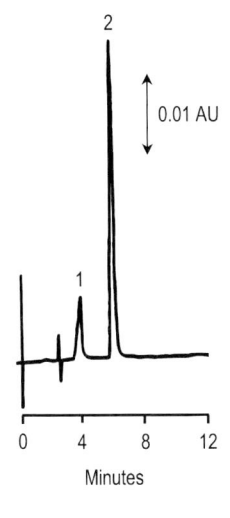

Fig. 9-100. Simultaneous analysis of chromium(III) and chromium(VI). – Separator column: IonPac AS7; eluant: 40 mmol/L $MgSO_4$ + 30 mmol/L $HClO_4$; flow rate: 0.5 mL/min; detection: photometry at 365 nm after post-chromatographic oxidation of chromium(III); derivatization reagent: 0.15 mol/L $K_2S_2O_8$ + 0.23 mmol/L $AgNO_3$; peaks: (1) chromium(III) and (2) chromium(VI); (taken from [165]).

Figure 9-101 shows the chromatogram of an electroplating sludge sample that was obtained after diluting the sample and treating it with a cation exchanger in the H$^+$ form. In addition to the major components fluoride, nitrate, and sulfate, small amounts of chloride and orthophosphate could also be detected.

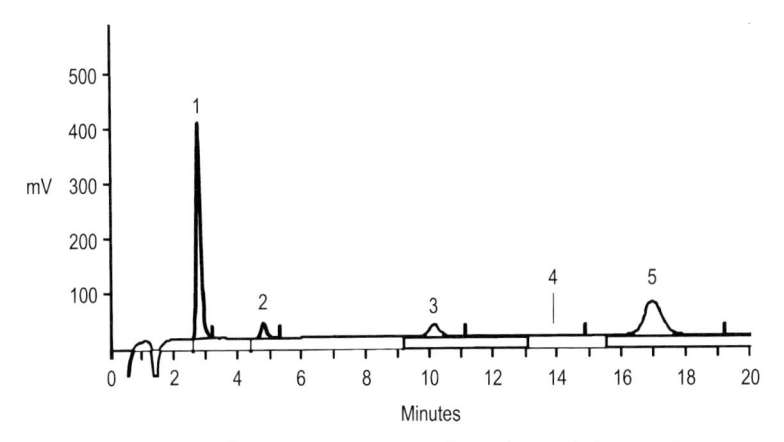

Fig. 9-101. Analysis of inorganic anions in an electroplating sludge sample. – Separator column: Metrosep Anion Dual 2; eluant: 1.3 mmol/L Na_2CO_3 + 2 mmol/L $NaHCO_3$; flow rate: 0.8 mL/min; detection: suppressed conductivity; injection: 20 µL sample (1:25 diluted); sample preparation: cation exchanger in the H$^+$ form; solute concentrations: 59.7 mg/L fluoride (1), 8.4 mg/L chloride (2), 24.1 mg/L nitrate (3), 2.0 mg/L orthophosphate (4), and 81.9 mg/L sulfate (5); (taken from [168]).

Weak inorganic acids such as borate and carbonate can be determined by means of ion-exclusion chromatography. In addition to chloride and sulfate, pure and alloyed nickel plating baths contain high concentrations of boric acid, which has a weak buffer effect. Conventional analysis methods for boric acid such as titrimetric determination are time-consuming and suffer from poor precision, because the titration end point is often difficult to recognize. The ion chromatographic analysis of boric acid in a nickel/iron plating bath is illustrated in Fig. 9-102. Again, the only sample preparation here was dilution with de-ionized water. Carbonate as an ingredient in gold plating baths that are used as buffer agents, for example, can be determined with a gross retention time of t_{ms} = 13.5 min under the same chromatographic conditions.

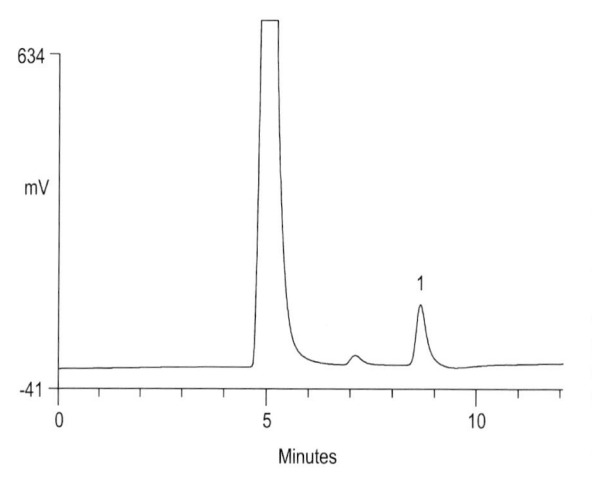

Fig. 9-102. Boric acid determination in a nickel/iron plating bath. — Separator column: IonPac ICE-AS1; eluant: 1 mmol/L octanesulfonic acid; flow rate: 1 mL/min; detection: suppressed conductivity; injection: 50 µL sample (1:1000 diluted); peaks: (1) borate.

9.4.2
Analysis of Metal Complexes

Traditionally, gold analysis in electroplating baths was carried out with conventional methods such as precipitation or titration techniques. However, these methods require time-consuming sample preparation to break the complex bond. Today, spectrometric methods such as atomic absorption spectrometry (AAS) or ICP-AES are usually applied, which are also somewhat problematic due to spectral interferences. Moreover, they do not allow speciation of individual oxidation states. This differentiation is important, because it is known from polarographic studies that gold(I) is oxidized to gold(III) in an acid gold bath [169]. Because trivalent gold cannot be deposited at the cathode, bath efficiency decreases with increasing gold(III) content [170]. Ion-pair chromatography is capable of determining the concentration ratio between gold(I) and gold(III). In the presence of an excess of cyanide, both gold species exist as anionic cyano complexes, $Au(CN)_2^-$ and $Au(CN)_4^-$. Owing to their stability, they may be separated by ion-pair chromatography and detected conductometrically. In Fig. 9-103 two chromatograms of an acid gold plating bath are compared; the samples were only diluted and filtrated prior to the measurement. The upper chromatogram in Fig. 9-103 represents a fresh bath, in which only gold(I) could be detected. The lower chromatogram was obtained from the same bath after nine months of operation. The decrease of the peak height for gold(I) in favor of the preceding gold(III) is significant. Due to the cyanide content in the mobile phase, this technique can also be used for analyzing gold baths that are not based on cyanide. Other precious metals such as silver, platinum, and palladium can also be analyzed in this way under modified chromatographic conditions (see also Fig. 6-23 in Section 6.4). Cobalt sulfate is added as a hardening agent to some plating baths. The cyano complex, $Co(CN)_6^{3-}$, that is formed during the plating process and that is ineffective as a hardening agent can be separated from the gold-cyano complexes (Fig. 9-104), because it elutes much later as a trivalent anion.

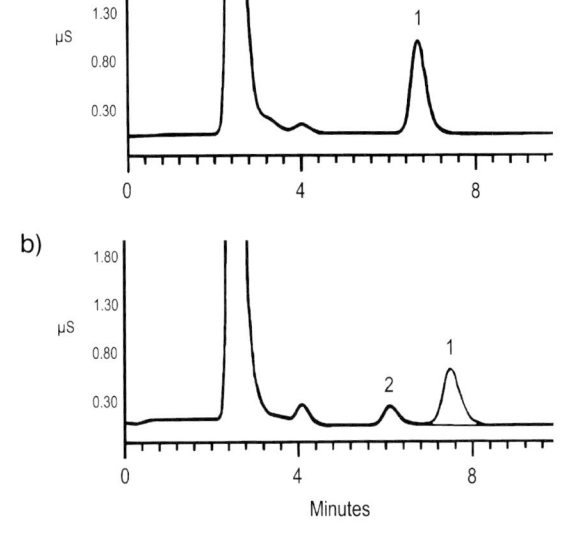

Fig. 9-103. Separation of gold(I) and gold(III) as cyano complexes in a gold plating bath. – Separator column: IonPac NS1; eluant: 2 mmol/L TBAOH + 5 mmol/L NaCN – methanol (40:60 v/v); flow rate: 1 mL/min; detection: suppressed conductivity; injection: 50 µL sample (1:1000 diluted); solute concentrations: (a) 9.46 mg/L gold(I) (1), (b) 6.65 mg/L gold(I) (1) and 2.86 mg/L gold(III) (2); (taken from [171]).

Metal-EDTA complexes may be separated by anion exchange chromatography (see Fig. 3-114 in Section 3.7.4). They are detected again via electrical conductivity measurements. In electroless copper baths, for example, it is possible to distinguish between free and complex-bound EDTA. Other bath constituents do not interfere with the determination of the copper-EDTA complex, $Cu(EDTA)^{2-}$.

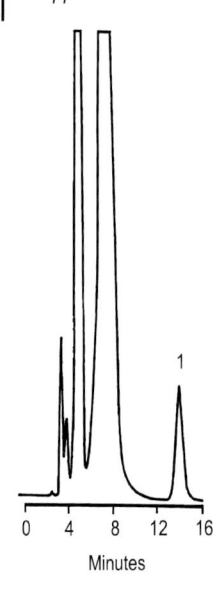

Fig. 9-104. Determination of the cobalt-cyano complex in a gold plating bath. — Separator column: IonPac NS1; eluant: 2 mmol/L TPAOH — acetonitrile (90:10 v/v); flow rate: 1 mL/min; detection: suppressed conductivity; injection: 50 µL sample (1:400 diluted); peak: (1) $Co(CN)_6^{3-}$.

9.4.3
Analysis of Organic Acids

Organic acids are usually separated by ion-exclusion chromatography and detected via their electrical conductivity. Applications are mainly in the field of copper and electroless nickel plating baths.

Normally, copper baths contain formaldehyde and a complexing agent such as tartaric acid. During the plating process, formaldehyde is oxidized to formic acid. Ion chromatography provides simultaneous detection of both acids. Therefore, the aging processes of the plating bath, which are indicated by increased formate content, can be immediately recognized. Figure 9-105 shows a chromatogram of organic acids in a zinc bath, which was obtained using an ion-exclusion phase and non-suppressed conductivity detection. The main emphasis was the determination of gluconic acid and salicylic acid. Again, in this case, the simultaneous detection capability of ion chromatography is impressive, even though total analysis time exceeds 30 minutes due to the high affinity of salicylic acid towards the stationary phase. As with almost all electroplating baths, the sample has to be strongly diluted as zinc electrolytes usually contain 300 g/L chloride and similar concentrations of potassium and zinc, which can be kept in the liquid form only by elevating the temperature to about 60 °C. In special cases, anion exchange chromatography can also be applied for separating organic acids in electroplating baths. Maquieira et al. [172] exemplified this with the determination of succinic acid in a zinc electrolyte.

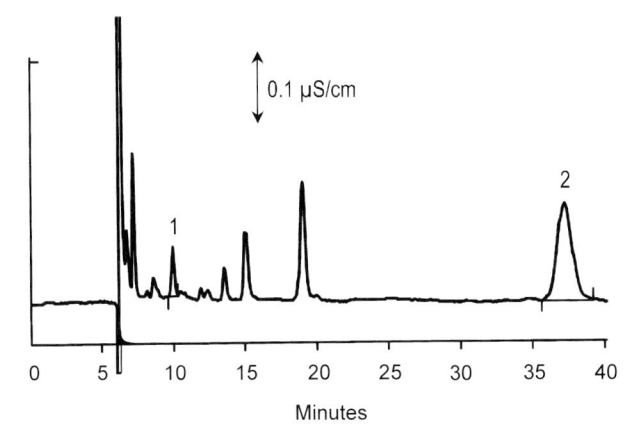

Fig. 9-105. Analysis of organic acids in a zinc bath. — Separator column: Metrosep Organic Acids; eluant: 0.5 mmol/L H_2SO_4 — aceton (85:15 v/v); flow rate: 0.5 mL/min; detection: non-suppressed conductivity; injection: 100 µL sample (1:1000 diluted); sample preparation: treatment with a cation exchanger in the H^+ form; solute concentrations: 4.5 g/L gluconic acid (1) and 6.1 g/L salicylic acid (2); (taken from [173]).

Ion-pair chromatography is capable of separating short-chain sulfonic acids such as allyl- and vinyl-sulfonic acid that are sometimes present in nickel baths. Conventional anion exchangers and ion-exclusion phases cannot be used because sulfonic acids elute as strong organic acids near the system void, making it impossible to separate them from electrolytic bath constituents. As short-chain sulfonic acids exhibit only a low hydrophobicity, the comparatively solvophobic tetrabutylammonium hydroxide is utilized as an ion-pair reagent. Detection is accomplished by measuring the electrical conductivity because olefinic sulfonic acids also do not possess suitable chromophores that enable a sensitive UV detection. The chromatogram of a vinyl- and allyl-sulfonic acid standard in Fig. 9-106 clearly illustrates that the retention behavior of these compounds resembles that of inorganic anions such as chloride and sulfate.

9.4.4.
Analysis of Inorganic Cations

While alkali- and alkaline-earth metals can also be rapidly and very sensitively detected by other instrumental analysis methods, the advantage of ion chromatography lies in its ability to simultaneously detect the ammonium ion. In copper pyrophosphate baths, for example, the addition of ammonia improves the plating evenness. However, because the ammonia concentration continuously decreases

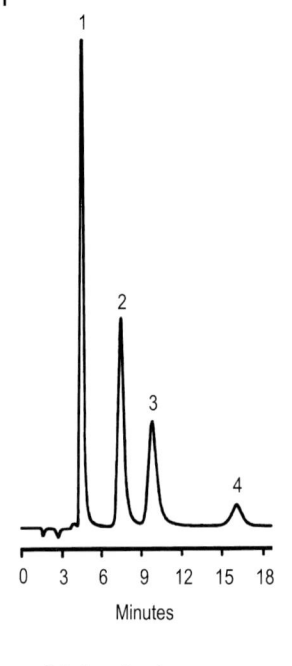

Fig. 9-106. Separation of short-chain olefinic sulfonic acids. — Separator column: IonPac NS1 (10-μm); eluant: 2 mmol/L TBAOH + 1 mmol/L Na_2CO_3 — acetonitrile (92:8 v/v); flow rate: 1 mL/min; detection: suppressed conductivity; injection volume: 50 μL; solute concentrations: chloride (1), 30 mg/L vinylsulfonic acid (2), 30 mg/L allylsulfonic acid (3), and sulfate (4).

at higher bath temperatures, it must be added to maintain optimal bath conditions. As seen in Fig. 9-107, ammonium can be determined quickly and reliably together with sodium and potassium after separation on a surface-sulfonated cation exchanger. If the bath also contains alkaline-earth metals, a weak acid cation exchanger would be used instead of the surface-sulfonated one, and the acid concentration in the mobile phase would be adjusted accordingly.

Ion chromatography provides an alternative to atom spectrometric methods for the determination of transition metals. Its main advantage is the simultaneity of the procedure and the ability to distinguish between different oxidation states. For example, the determination of iron(III), copper, and zinc in a chromic acid bath (Fig. 9-108) can be performed free of interferences despite the high chromium(VI) load. The two oxidation states of iron can be clearly distinguished via ion chromatography.

9.4.5
Analysis of Organic Additives

Organic additives such as brighteners, carriers, and levelers play an important role in regulating metal deposition and thus, have a significant influence on the plating quality. Usually, they consist of several components that affect the deposition in different ways. With high demands on the physical properties of the deposited metal, monitoring the concentrations of the individual components and the chemical composition of the bath is absolutely necessary. During the

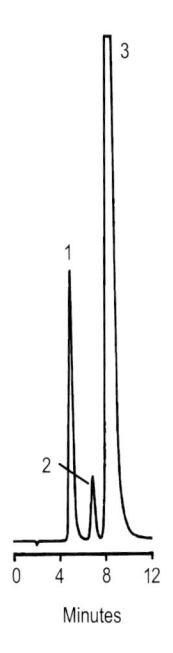

Fig. 9-107. Analysis of ammonium in a copper pyrophosphate bath. — Separator column: IonPac CS1; eluant: 5 mmol/L methanesulfonic acid; flow rate: 2.3 mL/min; detection: suppressed conductivity; injection: 50 µL sample (1:2500 diluted); peaks: (1) sodium, (2) ammonium, and (3) potassium.

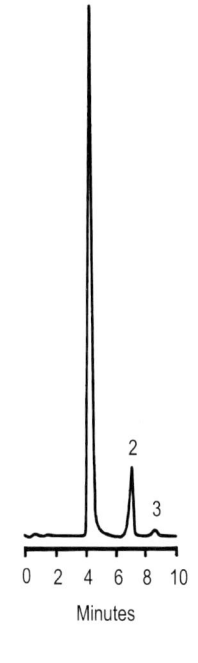

Fig. 9-108. Determination of iron(III), copper, and zinc in a chromic acid plating bath. — Separator column: IonPac CS5; eluant: 6 mmol/L pyridine-2,6-dicarboxylic acid + 8.6 mmol/L LiOH; flow rate: 1 mL/min; detection: photometry at 520 nm after reaction with PAR; injection: 50 µL sample (1:1000 diluted); solute concentrations: 3.7 g/L iron(III) (1), 1.3 g/L copper (2), and 0.17 g/L zinc (3).

plating process the concentration of the additives is subject to variations, as individual components can decompose electrochemically. If the concentrations of the additives change, the plating quality likewise changes. Although organic additives are extremely important for the plating process, almost nothing is found in the literature regarding the analysis of these compounds. Over many years, only some tests with the Hull probe were carried out, with which the total content of organic compounds can be estimated. However, the Hull probe does not provide quantitative data and cannot differentiate between individual organic additives. Towards the end of the 1970s, a polarographic technique known as Cyclic Voltammetric Stripping (CVS) was applied for analyzing organic additives [174]. The working principle of this method is based on an alternating deposition and stripping of copper on a rotating electrode surface. The current necessary for stripping deposited metal indicates the total concentration of organic additives. However, even CVS is subject to a number of limitations, as it also does not allow any statements on the concentration of individual organic components. Moreover, CVS is interfered with by degradation products and organic contaminants. CVS measurements are extremely sensitive to temperature changes and the rotary speed of the electrode.

Among the few chromatographic papers published in literature is the one by Reid [175], who describes the separation of organic additives in an acid copper bath utilizing a neutral polymer and UV detection. The biggest problem with determining individual organic components directly is the fact that, for competitive reasons, the manufacturers of such additives do not give any information regarding composition. Only in a few cases is the identity of certain additives known. Nickel/iron baths, for example, contain saccharin and surface-active compounds such as sodium lauryl sulfate. As was revealed in an investigation by Becker and Bolch [176], an appropriate concentration range has to be maintained to ensure constant plating quality. The determination of sodium lauryl sulfate is usually carried out by ion-pair chromatography using conductivity detection. The corresponding chromatogram in Fig. 9-109 reveals a baseline-resolved separation of this compound in a nickel/iron plating bath.

Three different techniques exist for the ion chromatographic determination of saccharin. Because saccharin is anionic at alkaline pH, anion exchange chromatography with conductivity or UV detection can be used (see Fig. 9-177 in Section 9.6). In ion-pair chromatography, tetramethylammonium hydroxide is used as an ion-pair reagent and the mobile phase is adjusted to pH 12 with NaOH. The respective chromatogram is illustrated in Fig. 9-110. Again, conductivity detection was used in this case. Last but not least, in the application note Number O-11 [177], Metrohm describes the separation of saccharin together with citric acid on an ion-exclusion phase with non-suppressed conductivity detection. The chromatogram in Fig. 9-111 reveals that saccharin can also be separated from all other bath constituents with this separation mode. However, the system peak that appears after 20 minutes significantly prolongs the total analysis time.

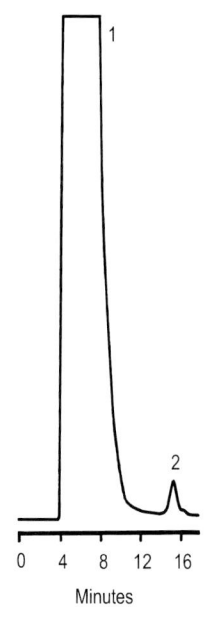

Fig. 9-109. Determination of sodium laurylsulfate in a nickel/iron plating bath. — Separator column: IonPac NS1 (10-µm); eluant: 10 mmol/L NH$_4$OH — acetonitrile (72:28 v/v); flow rate: 1 mL/min; detection: suppressed conductivity; injection: 50 µL sample (1:10 diluted); peaks: (1) chloride and sulfate, (2) sodium lauryl sulfate.

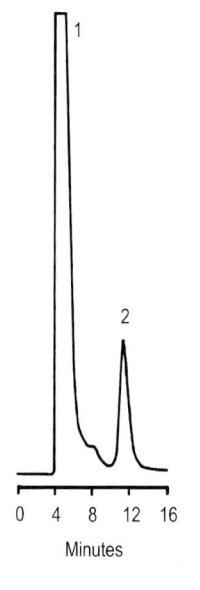

Fig. 9-110. Ion-pair chromatographic determination of saccharin in a nickel/iron bath. — Separator column: IonPac NS1 (10-µm); eluant: 2 mmol/L TMAOH (pH 12 with NaOH) — acetonitrile (95:5 v/v); flow rate: 1 mL/min; detection: suppressed conductivity; injection: 50 µL sample (1:50 diluted); peaks: (1) chloride and sulfate, (2) saccharin.

Apart from saccharin and sodium laurylsulfate, thiourea derivatives such as 2-imidazolidinthione and N,N'-diphenylthiourea are also of great importance. They are employed, for example, in electroless nickel and nickelborohydride plating baths as brighteners.

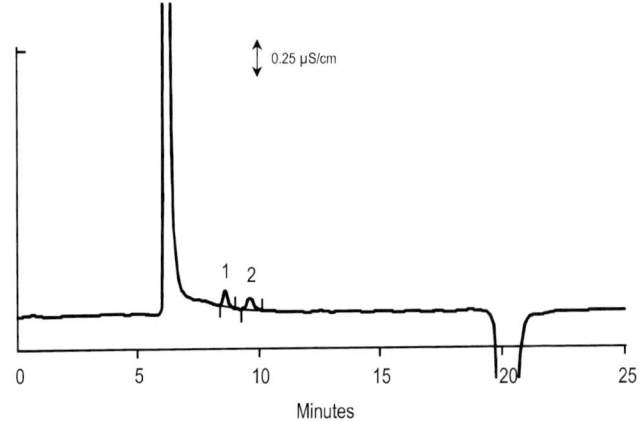

Fig. 9-111. Determination of saccharin in a nickel/iron bath by ion-exclusion chromatography. – Separator column: Metrosep Organic Acids; eluant: 0.5 mmol/L H_2SO_4 – acetone (85:15 v/v); flow rate: 0.5 mL/min; detection: non-suppressed conductivity; injection: 100 µL sample (1:1000 diluted); sample preparation: treatment with cation exchanger in the H^+ form; solute concentrations: 4.2 g/L citric acid (1) and 0.64 g/L saccharin (2); (taken from [177]).

To separate these basic substances, a neutral polymer is again suitable as a stationary phase; detection is performed by measuring light absorption of existing chromophores. The chromatographic conditions that have been developed for the elution of each compound differ only in the acetic acid and organic solvent content in the mobile phase. The corresponding chromatograms are shown in Fig. 9-112 and 9-113.

The separation of organic additives in acid copper plating baths on a conventional reversed-phase column with UV detection was published by Grushka et al. [178]. The characterization of brighteners in this kind of plating bath is extremely difficult due to the high electrolyte content. A typical acid copper bath contains 15 to 22.5 g/L copper salt, 170 to 230 g/L sulfuric acid, and 30 to 80 g/L chloride; the total content of organic additives is usually below 1%. To decrease the high concentrations of copper and sulfuric acid, Grushka et al. used a solid-phase extraction cartridge. Without this type of sample preparation, the intrinsic absorption of the copper salt would mask the UV absorption of the organic analyte. Additives commercially available in Europe are mostly derivatives of long-chain alkylsulfonates, which Grushka et al. characterized by means of ion-pair chro-

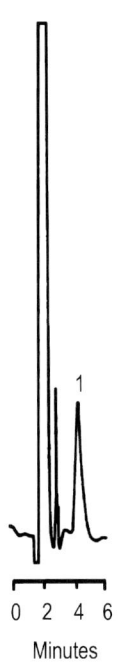

Fig. 9-112. Determination of 2-imidazolidinthione in an electroless nickel plating bath. — Separator column: IonPac NS1; eluant: acetic acid — methanol — water (1:5:94 v/v/v); flow rate: 0.8 mL/min; detection: UV (254 nm); injection: 50 µL sample with 3 mg/L of the relevant substance (1).

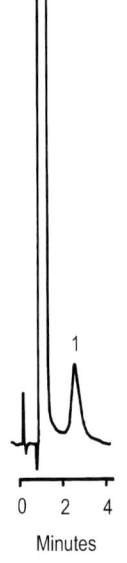

Fig. 9-113. Determination of N,N'-diphenylthiourea in an electroless nickelborohydride plating bath. — Separator column: IonPac NS1; eluant: acetic acid — methanol — acetonitrile — water (1:39:31:29 v/v/v/v); flow rate: 0.8 mL/min; detection: UV (254 nm); injection: 50 µL sample with 2 mg/L of the relevant substance (1).

matography using tetrabutylammonium hydroxide as an ion-pair reagent. Figure 9-114 shows the chromatogram of an active bath. Through comparison to a given standard, peaks 1 and 2 could be identified as components of the brightener. Peak 3 is a degradation product, which cannot be quantified for lack of information about its extinction coefficient. The third organic additive component, the carrier, is not visible because it is a non-chromophoric compound.

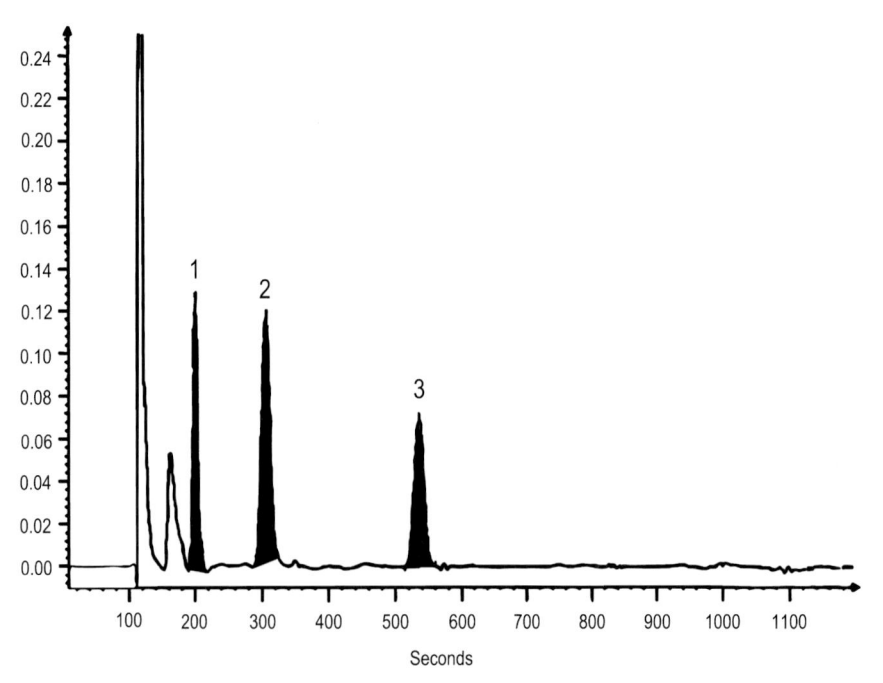

Fig. 9-114. Determination of organic additives in an acid copper bath after solid-phase extraction. – Separator column: 250 mm × 4.5 mm i.d. Lichrosorb RP-18; eluant: 3 mmol/L TBAOH – acetonitrile (91:9 v/v); flow rate: 1 mL/min; detection: UV (200 nm); peaks: (1) and (2) brighteners, (3) degradation product; (taken from [178]).

As an alternative to UV detection, Heberling et al. [179] applied pulsed amperometric detection on a gold working electrode for analyzing organic additives in Sel-Rex Cubath (Sel-Rex/OMI Corp.) and LeaRonal PC Gleam (Lea Ronal Corp.). The separations were carried out on a neutral DVB polymer with sulfuric acid/acetonitrile mixtures. As an example, Fig. 9-115 shows a chromatogram of an acid copper bath with a Sel-Rex Cubath additive that was obtained by direct injection of a 50-µL sample. The carrier component, actually a brightener, elutes after 5.5 minutes and the degradation product elutes after 3.5 minutes. The pulse sequence for the electrochemical detection consists of three potentials: $E1$: 0.85 V (t_1 = 480 ms), $E2$: 1.5 V (t_2 = 60 ms) and $E3$: -0.2 V (t_3 = 180 ms). The authors stated the precision for 89 subsequent analyses to be less than 4%

RSD with this detection method. Because the amperometric cell has to be manually cleaned only every one to two weeks, such an analysis technique can also be utilized for online bath monitoring.

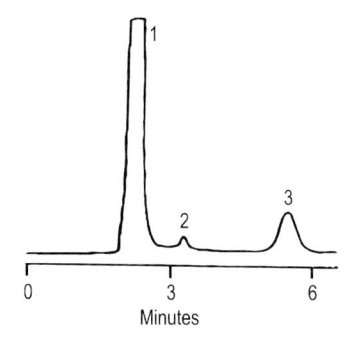

Fig. 9-115. Determination of organic additives in an acid copper bath with a Sel-Rex additive. — Separator column: IonPac NS1 (10-µm); eluant: 0.1 mol/L H_2SO_4 — acetonitrile (95:5 v/v); flow rate: 1 mL/min; detection: pulsed amperometry on a gold working electrode; injection: 50 µL sample; peaks: (1) copper sulfate, (2) degradation product, and (3) carrier component; (taken from [179]).

9.5 Ion Chromatography in the Detergent and Household Product Industry

9.5.1 Detergents

Detergents for household and industrial use consist of a large number of very dissimilar individual components. As a result of technical developments, and in response to ecological and economical restrictions, these products are constantly changing. Thus, it is not surprising that in this field, analytical methods are continually under further development and that product-related analytical methods come of age fairly rapidly.

Ionic detergent components like the following play a key role in the washing process:
- Surfactants
- Builders
- Bleaching agents
- Fillers and finishing materials

The determination of these compounds represents a classic analytical problem for the detergent industry. Conventional wet-chemical and instrumental methods for analyzing ionic detergent components embrace gravimetry for sulfate determination [180]; potentiometry for determining chloride, phosphate, and borate [181, 182]; complexometry for the quantification of NTA and EDTA [183]; and photometry again for sulfate determination [184]. Such methods are laborintensive and time-consuming. Therefore, they are being replaced by much faster and more sensitive ion chromatographic techniques [185].

Surfactants

Surfactants are the most important group of detergent components; they are constituents of all detergents. Generally, they are water-soluble, surface-active compounds that carry both a hydrophilic functional group and a long alkyl chain as a hydrophobic rest. These substances are generally categorized into the following surfactant groups:

- Anionic surfactants
- Cationic surfactants
- Non-ionic surfactants
- Amphoteric surfactants

Surfactants that can presently be analyzed by means of ion chromatography are summarized in Table 9-12. Until now, surfactants were analyzed by a two-phase titration [186, 187]. This technique is based on an equilibrium reaction between surfactant dyes and anionic/cationic surfactant salts and their distribution in a two-phase mixture of chloroform/water. Hyamine 1622 (diisobutyl-[2-(2-phenoxyethoxy)ethyl]-dimethyl-benzylammonium chloride) has proved to be a suitable titrant for anionic surfactants, whereas lauryl sulfate is used for cationic surfactants.

A chromatographic determination of the classes of compounds listed in Table 9-12 can be achieved using ion-pair chromatography in combination with conductivity or UV detection. The retention behavior of surface-active compounds has already been discussed in detail in Section 6.5. This section discusses a series of example chromatograms of various surfactant raw materials. For an unequivocal identification of surfactants in detergent products, however, a wet-chemical pre-separation into the individual surfactant types is necessary, because detergent formulations often contain a combination of surfactants such as anionic and non-ionic surfactants. The carbon-chain distribution can then be determined by ion-pair chromatography. If a crude material consists of several compound classes, a comparison of the chromatographic "fingerprints" can provide valuable information for raw material control.

Builders

The builders include complexing agents such as sodium triphosphate, ion exchangers such as Zeolite A, and washing alkalines such as sodium carbonate and sodium silicate.

Inorganic and organic complexing agents are of central importance in the course of the washing process, because magnesium and calcium ions stemming from water, dirt, or textiles are complexed with their help. At least in central Europe, sodium triphosphate is no longer used for environmental reasons, as high consumption contributed to eutrophication of aquatic systems. In the so-called phosphorus-free detergents, sodium triphosphate has been replaced by Zeolith A and polyphosphonic acids.

Table 9-12. Summary of surfactants that can be analyzed by ion chromatography.

Formula		Chemical name
R-CH_2-COONa	R = C_{10} to C_{16}	Soap
R-C_6H_4-SO_3Na	R = C_{11} to C_{13}	Alkylbenzene sulfonate
R-CH-SO_3Na | R′	R, R′ = C_{10} to C_{17}	Alkane sulfonate
R-CH_2-CH=CH-$(CH_2)_n$-CH_2-SO_3Na	R = C_{10} to C_{14} n = 1 to 2	α-Olefin sulfonate
R-CH_2-CH-$(CH_2)_n$-CH_2-SO_3Na | OH	R = C_9 to C_{13} n = 1 to 2	Hydroxyalkane sulfonate
R-CH-COOCH_3 | SO_3Na	R = C_{14} to C_{16}	α-Sulfo-fatty acid methylester
R-CH_2-O-SO_3Na	R = C_{11} to C_{17}	Alkyl sulfate
R-CH-O-$(C_2H_4O)_2$-SO_3Na | R′	a) R′ = H R = C_{10} to C_{13} b) R, R′ = C_{10} to C_{14}	Alkylether sulfate a) Fatty alcohol ethersulfate b) sec. Alkyl ether sulfate
R\ \ \ /CH_3 N$^+$ Cl^- R′/ \ CH_3	R = C_{16} to C_{18}	Dialkyl-dimethylammonium chloride
R-C⟨N-CH_2 / CH_2⟩N$^+$ $CH_3OSO_3^-$ H_3C CH_2-CH_2-NH-CO-R	R = C_{16} to C_{18}	Imidazolinium salts
C_6H_5-CH_2-N$^+$-$(CH_3)_2$ Cl^- | R	R = C_8 to C_{18}	Alkyl-dimethyl-benzyl-ammonium chloride

In the past, both sodium triphosphate and polyphosphonic acids were determined with time-consuming ion-exchange chromatographic techniques [188]. Owing to their polyvalent character, they were not amenable to ion chromatography. However, the analysis of these compounds became possible with an IonPac AS7 latexed anion exchanger in combination with a specific post-column derivatization [189, 190]. The elution of condensed phosphates in finished products requires the use of 0.07 mol/L nitric acid. As revealed by Fig. 9-116, sulfate, which is added to the washing powder as a formulating agent, can also be detected under these chromatographic conditions. Calibration for the quantification of triphosphate is difficult, because commercial triphosphate also contains noticeable amounts of diphosphate. In the absence of a high-purity triphosphate, one can use the phosphorus-specific detection method developed by Vaeth et al. [191], in which di- and tri-phosphate are hydrolyzed to orthophosphate. The latter is detected photometrically after reaction with sodium molybdate (see Section

3.8.2). This method is extremely well suited for determining the degree of phosphate conservation. Today, condensed phosphates can be separated together with mineral acids on novel hydroxide-selective stationary phases with a hydroxide gradient, and detected by suppressed conductivity.

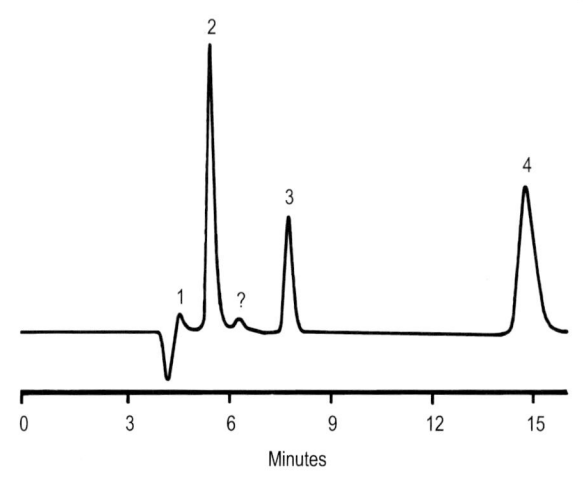

Fig. 9-116. Analysis of condensed phosphates in a detergent. – Separator column: IonPac AS7; eluant: 0.07 mol/L HNO_3; flow rate: 0.5 mL/min; detection: photometry at 330 nm after reaction with ferric nitrate; injection: 50 µL of a 0.1% washing powder solution; peaks: (1) orthophosphate, (2) pyrophosphate, (3) sulfate, and (4) triphosphate.

Figure 9-117 shows the chromatogram of a NTA-containing detergent that was obtained with 0.03 mol/L nitric acid. Under these chromatographic conditions, polyphosphonic acids, which are often present in detergents at low concentrations, can also be separated. Here, the polyphosphonic acid contained in the

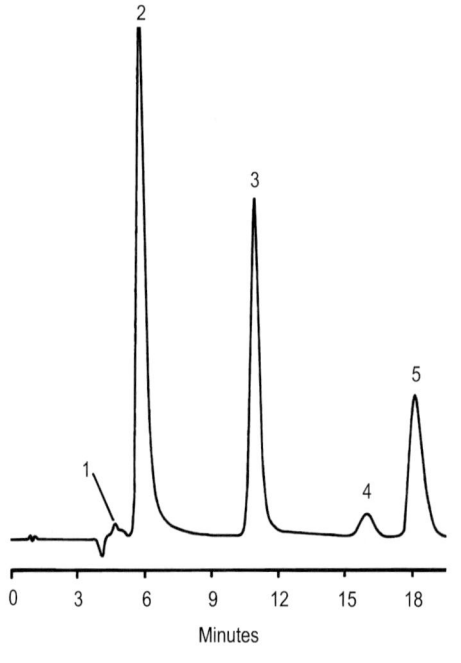

Fig. 9-117. Analysis of inorganic and organic complexing agents in a detergent. – Separator column: IonPac AS7; eluant: 0.03 mol/L HNO_3; flow rate and detection: see Fig. 9-116; injection: 50 µL of a 0.2% washing powder solution; peaks: (1) orthophosphate, (2) NTA, (3) pyrophosphate, (4) polyphosphonate, and (5) sulfate.

detergent sample elutes in the sulfate retention range. Triphosphate has a very high retention time under these conditions, thus the separator column must occasionally be flushed with concentrated nitric acid. It should be noted that the analysis with conventional methods of NTA in finished product is impossible in presence of other complexing agents such as citric acid.

Carbonate and silicate can be analyzed ion chromatographically via ion-exclusion chromatography. While carbonate is detected by electrical conductivity (see Fig. 9-118), silicate is detected photometrically – as described above – after derivatization with sodium molybdate.

Bleaching Agents

In an alkaline medium, the perhydroxide anion of the peroxide bleach [192], predominant in Europe, is formed as an active intermediate from hydrogen peroxide:

$$H_2O_2 + OH^- \rightleftharpoons H_2O + HO_2^- \tag{320}$$

The hydrogen peroxide is provided by sodium perborate, which exists as peroxoborate in crystalline state.

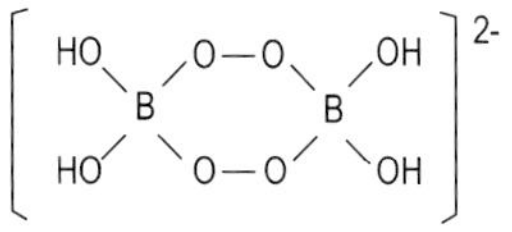

Peroxoborate Anion

In aqueous solution, the peroxoborate anion hydrolyzes to form hydrogen peroxide. To obtain a good bleaching process at temperatures below 60 °C, so-called bleaching activators are used. Together with hydrogen peroxide they form organic peracid intermediates at pH values between 9 and 12. Organic peracid intermediates develop a good bleaching effect at low temperatures. Tetraacetylethylene diamine (TAED), preferred in this regard, reacts as shown in the following scheme:

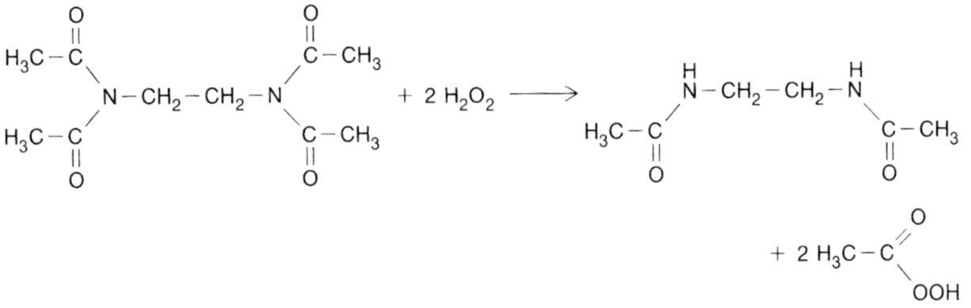

$$(321)$$

When stored for some time, TAED decomposes as acetic acid is lost. As illustrated in Fig. 9-118, perborate and acetic acid can be easily and rapidly determined via ion chromatography in the finished detergent product. Carbonate analysis in the same run is possible, too.

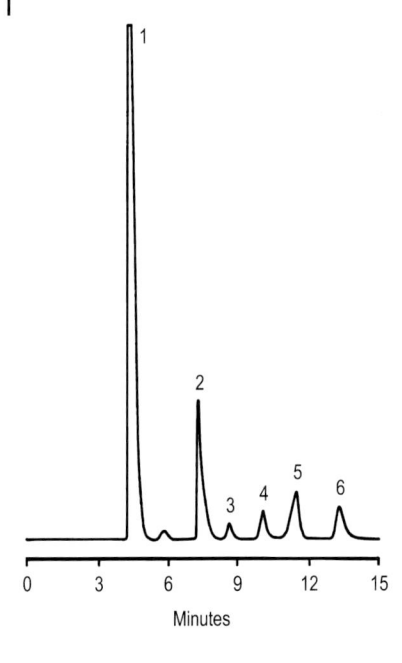

Fig. 9-118. Analysis of sodium perborate by ion-exclusion chromatography. — Separator column: IonPac ICE-AS1; eluant: 1.1 mmol/L octanesulfonic acid; flow rate: 1 mL/min; detection: suppressed conductivity; injection: 50 µL of a 0.05% detergent solution; peaks: (1) sulfate, (2) perborate, (3) H_2O_2 interference, (4) acetate, (5) NTA, and (6) carbonate.

The effort required to analyze perborate, NTA, carbonate, and acetate by ion chromatography takes about 0.8 hours per sample, including the necessary calibrations [193]. In comparison, the effective working time for conventional borate analysis via the extraction method alone is about 2.5 hours; to determine all the above components requires about 5.8 hours [193]. This example demonstrates the significant cost saving brought about by the introduction of ion chromatographic techniques.

Fillers and Finishing Materials
Sodium sulfate is generally used as a filler in powdered detergents. Sulfate can be analyzed by anion exchange chromatography together in a single run with chloride and orthophosphate (see Fig. 9-119). Orthophosphate, via the diphosphate anion, represents the degradation product of triphosphate. To demonstrate the superiority of ion chromatography over wet-chemical methods again in this case, the two sets of working steps for sulfate analysis are contrasted in Table 9-13.

The time expenditure per sample is about 24 hours for the wet-chemical procedure, with 2.5 hours being the actual working time. In contrast, ion chromatography requires only 20 to 25 minutes, including calibration time, for each sulfate determination [193]. Furthermore, ion chromatography has the advantage that short-chain alkyl sulfonates and sulfates do not interfere with the sulfate determination, and chloride and orthophosphate can be detected in the same run.

9.5 Ion Chromatography in the Detergent and Household Product Industry

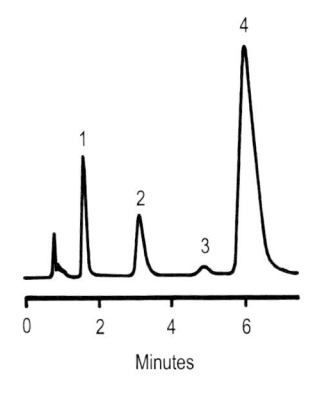

Fig. 9-119. Analysis of inorganic anions in a detergent. — Separator column: IonPac AS3; eluant: 2.8 mmol/L $NaHCO_3$ + 2.2 mmol/L Na_2CO_3; flow rate: 2.3 mL/min; detection: suppressed conductivity; injection: 50 µL of a 0.1% detergent solution (1:5 diluted); solute concentrations: 4 mg/L chloride (1), 7 mg/L orthophosphate (2), nitrate (3), and 45 mg/L sulfate (4).

Table 9-13. Comparison of ion chromatography and wet chemistry for sulfate analysis.

	Sulfate Analysis
Ion Chromatography	Wet Chemistry
1. Preparation of the sample solution	1. Preparation of the sample solution
2. Injection of the sample through a membrane filter	2. Acidification of the sample solution and heating with active carbon for removal of surfactants
3. Evaluation of the chromatogram	3. Filtration
	4. Addition of a precipitant
	5. Precipitation of sulfate
	6. Crystallization of the precipitate over night
	7. Filtration and rinsing of the precipitate
	8. Ashing of the filter
	9. Glowing of the residue
	10. Weighing and evaluation

The finishing of liquid detergents and cleansers typically employs short-chain alkylbenzene sulfonates such as toluene sulfonate or cumene sulfonate, which, because of their hydrotropic properties, ensure the solubility of other detergent components in an aqueous environment. Analysis of hydrotropic compounds is performed by ion-pair chromatography. The compounds are eluted in the order of increasing alkyl substitution. As demonstrated in Fig. 9-120, these compounds can be determined directly in the finished product without extensive sample preparation. The analyte samples are only diluted with de-ionized water and membrane filtered. As aromatic sulfonates may be detected both by their electrical conductivity and their UV absorption, the choice of suitable detection method mainly depends on the type of matrix.

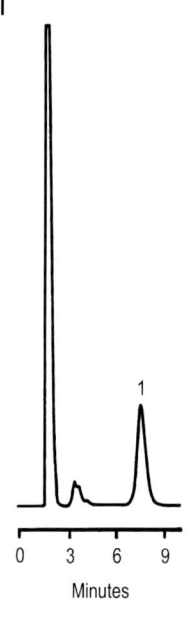

Fig. 9-120. Separation of cumene sulfonate in a cleanser. — Separator column: IonPac NS1 (10-µm); eluant: 10 mmol/L NH$_4$OH — acetonitrile (91:9 v/v); flow rate: 1 mL/min; detection: suppressed conductivity; injection: 50 µL of a 0.2% sample solution; peak: (1) cumene sulfonate.

9.5.2
Household Products

As an example of household products and industrial cleansing agents, Fig. 9-121 shows the separation of gluconic acid and citric acid as key components of a weakly basic cleansing agent. Both compounds are hydroxycarboxylic acids, thus separation was performed by ion-exclusion chromatography on IonPac ICE-AS5. As seen in Fig. 9-121, this stationary phase exhibits a high selectivity for the above-mentioned organic acids. Meanwhile, IonPac ICE-AS5 has been replaced by IonPac ICE-AS6, which exhibits a similar selectivity and which is operated with the same eluant at slightly lower concentration (0.4 mmol/L). Detection was carried out by suppressed conductivity.

Triethanolamine, which acts as a base in this cleansing agent, can be separated on a neutral DVB-based polymer. An aqueous sodium hydroxide solution is employed as an eluant to prevent the protonation of the nitrogen functional group and as a prerequisite for the subsequent pulsed amperometric detection. Like carbohydrates, alkanolamines can also be oxidized on a gold working electrode at pH 13. In contrast to the more common conductivity detection, this detection method is much more specific and sensitive. This is very important for this particular example, as it allows the sample to be heavily diluted, so that the sample matrix does not interfere. Figure 9-122 illustrates the chromatogram of the 1000-fold diluted sample.

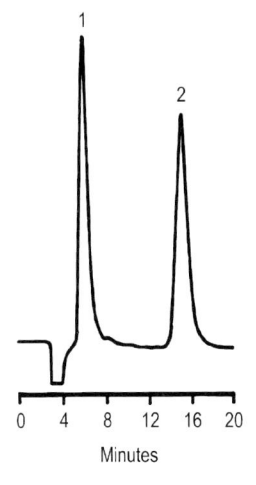

Fig. 9-121. Separation of gluconic acid and citric acid in a weakly basic cleansing agent. — Separator column: IonPac ICE-AS5; eluant: 1.6 mmol/L perfluorobutyric acid; flow rate: 0.5 mL/min; detection: suppressed conductivity; injection: 50 µL sample (1:1000 diluted); solute concentrations: 309 g/L gluconic acid (1) and 50 g/L citric acid (2).

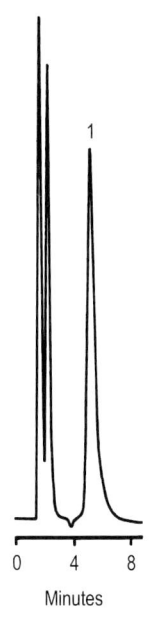

Fig. 9-122. Determination of triethanolamine in a weakly basic cleansing agent. — Separator column: IonPac NS1 (10-µm); eluant: 0.1 mol/L NaOH; flow rate: 1 mL/min; detection: pulsed amperometry on a gold working electrode; injection: 50 µL sample (1:1000 diluted); solute concentration: 157 g/L triethanolamine (1).

Often, triethanolamine in cleansing agents is substituted by hydroxylamine, which acts as a base. After protonation, hydroxylamine can be separated on a low-capacity cation exchanger in an acidic medium. Because hydroxylamine is a weak base, detection is somewhat difficult. If detection needs to be specific, again amperometry is the only sufficient method. Otherwise, the less specific and less sensitive non-suppressed conductivity detection can be used. No signal is obtained upon application of suppressed conductivity detection, as the product of the suppressor reaction is not sufficiently dissociated. Unlike ammonium ions, hydroxylammonium ions do not react with o-phthaldialdehyde, thus fluorescence detection is also not feasible. However, hydroxylamine can be easily oxidized at a platinum working electrode at potentials around +0.8 V; this provides the required selectivity and sensitivity. The chromatogram of a 2000-fold diluted sample in Fig. 9-123 almost looks like that of a standard.

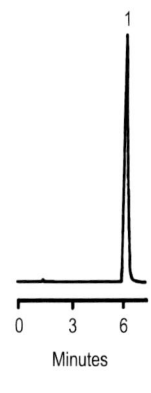

Fig. 9-123. Determination of hydroxylamine in a weakly basic cleansing agent. — Separator column: IonPac CS3; eluant: 30 mmol/L HCl; flow rate: 1 mL/min; detection: D.C. amperometry on a platinum working electrode; oxidation potential: +0.8 V; injection: 50 µL sample (1:2000 diluted); peak: (1) hydroxylamine.

Besides low-molecular weight organic acids and bases, cleansing agents also contain surfactants, which are usually analyzed by ion-pair chromatography on a non-functionalized DVB polymer. As most of the surfactant raw materials are multi-component mixtures, surfactant characterization by means of chromatography is based on separations according to chain length. Examples of anionic and cationic surfactant separations can be found in Chapter 6. If those raw materials are to be determined in finished products, the chain length distribution is not very helpful due to the resulting long analysis time. On the other hand, liquid chromatography is too good a separation technique to give only one peak for a surfactant raw material. Therefore, chromatographic conditions are optimized so that the surfactant is completely separated from the matrix, without separating all individual components to baseline. For quantification, the peak areas of the resulting pattern are summed up. For example, Fig. 9-124 illustrates the analysis of alkyl sulfonates in a cleansing agent using ammonium hydroxide as an ion-pair reagent. To rapidly elute the various components of the surfactant, a rather steep acetonitrile gradient was applied. Because the detection of the non-chromophoric alkyl sulfonates was carried out by suppressed conductivity, the increasing organic solvent content in the mobile phase leads to a negative baseline drift during the chromatographic run.

As an example of cationic surfactant analysis in finished products, Fig. 9-125 shows the separation of alkylbenzyl-dimethyl-ammonium compounds in a toilet cleaner. As can be seen from the respective chromatogram, this raw material basically consists of two species, the C_{12}- and the C_{14}-component. Thus, chromatographic separation is much simpler than shown above for alkyl sulfonates. Because these long-chain quaternary ammonium compounds exhibit a marked hydrophobicity, surface-inactive hydrochloric acid is used as an ion-pair reagent.

Due to the aromatic character of the analytes, both suppressed conductivity and UV detection at low wavelength are suitable detection methods; the two methods are compared in Fig. 9-125.

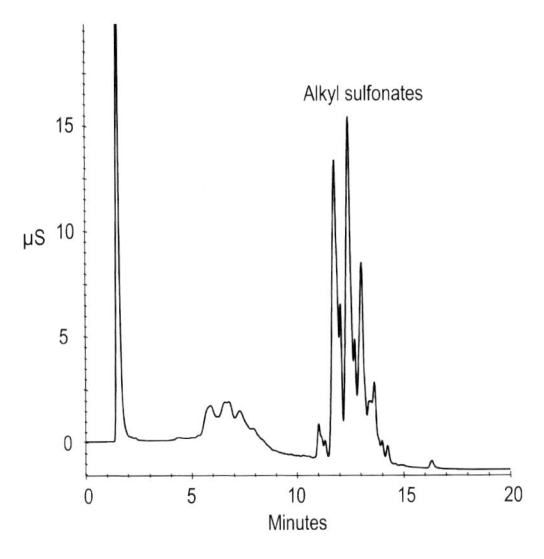

Fig. 9-124. Determination of alkyl sulfonates in a cleansing agent. — Separator column: IonPac NS1 (10-μm); eluant: 4 mmol/L NH_4OH — acetonitrile; gradient: linear, 0% to 50% acetonitrile in 15 min; flow rate: 1 mL/min; detection: suppressed conductivity; injection: 50 μL of a cleansing agent; (courtesy of Henkel KGaA Düsseldorf, Germany).

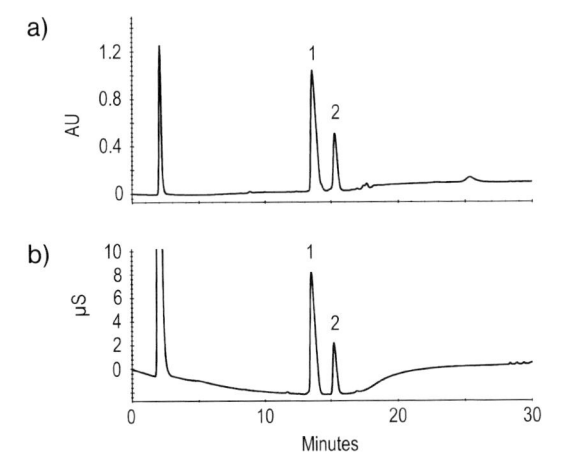

Fig. 9-125. Determination of alkylbenzyl-dimethyl-ammonium compounds in a toilet cleaner. — Separator column: IonPac NS1 (10-μm); eluant: 8 mmol/L HCl — acetonitrile; gradient: linear, 0% to 90% acetonitrile in 15 min; flow rate: 1 mL/min; detection: (a) UV (205 nm), (b) suppressed conductivity; injection: 50 μL of a toilet cleaner sample; peaks: (1) dodecylbenzyl-dimethyl-ammonium salt and (2) tetra-decylbenzyl-dimethyl-ammonium salt (courtesy of Henkel KGaA Düsseldorf, Germany).

The high selectivity of pulsed amperometric detection is exemplified by the determination of glycerol in soap. Due to its low volatility, glycerol cannot be separated by GC. Separation by ion-exclusion chromatography is straightforward, but there is no alternative to pulsed amperometry for sensitive detection of glycerol because of its non-chromophoric structure. Because glycerol is separated at low pH, a platinum working electrode is used for its oxidation. Sample preparation in this case is very easy. The soap sample is dissolved in de-ionized water and the resulting solution is diluted and membrane filtrated. Due to the high selectivity of pulsed amperometry the chromatogram in Fig. 9-126 only shows the glycerol peak eluting at approximately 10 minutes. Nevertheless, total analysis time is a little longer because of the negative peak appearing in the chromatogram after 15 minutes, which is attributed to the reduction of oxygen in the sample.

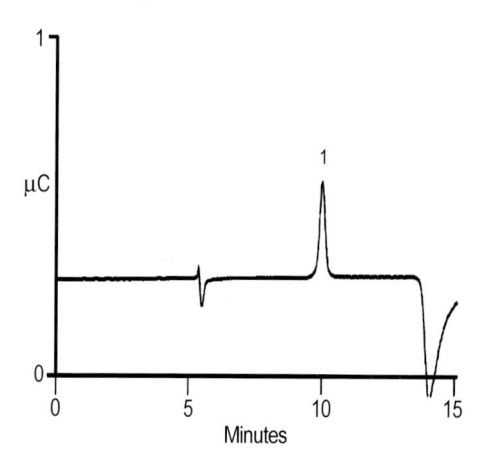

Fig. 9-126. Determination of glycerol in soap. — Separator column: IonPac ICE-AS1; eluent: 50 mmol/L HClO$_4$; flow rate: 1 mL/min; detection: pulsed amperometry on a platinum electrode; injection: 50 µL sample (1:500 diluted); peak: (1) 6.2 mg/L glycerol.

The composition of dishwashing agents very much depends on their application. Condensed inorganic phosphates are usually a constituent of products for automatic dishwashers and can be analyzed by anion exchange chromatography and suppressed conductivity detection. The most suitable stationary phase is the hydroxide-selective IonPac AS16 latexed anion exchanger, which allows rapid elution of polyphosphates with a purely aqueous hydroxide eluent. A chromatogram of a dishwashing agent sample diluted with de-ionized water is shown in Fig. 9-127; tripolyphosphate was detected as a major constituent.

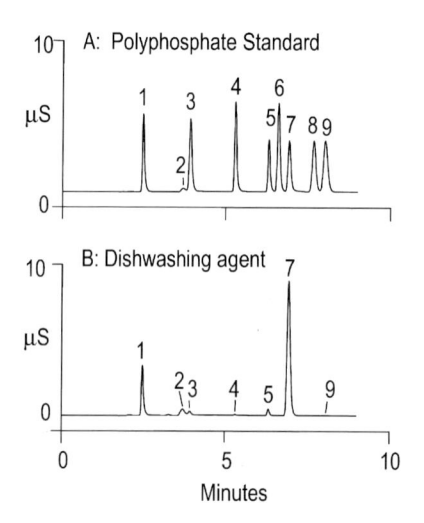

Fig. 9-127. Separation of polyphosphates in a dishwashing agent. — Separator column: IonPac AS16 with guard; column temperature: 30 °C; eluent: KOH (EG40); gradient: 25 mmol/L for 1.7 min to 65 mmol/L in 2.5 min; flow rate: 1.5 mL/min; injection volume: 10 µL; solute concentrations: (A) 3 mg/L chloride (1), carbonate (2), 5 mg/L sulphate (3), 10 mg/L orthophosphate (4), 10 mg/L pyrophosphate (5), 10 mg/L trimetaphosphate (6), 10 mg/L tripolyphosphate (7), 10 mg/L tetrametaphosphate (8), and 10 mg/L tetrapolyphosphate (9), (B) 300 mg/L sample.

Anionic surfactants are a common constituent of all dishwashing agents, whether they are used for automatic dishwashers or for manual dishwashing. Figure 9-128 illustrates the analysis of fatty alcohol ether sulfates (FAES) derived from coco fatty acids (lauric and myristic acid) in such a matrix. As with the alkyl sulfonates in industrial cleansing agents described above, coco-FAES is a multi-component raw material (see Fig. 6-41 in Section 6.5). Again, a rather steep acetonitrile gradient was applied to rapidly elute the various components of this surfactant. The peak areas of the resulting pattern were summed up for quantification.

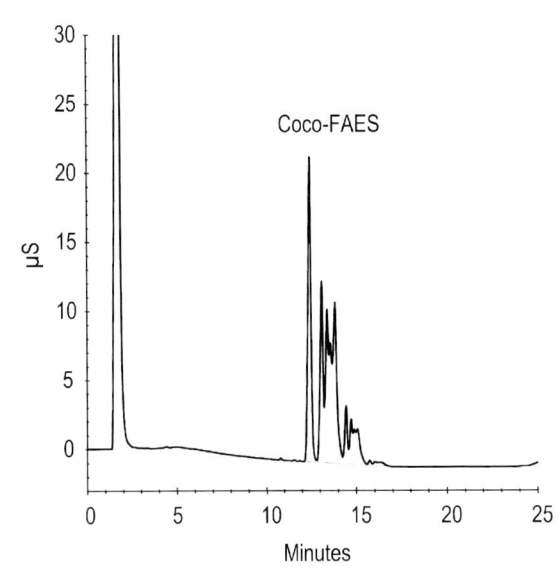

Fig. 9-128. Determination of fatty alcohol ether sulfates in a dishwashing agent. — Separator column: IonPac NS1 (10-µm); eluant: 4 mmol/L NH_4OH – acetonitrile; gradient: 0% to 50% acetonitrile in 15 min; flow rate: 1 mL/min; detection: suppressed conductivity; injection: 50 µL of a diluted dishwashing agent containing coco-FAES; (courtesy of Henkel KGaA, Düsseldorf, Germany).

A characteristic example in the field of cosmetics is the separation and determination of anionic surfactants in shampoos. The only required sample preparation step is the dilution of the product to be analyzed in de-ionized water and subsequent membrane filtration (0.45 µm) prior to injection. Figure 9-129 illustrates the gradient elution of anionic surfactants in a commercial shampoo with simultaneous non-suppressed conductivity and UV detection. While the peak pattern in the UV trace indicates the presence of alkylbenzene sulfonate (ABS) (see also Fig. 6-42 in Section 6.5), the characteristic peak profile for alkyl-ether sulfates (see Fig. 6-41 in Section 6.5) is evident in the conductivity trace after visual subtraction of the ABS signal. Therefore, ion chromatography provides a way to unequivocally identify surfactant raw materials in finished cosmetic products without time-consuming sample preparation.

This not only applies to shampoos but also to toothpaste, in which anionic surfactants can also be identified, depending on the product. The chromatogram shown in Fig. 9-130 clearly demonstrates that the method developed for the analysis of anionic surfactants in shampoos is also suited for the investigation of toothpaste [194]. To prepare the sample, the weighed amount is suspended in de-ionized water and the extract is membrane filtered (0.45 µm) prior to injection. In the present case, lauryl sulfate was clearly identified as a surfactant component. The simultaneous application of UV detection also enables the detection of other organic constituents in the same run, the chemical structure of which cannot be clarified without knowing the basic formulation. Inorganic constituents such as monofluorophosphate (MFP), which is included in many formulations, can be directly determined by anion exchange chromatography. Besides MFP, orthophosphate as its degradation product as well as chloride and sulfate can be analyzed simultaneously under isocratic conditions with a carbonate-selective anion exchanger (Fig. 9-131). However, shorter analysis times, higher sensitivity, and better separation between the inorganic constituents and the various organic acids present in the sample are obtained with a gradient elution technique using a modern, low-capacity IonPac AS17 anion exchanger with a hydroxide concentration gradient (Fig. 9-132). In both cases, sample preparation is performed as described above. In addition, the extract is passed through an OnGuard-RP cartridge prior to injection to remove non-ionic organics and especially the titanium dioxide colloid.

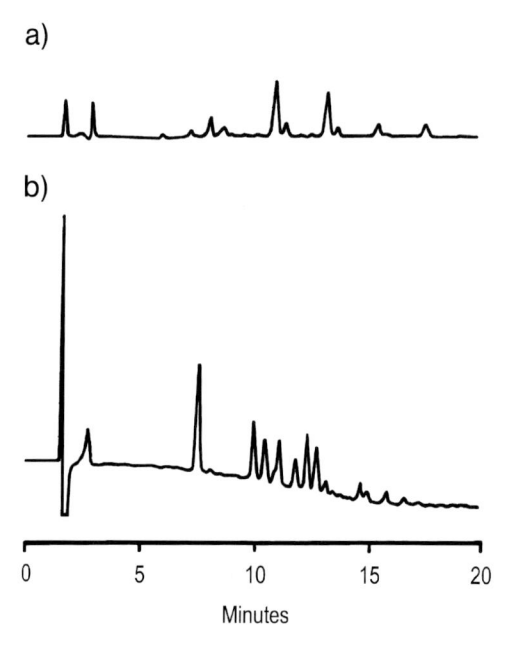

Fig. 9-129. Gradient elution of anionic surfactants in a shampoo. — Separator column: Hypersil 5 MOS; eluant: (a) 1 mmol/L NaOAc, (b) acetonitrile; gradient: linear, 31% B to 41% B in 10 min; flow rate: 1 mL/min; detection: (a) UV (254 nm), (b) non-suppressed conductivity; injection: 50 µL of a 0.2% solution.

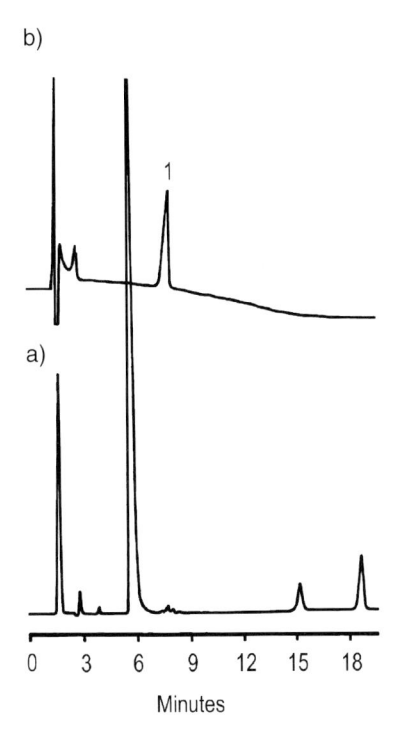

Fig. 9-130. Analysis of lauryl sulfate and other organic constituents in toothpaste. — Separator column: Hypersil 5 MOS; chromatographic conditions: see Fig. 9-129; detection: (a) non-suppressed conductivity, (b) UV (254 nm); injection: 50 µL of a 0.5 % solution; peak: (1) lauryl sulfate; (taken from [194]).

9.6
Ion Chromatography in the Food and Beverage Industry

It is the task of the food and beverage industry to provide the customer with a tasty and healthy product whose nutritive value and storage quality are well defined. Manufacturing and monitoring processes have been automated to such an extent that only a few, simple quality tests are required. However, in some cases it is still necessary to carry out costly analyses for quality assurance purposes. In these cases, ion chromatography permits a rapid and precise determination of both inorganic and organic anions and cations even in complex matrices.

Ion chromatography is increasingly being adopted by many test and research laboratories in the food and beverage industry, because minimal sample preparation is required for analyses utilizing this method [195]. Often, sample extraction with de-ionized water and membrane filtration (0.45 µm) are completely sufficient for sample preparation. The reason ion chromatography easily deals with complex matrices lies not only in the stability and resistance to fouling of the stationary phases used, but also in the sensitivity and specificity of the detection methods employed. A survey of the applicability of ion chromatography in the food and beverage industry is given in Table 9-14.

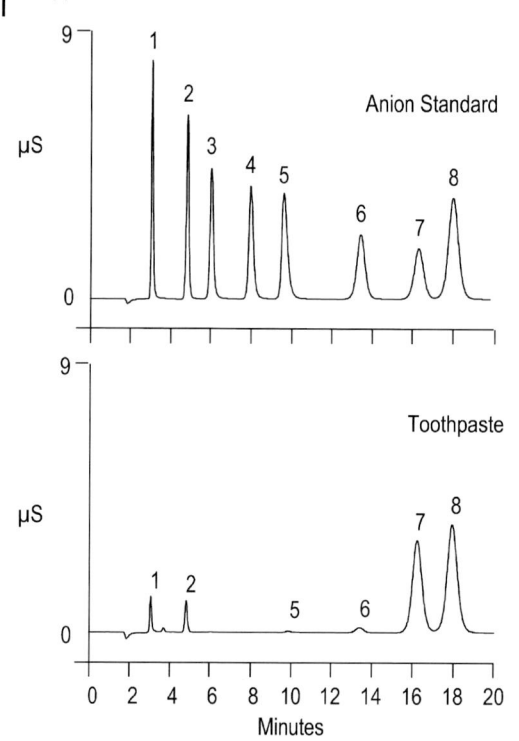

Fig. 9-131. Isocratic analysis of monofluorophosphate in toothpaste. — Separator column: IonPac AS14; eluant: 2.5 mmol/L NaHCO$_3$ + 2 mmol/L Na$_2$CO$_3$; flow rate: 1 mL/min; detection: suppressed conductivity; injection volume: 10 µL; solute concentrations in the anion standard: 5 mg/L fluoride (1), 10 mg/L chloride (2), 15 mg/L nitrite (3), 25 mg/L bromide (4), 25 mg/L nitrate (5), 40 mg/L orthophosphate (6), 50 mg/L monofluorophosphate (7), and 30 mg/L sulfate (8).

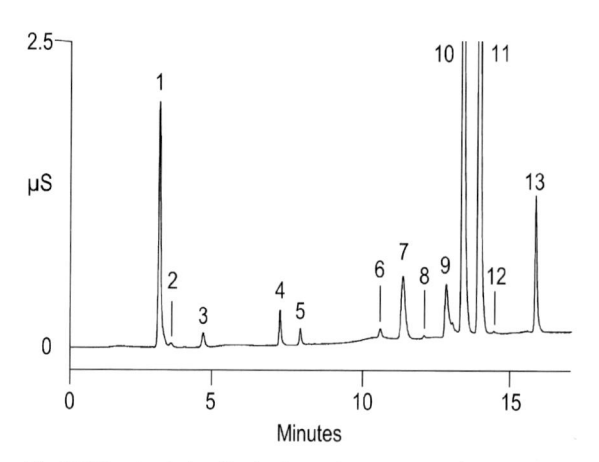

Fig. 9-132. Analysis of anionic toothpaste constituents using a gradient elution technique. — Separator column: IonPac AS17; column temperature: 30 °C; eluant: KOH (EG40); gradient: 1 mmol/L isocratic for 3 min, then linear to 12 mmol/L in 7 min, then to 35 mmol/L in 4 min; flow rate: 1 mL/min; detection: suppressed conductivity; injection volume: 10 µL; sample: 0.5 % toothpaste solution: peaks: (1) fluoride, (2) acetate, (3) formate, (4) chloride, (5) nitrite, (6) nitrate, (7) benzoate, (8) unknown, (9) carbonate, (10) monofluorophosphate, (11) sulfate, (12) oxalate, and (13) orthophosphate.

Table 9-14. Application areas and typical analytical examples in the food and beverage industry.

Application area	Analytical example
Milk products	Determination of iodide in whole milk; chloride and/or sodium in butter; lactate, pyruvate, and citrate in cheese
Meat processing	Determination of the nitrite/nitrate ratio in meat products; nitrate in the water being used
Beverages	Determination of inorganic anions and cations in the water being used; in sweeteners and flavours and in the finished products; organic acids and carbohydrates in beer, wine, and juice
Canned food	Determination of chloride, nitrate, sodium, organic acids, and transition metals in canned fruit and canned vegetables, spices, vinegar, and fish
Baby food	Determination of iodide, choline, and transition metals
Cereal products	Determination of bromate and propionate in bakery products; iron(II)/iron(III) ratio
Fats, oils, carbohydrates and flavours	Determination of fatty acids and carbohydrates in corn syrup

9.6.1
Beverages

One of the key applications of ion chromatography in the food and beverage industry is the analysis of inorganic and organic anions and carbohydrates in beverages of all kinds [196, 197]. This includes predominantly the investigation of wine, beer, fruit juices, various refreshers, coffee and tea. In addition to inorganic anions, all these beverages contain various organic acids whose retention behavior is very similar to that of inorganic anions. Thus, modern anion exchangers of any type do not allow a baseline-resolved separation of all inorganic and organic anions under isocratic conditions. In special cases, inorganic anions such as chloride, nitrate, orthophosphate, and sulfate together with dicarboxylic acids such as malic acid and tartaric acid may be separated to baseline if two anion exchangers with different selectivities are combined. Detection is performed by suppressed conductivity utilizing a continuously regenerated membrane suppressor. Figure 9-133 shows the application of this method to the analysis of an old port wine. When evaluating the orthophosphate results, it should be noted that only free orthophosphate, but not the total phosphate is detected via ion chromatography. As some phosphate is bound by calcium, especially in fruit juices, the phosphate content determined via ion chromatography is much lower than that obtained with conventional methods after sulfuric acid digestion. Conversely, this also applies to the calcium analysis, if the samples are not treated with an acid. Nitrate, eluting shortly after orthophosphate, is usually present in very low concentrations, but may be recorded via simultaneous UV detection at 215 nm in the same run without any problems.

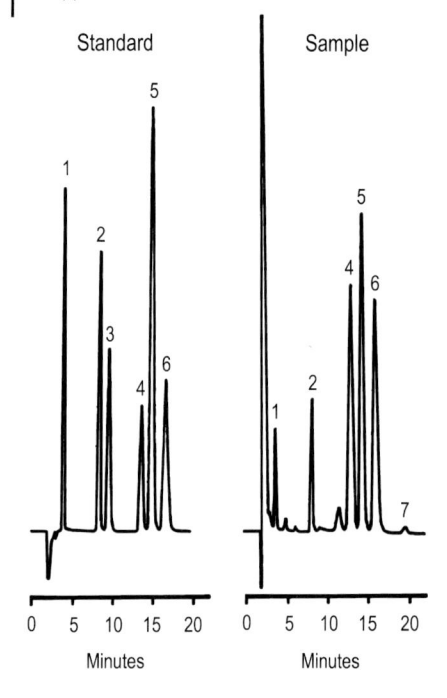

Fig. 9-133. Analysis of inorganic and organic anions in an old port wine. — Separator columns: IonPac AS4 + AS4A(-SC); eluant: 2.8 mmol/L NaHCO$_3$ + 2.2 mmol/L Na$_2$CO$_3$; flow rate: 1.5 mL/min; detection: suppressed conductivity; injection: (A) 50 µL of an anion standard with 2 mg/L chloride (1), 10 mg/L orthophosphate (2), 10 mg/L nitrate (3), 10 mg/L malate (4), 10 mg/L sulfate (5), and 10 mg/L tartrate (6), (B) 50 µL sample (Messias Port, 10 years old, diluted 1:50); additional peak: (7) oxalate.

This method is not applicable to the analysis of citrus juices, because the citric acid that is present at high concentrations in these products is strongly retained under the given chromatographic conditions. After repeated injections, this results in a marked reduction of the ion-exchange capacity. Therefore, a gradient technique with a sodium hydroxide eluant on suitable stationary phases must be used for analyzing such samples. First attempts are found in the paper by Sarzanini et al. [198], in which the separation of major and minor components (citric acid, malic acid, tartaric acid as well as galacturonic acid, isocitric acid, lactic acid, and acetic acid) in various fruit juices on an OmniPac PAX-500 (see Section 6.7) multimode phase is described. However, the authors used a very complex tertiary gradient based on NaOH, methanol, and ethanol. Although the above-mentioned organic acids are separated with this mixture, pronounced positive and negative baseline drifts and a significant band broadening for the major components result, which can be attributed to the relatively low ion-exchange capacity of this separator column. Moreover, co-elution between sulfate and ascorbic acid is observed under the given chromatographic conditions. The main emphasis of this paper was the characterization of orange juices from the various provenances, which included not only the categorization of citrate-, isocitrate-, and malate contents, but also the citrate/malate- and citrate/isocitrate ratios. According to their investigations, orange juices contain vastly different amounts of citrate and isocitrate depending on the origin. In orange juices from Italy, citrate and isocitrate concentrations are the highest, followed by juices from Argentine, Brazil, Uruguay, Cuba, Florida, and

California. In orange juices from Mediterranean countries such as Morocco, Egypt, and Israel citrate and isocitrate concentrations are the lowest. On the other hand, malate concentrations do not differ much.

The introduction of hydroxide-selective anion exchangers greatly simplified gradient elution of organic acids. Out of all the hydroxide-selective anion exchangers available today, IonPac AS11 is the most suitable for fruit juice analysis. Meanwhile, this stationary phase is also available in a high-capacity format (IonPac AS11-HC, see Section 3.4.2). For the first time, the above-mentioned organic acids could be separated together with the most important mineral acids using this column and a purely aqueous hydroxide gradient [199]. Figure 9-134 exemplifies this with the analysis of an apple juice concentrate containing significant amounts of quinic acid, malic acid, and citric acid. Traces of nitrate can be analyzed in the same run; the minimum detection limit is 0.5 mg/L in this matrix. Because the various anions are present at largely different concentrations, different dilutions of the fruit juices have to be chromatographed for quantitative analysis of major and minor components when using low-capacity anion exchangers. Although analysis times are significantly higher when using high-capacity columns, separate runs for analyzing major and minor components are usually unnecessary.

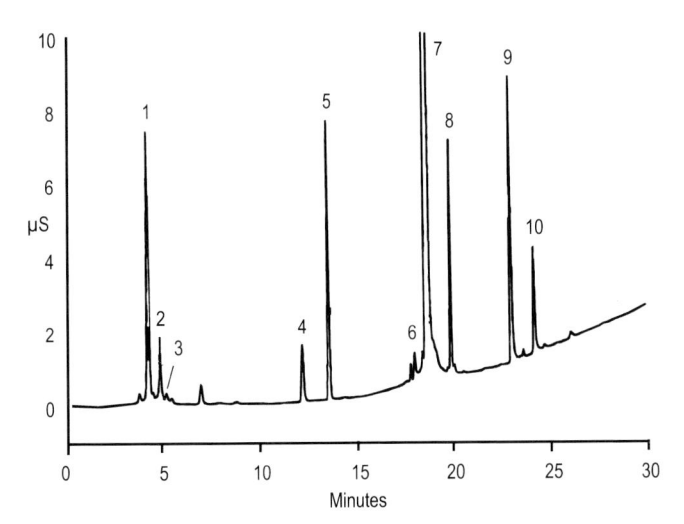

Fig. 9-134. Gradient elution of inorganic and organic anions in an apple juice concentrate. — Separator column: IonPac AS11; eluant: NaOH; gradient: linear, 0.3 mmol/L to 50 mmol/L; flow rate: 1 mL/min; detection: suppressed conductivity; injection: 25 µL of a 1:250 diluted sample; peaks: (1) quinate, (2) lactate, (3) acetate, (4) galacturonate, (5) chloride, (6) nitrate, (7) malate, (8) sulfate, (9) orthophosphate, and (10) citrate; (taken from [200]).

As an alternative to anion exchange chromatography, ion-exclusion chromatography is also used for determining organic acids in fruit juices. However, due to the limited resolution of conventional ion-exclusion phases, co-elution phenomena are often observed, so that caution is advised in quantitative analysis. As an example, Fig. 9-135 shows the chromatogram of a freshly squeezed Italian grape juice utilizing UV detection at 210 nm [201]. According to the column manufacturer, the column has to be operated at elevated temperature to obtain maximum peak efficiency.

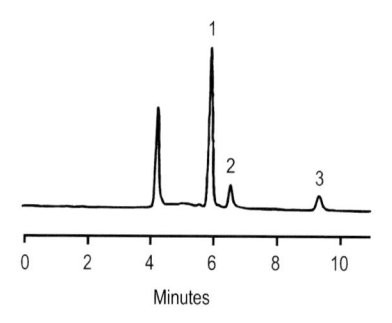

Fig. 9-135. Ion-exclusion chromatography of organic acids in a freshly squeezed grape juice utilizing UV detection. – Separator column: Shimpack IE; column temperature: 60 °C; eluant: 5 mmol/L H_2SO_4; detection: UV (210 nm); sample: freshly squeezed Italian grape juice; peaks: (1) tartrate, (2) malate, and (3) fumarate; (taken from [201]).

An interesting application of ion-exclusion chromatography in the field of fruit juice analysis is the analysis of lactic acid and acetic acid, whose concentrations are regarded as markers for product quality [202]. The AIJN (Association of the Industry of Juices and Nectars from Fruits and Vegetables of the European Economic Community) suggested the following thresholds for these two acids in fruit juices: 0.5 g/kg lactic acid and 0.4 g/kg acetic acid [203]. If these values are exceeded, microbial growth is more than likely. Such growths can be caused by defective processing of the fruits or by using bad fruits. Lactic acid and acetic acid both are formed as degradation products of carbohydrates. Recent investigations revealed that the formation of fumaric acid is also caused by microorganisms, especially by *Rhizopus stolonifer* [204]. However, elevated levels of fumaric acid can also occur after adding synthetic malic acid, which supposes a possible adulteration of the juice [205]. The threshold for fumaric acid in apple juice has been fixed by the AIJN to be 5 mg/kg. While traditional enzymatic analysis is very time-consuming because it requires individual determinations of these three acids, lactic acid, acetic acid, and fumaric acid can be analyzed by ion-exclusion chromatography in the same run. Because very low concentrations of these acids have to be detected, lactic acid and acetic acid are detected conductometrically and fumaric acid by measuring the light absorption at 207 nm. These two detection systems can be used in series. Incubation experiments with *Rhizopus stolonifer* on pear and peach nectars showed significantly increased amounts of lactic acid and fumaric acid and marginally increased amounts of acetic acid over a period of several days. Figure 9-136 shows the chromatograms of a pear nectar sample obtained with both detection systems prior to and after incubation with *Rhizopus stolonifer*. Sample preparation included a dilution of the nectar with de-ionized water and subsequent filtration with a 0.45-μm filter.

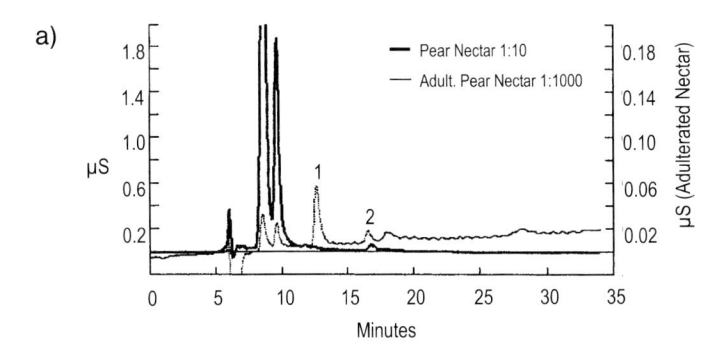

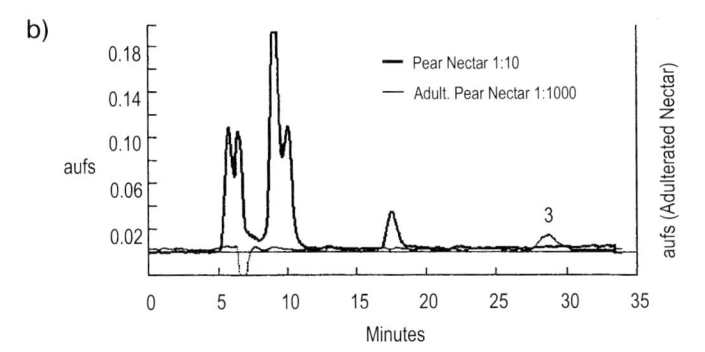

Fig. 9-136. Ion-exclusion chromatographic separation of lactic acid, acetic acid, and fumaric acid in a pear nectar prior to and after incubation with *Rhizopus stolonifer*. — Separator column: IonPac ICE-AS6; eluant: 0.5 mmol/L perfluorobutyric acid; flow rate: 1 mL/min; detection: (a) suppressed conductivity, (b) UV (207 nm); sample: pear nectar diluted 1:10 with de-ionized water, incubated sample diluted 1:1000; peaks: (1) lactic acid, (2) acetic acid (3) fumaric acid; (taken from [202]).

Ion chromatography can be regarded as an alternative to atomic absorption spectrometry for the analysis of alkali- and alkaline-earth metals in fruit juices, although AAS is still recommended by the IFFJU (International Federation of Fruit Juices Producers) as the official method. The broad distribution of ion chromatography, its versatile applications, and its speed and high precision are the most important attributes of this method regarding its applicability to the cation analysis of fruit juices. Moreover, mono- and di-valent cations can be separated in the same run under isocratic conditions, even though the concentration ratio between sodium and potassium may be quite substantial. (Potassium concentrations up to 3000 mg/kg are no exception, while sodium concentrations are usually around 10 mg/kg.) It is important to note that only dissolved calcium is detected by ion chromatography. Comparative investigations of juices with IC and AAS by Trifirò et al. [206] clearly showed that calcium bound to pectins does not go into solution after sample acidification. On the other hand, if the samples are ashed and the residue is taken up with dilute acid, IC and AAS results for this parameter are in good agreement. Trifirò et al. recommended an IonPac CS12A cation exchanger for the separation of alkali- and alkaline-earth metals in fruit juices, which according to their opinion shows good resistance towards polyphenols. Typical cation chromatograms for a variety of fruit juices and fruit purées are shown in Fig. 9-137. All samples were diluted 1:50 with de-ionized water and filtrated (0.45 µm).

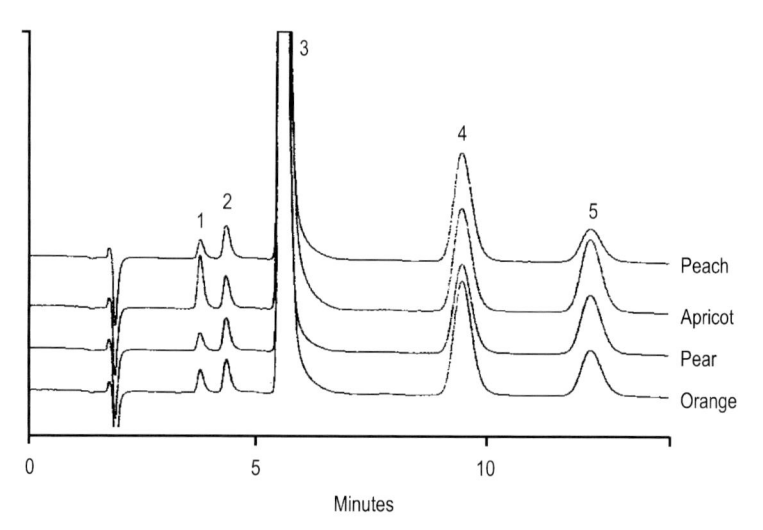

Fig. 9-137. Separation of alkali- and alkaline-earth metals in various fruit juices and fruit purées. — Separator column: IonPac CS12A; eluant: 20 mmol/L methanesulfonic acid; flow rate: 1 mL/min; detection: suppressed conductivity; samples: various fruit juices and fruit purées diluted 1:50 with de-ionized water; peaks: (1) sodium, (2) ammonium, (3) potassium, (4) magnesium, and (5) calcium; (taken from [206]).

An important application area of ion chromatography in the characterization of fruit juices is the analysis of carbohydrates. Traditionally, sorbitol and the main sugars such as glucose, fructose, and sucrose were determined with enzymatic methods, which are usually very sensitive and specific. Interferences caused by the matrix are rare; however, they are possible when determining sucrose. The principle of the sucrose determination is based on the determination of the glucose content prior to and after enzymatic hydrolysis of sucrose. Because fruit juices often contain large amounts of glucose in the presence of small amounts of sucrose, precise quantification of small sucrose concentrations is only possible when the glucose has first been degraded. Using ion chromatography, small amounts of sucrose in the presence of large amounts of glucose can be determined without any problem. Interference in enzymatic analysis is also observed with colored solutions. Enzymatic determination of sorbitol is especially interfered with by the fruit-owned anthocyanins in colored juices. These interferences do not occur in anion exchange chromatography with pulsed amperometric detection. Due to the high sensitivity of this detection method, the samples to be analyzed can be strongly diluted, so matrix interferences are usually not observed. For example, the determination of sorbitol in an apple juice was carried out by direct injection of the sample diluted 1:500 with de-ionized water (Fig. 9-138). Comparing IC measurements with enzymatic techniques by Baumgärtner [207] (see Table 9-15) yields a good agreement between both methods; this was also for the other main components, glucose and fructose, which can be determined together with sucrose in the same run.

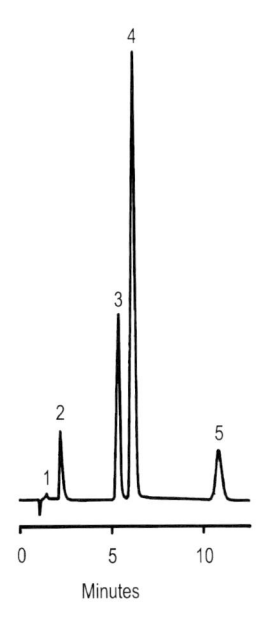

Fig. 9-138. Analysis of carbohydrates in an apple juice. – Separator column: CarboPac PA1; eluant: 0.1 mol/L NaOH; flow rate: 1 mL/min; detection: pulsed amperometry on a gold working electrode; injection: 50 µL sample (diluted 1:500); peaks: (1) inositol, (2) sorbitol, (3) glucose, (4) fructose, and (5) sucrose; (taken from [207]).

Table 9-15. Comparison of the results obtained with IC and enzymatic techniques for carbohydrate analysis in apple juice; (taken from [207]).

Carbohydrate	Ion Chromatography [g/L]	Enzymatic Analysis [g/L]
Sorbitol	5.11	5.22
Glucose	24.7	24.5
Fructose	62.7	63.3

In routine analysis, it is possible to quickly determine individual carbohydrates; total carbohydrate is determined by summing up the individual ones. Sugar alcohols and carbohydrates that are not of natural origin are recognized as adulterations. Thus, a quick examination of the authenticity of fruit- and vegetable juices can easily be carried out. In addition to sorbitol, inositol, xylitol, and mannitol are also used for characterizing juices [208]. Strawberry, blackberry, and blueberry juices, for example, contain significant amounts of *myo*-inositol and mannitol in addition to sorbitol. Black currant juice contains around 1 g/L inositol.

Sometimes, the analysis of carbohydrates in fruit juices reaches its limit. The identification of fruit juice adulteration by adding water and partly inverted sucrose represents an analytical problem that has been inadequately solved. The difficulties in proving an addition of partly inverted sucrose to orange juice, for example, are attributed to the fact that no additional compounds are introduced into the juice by this adulteration. Because partly inverted sucrose shows the same ratio between glucose, fructose, and sucrose as orange juice, the addition of an inverted sugar cannot be recognized by analyzing the main sugars. However, inverted sugar contains various oligosaccharides that are primarily formed during the inverting (acid hydrolysis) of beet sugar. Those oligosaccharides, which also exist in natural orange juice at very low concentrations, are introduced into the orange juice when it is adulterated with inverted sugar.

In 1989, Low et al. [209] developed a method based on anion exchange chromatography and pulsed amperometric detection, with which the addition of inverted sugar in orange juice can be identified via the characteristic oligosaccharide pattern. For this analysis, the orange juices and orange juice concentrates have to be prepared accordingly. After adjusting to an extract content of 5.6 Bx, interfering components such as organic acids, peptides, and dyes are removed via treatment with anion- and cation exchange resins. The sample thus prepared is directed through a reversed-phase extraction cartridge prior to injection. The addition of an internal standard such as maltotetraose is vitally important for quantification and compilation of a data bank because the detector response depends on the degree of electrode fouling. A major disadvantage of the Low method, which recommends CarboPac PA1 as a stationary phase for oligosaccharide separation, is the long analysis time of three hours per sample. Moreover, identifying a sugar addition by comparing the chromatograms instead of by a quantitative evaluation is not especially feasible. Wiesenberger et al. [210] optimized the Low method in terms of quantitative evaluation and standardization, so that an addition of partly inverted sucrose down to 5% (w/w) can still be recognized. As a further improvement, the authors used a CarboPac PA-100 instead of a CarboPac PA1. The CarboPac PA-100 was especially developed for oligosaccharide separations. Finally, Trotzer et al. [208] refined this method to develop the technique currently in use. Because the minor sugars in the sample only represent a very small amount of the total sugar content, complete separation with high chromatographic resolution and reproducible retention times between the minor sugars and the excess of mono- and di-saccharides is a prerequisite for the quantitative analysis of minor sugars. Therefore, Trotzer et al. used two CarboPac PA-100 columns in series. Six characteristic minor sugars were used to evaluate the addition of inverted sugar to orange juices and orange juice concentrates (Fig. 9-139). The component marked with "A" does not exist at all or only at very low concentration in inverted sugar. Thus, the addition of inverted sugar to orange juice does not necessarily lead to an increase of this signal. The other six minor sugars show significantly larger signals when inverted sugar was added (Fig. 9-140). For precise quantification of an adulteration, it has proved to be of advantage to sum up the peak areas of these six peaks, which corresponds on the average to 90 mg/L maltotetraose in authentic orange juices and orange juice concentrates with a standard deviation of 10%. Based on this data, an adulteration is considered to be proven if the summed amount of the six minor sugars exceeds 120 mg/L maltotetraose. Thus, the safety range corresponds to the threefold standard deviation of the mean content of an authentic sample.

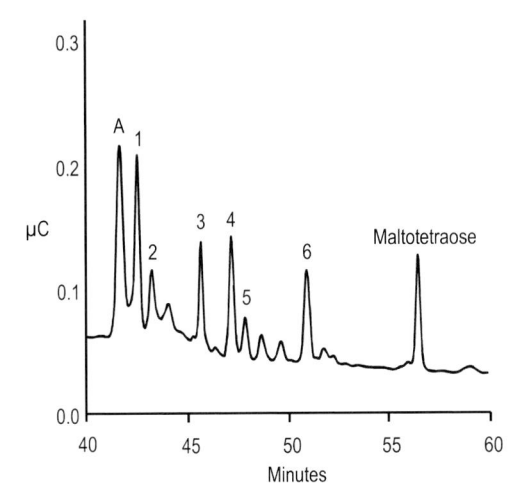

Fig. 9-139. Oligosaccharide peak pattern of an authentic orange juice. – Separator columns: 2 CarboPac PA-100, eluant: NaOH/NaOAc gradient; flow rate: 0.7 mL/min; detection: pulsed amperometry on a gold working electrode; injection: 25 µL sample; sample preparation according to Low; peaks (1) to (6) are minor sugars, whose signals increase with the addition of inverted sugar; (taken from [208]).

It is very important to standardize the handling of the method, which is also suitable for identifying adulterations in grapefruit juices and grapefruit juice concentrates. Even small changes in the chromatographic conditions cause retention time shifts and changes in the peak areas, so that a comparison to data from the data bank is invalid. The method has also been criticised because false-positive results are supposedly obtained for concentrates that are exposed to a heating process. This is attributed to the fact that when heating a concentrated sugar solution in the presence of citric acid, the same oligosaccharides are formed that are introduced into the juice by adulteration with inverted sugar

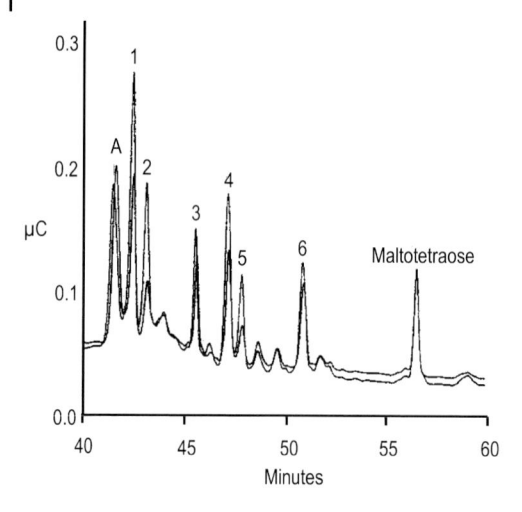

Fig. 9-140. Overlay of chromatograms of an authentic orange juice and an orange juice adulterated with 10% of a partly inverted sucrose. – Chromatographic conditions: see Fig. 9-139; sample preparation according to Low; peaks (1) to (6) are increased by the addition of inverted sugar; (taken from [209]).

[211]. However, further investigations by Trotzer et al. showed that drastic conditions are necessary to form these oligosaccharides at concentrations which normally result from the addition of inverted sugar.

An adulteration of citrus juices via addition of glucose syrup can be identified by determining the content of maltose, as glucose from starch hydrolysates contains small amounts of maltose, maltotriose, and maltotetraose as degradation products of starch hydrolysis. In the past, these sugars were separated as borate complexes on strong basic anion exchangers and detected photometrically after derivatization with orcinol-sulfuric acid. However, this reaction only occurs at elevated temperature. Although six different borate buffers are used for this method, trimers and tetramers are poorly separated; higher homologues are not separated at all. In contrast, using the Low method modified by Trotzer et al., these three sugars can be detected in the matrix of an orange juice without any problem. In comparison with the oligosaccharide separation described above only the gradient conditions have to be changed. Figure 9-141 shows the corresponding sector of two chromatograms as an overlay; the bottom one represents an authentic orange juice and the top one an orange juice adulterated with maltose. It proves that maltose and maltotriose can be detected as additional peaks in the oligosaccharide pattern. The method, of course, is only applicable to fruit juices that naturally do not contain any maltose. The addition of less than 1% sugar can be identified.

In some countries, ion chromatography is used for characterizing beer and for monitoring the brewing process [212]. The analysis of organic acids, carbohydrates, sulfite, ascorbic acid, and ethanol is of primary interest. Monitoring the water used in the brewing process for inorganic anions and cations is also an important application. The determination of alcohols such as ethanol and glycerol is usually performed with an ion-exclusion phase and refractive index detection. A typical example is the determination of ethanol in an alcohol-free beer on an Aminex HPX 87H stationary phase, which is illustrated in Fig. 9-142. While sugars eluting close to the system void are not separated or are incompletely separated, lactic acid, glycerol, and ethanol can be detected in this matrix without any problem [213]. For this purpose, the sample to be analyzed is degassed and diluted 1:10 with the eluant.

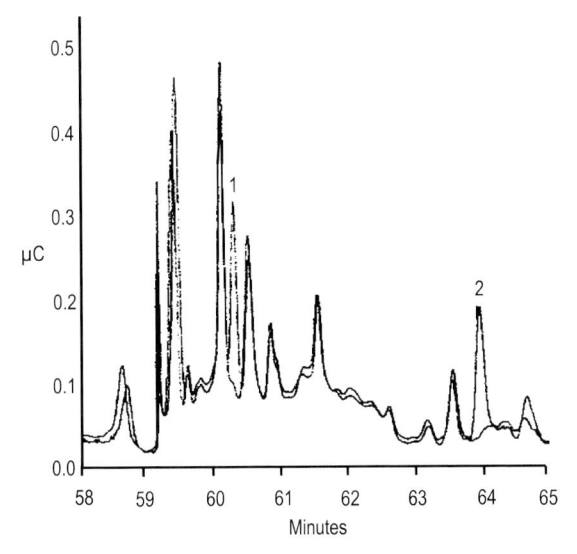

Fig. 9-141. Overlay of chromatograms of an authentic orange juice and an orange juice adulterated with maltose. — Separator columns: 2 CarboPac PA-100; eluant: NaOH/NaOAc gradient; flow rate: 0.7 mL/min; detection: pulsed amperometry on a gold working electrode; injection: 25 µL sample; sample preparation according to Low; peaks: (1) maltose and (2) maltotriose; (taken from [208]).

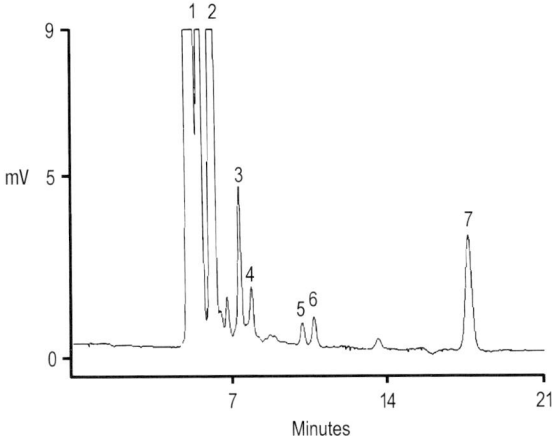

Fig. 9-142. Determination of ethanol in an alcohol-free beer by ion-exclusion chromatography. — Separator column: 300 mm × 7.8 mm i.d. Aminex HPX 87H; column temperature: 65 °C; eluant: 5 mmol/L H_2SO_4; flow rate: 0.6 mL/min; detection: RI; sample preparation: dilution 1:10 with eluant and degassing; peaks: (1) maltotriose, (2) maltose + sucrose, (3) glucose, (4) fructose, (5) lactic acid, (6) glycerol, and (7) ethanol; (taken from [212]).

For the analysis of organic acids in beer, the literature describes separation on ion exchangers utilizing UV detection [213, 214]. However, detection via measuring the light absorption at 210 nm is interfered with by other matrix components such as phenols, bitter components, and fructose, which also absorb at this wavelength. To avoid extensive sample preparation, the application of conductivity detection is recommended. As an example, Fig. 9-143 shows the separation of organic acids in a dark lager beer on IonPac ICE-AS6. Even though the resolution on an anion exchanger under gradient conditions is somewhat larger, at least the main components can be determined by ion-exclusion chromatography, which requires significantly less analysis time because of the isocratic conditions. Sample preparation in this case is restricted to a simple dilution with de-ionized water.

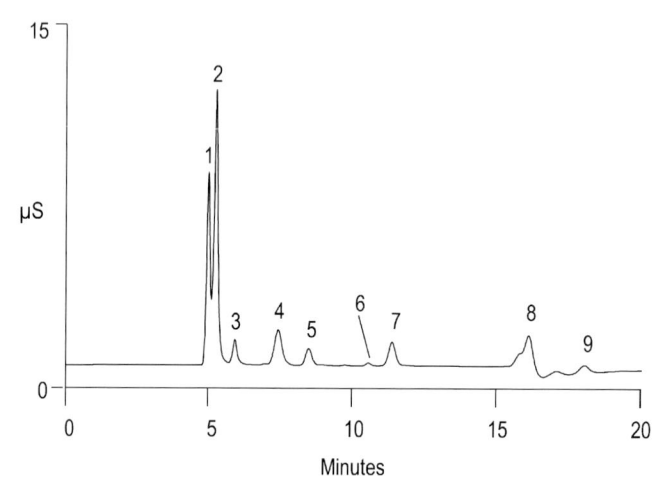

Fig. 9-143. Separation of organic acids in beer by ion-exclusion chromatography. – Separator column: IonPac ICE-AS6; eluant: 0.4 mmol/L perfluorobutyric acid; flow rate: 1 mL/min; detection: suppressed conductivity; injection volume: 50 µL; sample: dark lager beer, diluted 1:30 with de-ionized water; peaks: (1) and (2) mineral acids, (3) maleic acid, (4) citric acid, (5) malic acid, (6) formic acid, (7) lactic acid, (8) carbonate, and (9) succinic acid.

For monitoring the brewing process in terms of carbohydrates, anion exchange chromatography with pulsed amperometric detection is the method of choice [215]. Sugars such as glucose, maltose, and oligosaccharides, which are formed by enzymes from barley during the mash process, can be analyzed with a NaOH/NaOAc gradient within 30 minutes. After the treatment with hops and the addition of yeast, the smaller saccharides are fermented to form alcohol. Saccharides larger than DP_3 are not fermented, but contribute to the caloric value, the taste, and the stability of the foam. The three chromatograms in Fig. 9-144 characterize the individual steps of the brewing process with the respective carbohydrate pattern.

As revealed in the first chromatogram, only traces of oligosaccharides, in addition to the three fermentable sugars, can be detected after the mash. In the wort (Chromatogram B), the oligosaccharide fraction is significantly larger and it is highest in the finished product (chromatogram C). The decrease in glucose and maltose concentration in favour of ethanol can also be monitored in this way.

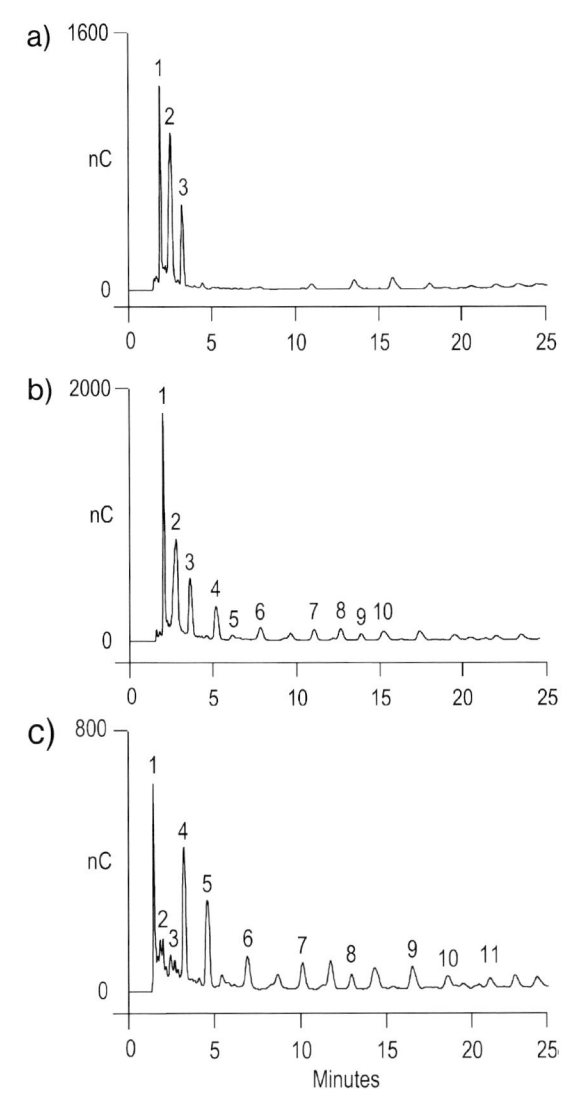

Fig. 9-144. Characterization of the brewing process via carbohydrate screening. (a) after the mash, (b) in the wort, and (c) in the finished product. – Separator column: CarboPac PA-100; eluant: NaOH/NaOAc gradient; flow rate: 1 mL/min; detection: pulsed amperometry on a gold working electrode; injection volume: 10 µL; peaks: (1) ethanol, (2) glucose, (3) maltose, (4) maltotriose, (5) maltotetraose, (6) maltopentaose, (7) maltohexaose, (8) maltoheptaose, (9) maltooctaose, (10) maltononaose, and (11) maltodecaose.

In the field of wine analysis within the European Union, the analytical methods are defined to a large extent by VO (EWG) No. 2676/90 [216]. Therefore, wet-chemical and enzymatic techniques still play an important role. However, the liquid chromatographic determination of the main sugars such as glucose, fructose, and sucrose together with glycerol and ethanol gains in importance due to its potential for automation. Separation is performed on a calcium-loaded ion-exclusion phase with pure de-ionized water as an eluant; refractive index detection is applied [217]. Thus, carbohydrate separation is based on the formation of different coordination complexes with metal cations, whose stability depends on the steric constitution of the hydroxide groups on the carbohydrate skeleton. A potential problem of such separations is the interference of organic acids, which — depending on the type of column — have to be removed via strong basic anion exchangers [218]. With the Polyspher CHCA (Merck, Darmstadt, Germany), on the other hand, quantitative analysis of the above-mentioned sugars and alcohols is possible without a preceding separation of organic acids. But even this separator column requires an operating temperature of 85 °C. The chromatograms of a dry white wine and a corresponding standard in a wine matrix are illustrated in Fig. 9-145.

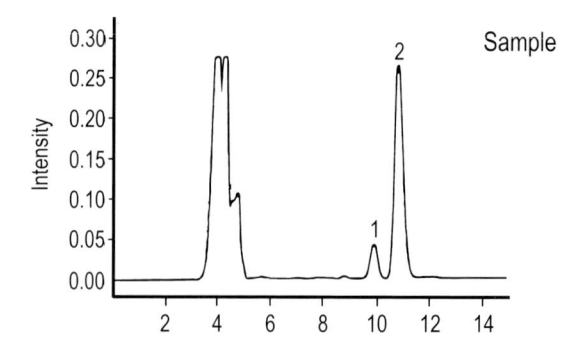

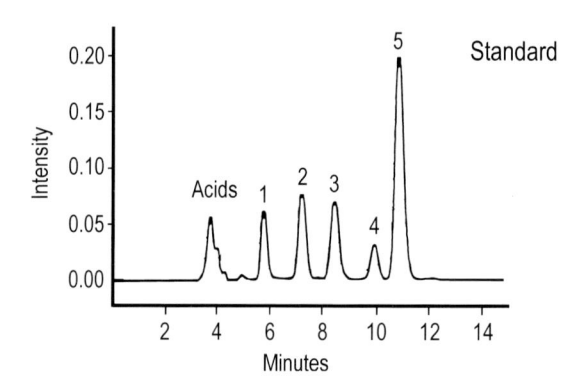

Fig. 9-145. Ion-exclusion chromatography of carbohydrates and alcohols in white wine. — Separator column: RT 300-6.5 Polyspher CHCA; column temperature: 85 °C; eluant: de-ionized water; flow rate: 0.6 mL/min; detection: RI; injection volume: 20 µL; sample: dry white wine with a total sugar content of 0.1 g/L, spiked with tartaric acid, lactic acid, and malic acid to yield 17.6 g/L total acid; peaks: (1) glycerol and (2) ethanol; standard: (1) sucrose, (2) glucose, (3) fructose, (4) glycerol, and (5) ethanol; (taken from [218]).

Ohs [219] and Baumgärtner [207] made important investigations using the HPAEC-PAD method into the analysis of carbohydrates in wine, in which small quantities of sorbitol, rhamnose, arabinose and, to some extent, trehalose are present in addition to the main components glucose and fructose. An example chromatogram in Fig. 9-146 demonstrates that the separation of the above-mentioned sugars is possible in spite of the high concentration differences between main and minor components. Moreover, all wines that have been investigated by Baumgärtner have a common characteristic peak pattern in the oligosaccharide region.

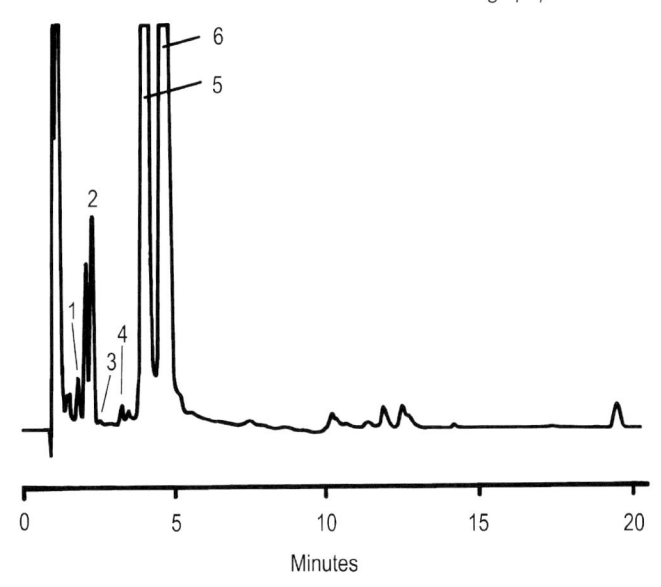

Fig. 9-146. Gradient elution of carbohydrates in wine with the HPAEC-PAD method. — Separator column: CarboPac PA1; eluant: (A) 0.1 mol/L NaOH, (B) 0.1 mol/L NaOH + 0.5 mol/L NaOAC; gradient: 100% A isocratic for 4 min, then linearly to 100% B in 40 min; flow rate: 1 mL/min; detection: pulsed amperometry on a gold working electrode; injection: 50 µL sample (Wallhäuser Pfarrgarten, Riesling 1988, diluted 1:50); peaks: (1) sorbitol, (2) trehalose, (3) rhamnose, (4) arabinose, (5) glucose, and (6) fructose; (taken from [207]).

The spectrum of organic acids in wine is extremely complex and represents a challenge for the ion chromatographic analysis due in part to large concentration differences. Thus, isocratic elution on an anion exchanger with a carbonate/bicarbonate eluant as described in an application note by Metrohm [220] is completely useless, because peak purity of inorganic and organic acids is not ensured. Even conventional hydroxide-selective anion exchangers suitable for gradient elution do not possess the required ion-exchange capacity to obtain acceptable resolution among early eluting monocarboxylic acids. This problem could be solved, however, with the introduction of high-capacity anion exchangers. Figure 9-147 shows an example chromatogram that was obtained with a high-capacity IonPac AS11-HC anion exchanger. The 100% solvent compatibility of this stationary phase allows the use of methanol for optimizing selectivity, especially in the retention range of tartaric acid. Optimum separation is obtained — as in the present case — by increasing and decreasing the methanol content during the gradient run. The only disadvantage is the long analysis time of about 40 minutes, which is typical for hydroxide-selective high-capacity anion exchangers. Ion-exclusion chromatography is also unsuited for such a complex analytical problem, unless only the main components are to be analyzed. On the other hand, in combination with an amperometric detector, ion-exclusion chromatography is able to successfully determine sulfite. As can be seen from the chromatogram in Fig. 9-148, sulfite can easily be oxidized on a platinum electrode at an oxidation potential of +0.6 V. Although not quantified in the present case, glycerol and ethanol may be determined, too, under these chromatographic conditions.

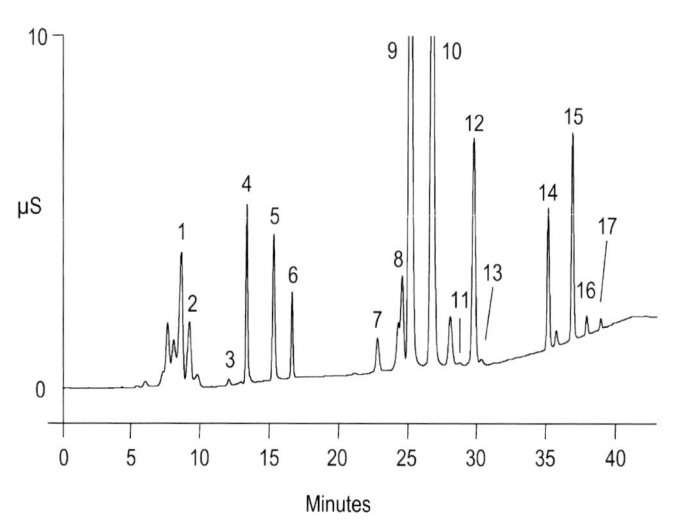

Fig. 9-147. Gradient elution of inorganic anions and organic acids in wine. – Separator column: IonPac AS11-HC with guard; column temperature: 30 °C; eluant: NaOH – methanol; gradient: NaOH: 1 mmol/L isocratic for 8 min, then linearly to 30 mmol/L in 20 min, then to 60 mmol/L in 10 min, methanol: 0% to 20% from 8 to 18 min, 20% from 18 to 28 min, 20% to 0% from 28 to 38 min; flow rate: 1.5 mL/min; injection volume: 10 µL; detection: suppressed conductivity; sample: wine diluted 1:10 with de-ionized water, sample preparation via OnGuard-RP; peaks: (1) lactate, (2) acetate, (3) formate, (4) pyruvate, (5) galacturonate, (6) chloride, (7) nitrate, (8) succinate (9) malate, (10) tartrate, (11) fumarate, (12) sulfate, (13) oxalate, (14) orthophosphate, (15) citrate, (16) isocitrate, and (17) cis-aconitate.

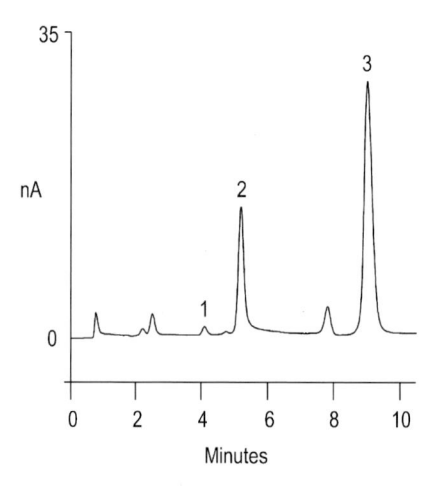

Fig. 9-148. Determination of sulfite in white wine utilizing electrochemical detection. – Separator column: IonPac ICE-AS1; eluant: 5 mmol/L H_2SO_4; flow rate: 1 mL/min; injection volume: 50 µL; detection: DC amperometry on a platinum working electrode; sample: white wine diluted 1:10 with de-ionized water; peaks: (1) glycerol, (2) 1.5 mg/L sulfite (the sulfite content in the undiluted sample was 15 mg/L) and (3) ethanol.

The analysis of alkali- and alkaline-earth metals in wine, which is usually only diluted with de-ionized water and filtrated prior to injection, is relatively easy. An example is shown in Fig. 9-149. If low analyte concentrations do not allow sample dilution, a pre-treatment with a polyvinylpyrrolidone resin (OnGuard-P) is recommended to remove dyes, polyphenols, anthocyanines, and other molecular organic compounds which could block the ion-exchange sites of the separator column.

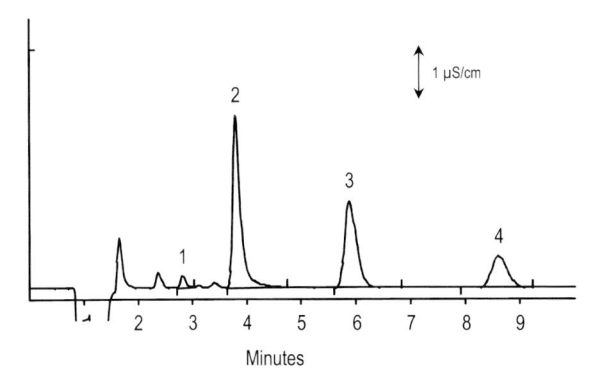

Fig. 9-149. Determination of alkali- and alkaline-earth metals in a red wine. — Separator column: Metrosep Cation 1-2; eluant: 4 mmol/L tartaric acid + 1 mmol/L pyridine-2,6-dicarboxylic acid; flow rate: 1 mL/min; injection volume: 10 µL; detection: non-suppressed conductivity; sample: Italian red wine diluted 1:50 with de-ionized water; solute concentrations: 9.8 mg/L sodium (1), 822 mg/L potassium (2), 306 mg/L calcium (3), and 72.2 mg/L magnesium; (taken from [221]).

In contrast, transition metal analysis in wine is not a trivial matter. Metals such as aluminum, cobalt, iron, or nickel impact the taste of wine and lead to unwanted clouding [222], and health organizations impose strict thresholds for toxic elements such as lead, cadmium, copper, and zinc. Thus, there is a great demand for a fast, simple, and sensitive multi-element method for determining transition metals in wine. In the wine matrix, some of the metals are complexed. Therefore, the organic matrix has to be eliminated. Widely distributed sample preparation techniques are dry ashing and wet digests, neither of which is particularly suited for trace analysis. Alternatively, microwave digests are promising because the consumption of chemicals and sample preparation time can be lowered significantly with this technique.

In their recent paper, Buldini et al. [223] describe UV photolysis as a new digestion method for wine that is characterized by ease-of-use and excellent reliability. The most important part of this digestion apparatus (705 UV, Metrohm, Herisau, Switzerland) is a 500-W air- and water-cooled high pressure lamp. Photolysis is performed in the presence of hydrogen peroxide as a source for $OH^{\bullet}$-radicals, which accelerate the degradation of organic matrix components. Water and oxygen as degradation products of hydrogen peroxide do not interfere with the subsequent metal determination. Depending on the type of wine to be analyzed, the photolysis time can be lowered by increasing the hydrogen peroxide concentration. In general, the photolysis time for wine at constant hydrogen peroxide concentration increases with increasing TOC content. For the subsequent ion chromatographic determination of transition metals, a sample that has a pH of 9 after photolysis has to be buffered to pH 5 to 6. The sample thus prepared can then be investigated for its transition metal content using an ion exchanger with a defined anion- and cation exchange capacity. Depending on the type of analyte, oxalic acid or pyridine-2,6-dicarboxylic acid (PDCA) are used as complexing agents. Alkali- and alkaline-earth metals up to concentration differences of 1000:1 do not interfere with the transition metal determination due to the selective detection via post-column derivatization with PAR (see Section 4.5.3). Manganese cannot be quantitated under these conditions, because it is partly oxidized under the influence of UV light. An overlay of chromatograms for different wines using an oxalic acid eluant is illustrated in Fig. 9-150.

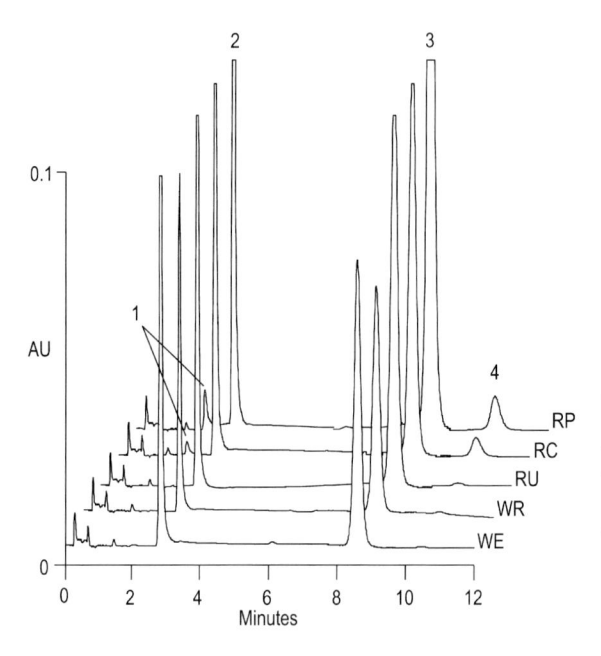

Fig. 9-150. Isocratic separation of transition metals in various wines after UV photolysis. – Separator column: IonPac CS5A; eluant: 80 mmol/L oxalic acid + 50 mmol/L KOH + 100 mmol/L TMAOH, pH 4.7; flow rate: 1.2 mL/min; detection: photometry at 530 nm after derivatization with PAR; injection volume: 150 µL; samples: 1:10 diluted wines, WE white wine Emilia, WR white wine Romagna, RU red wine Umbria, RC red wine Campania, RP red wine Piemont; peaks: (1) lead, (2) copper, (3) zinc, and (4) nickel; (taken from [223]).

In comparison to fruit juices, beer, and wine, the matrix of refreshers does not pose any problem for their ion chromatographic characterization. In almost all cases, sample preparation is restricted to dilution with de-ionized water, degassing (for CO_2-containing refreshers), and filtration (0.45 µm). This type of samples is exemplified in Figures 9-151 and 9-152, which show the separations of inorganic anions and cations in Coke. Cation analysis was performed with a small step gradient to obtain optimum resolution between sodium and ammonium in the shortest analysis time possible.

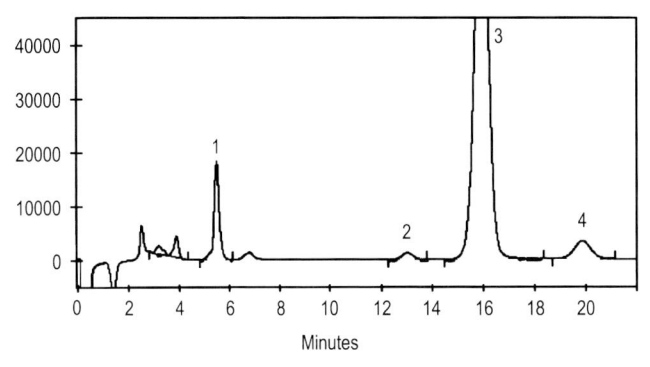

9.6 Ion Chromatography in the Food and Beverage Industry | 731

Fig. 9-151. Separation of inorganic anions in a Coke. – Separator column: Metrosep Anion Dual 2; eluant: 1.3 mmol/L Na_2CO_3 + 2 mmol/L $NaHCO_3$; flow rate: 0.8 mL/min; detection: suppressed conductivity; injection volume: 20 µL; sample: Coke diluted 1:5 with de-ionized water; solute concentrations: 15.2 mg/L chloride (1), 3.9 mg/L nitrate (2), 506 mg/L orthophosphate (3), and 11.8 mg/L sulfate (4); (taken from [224]).

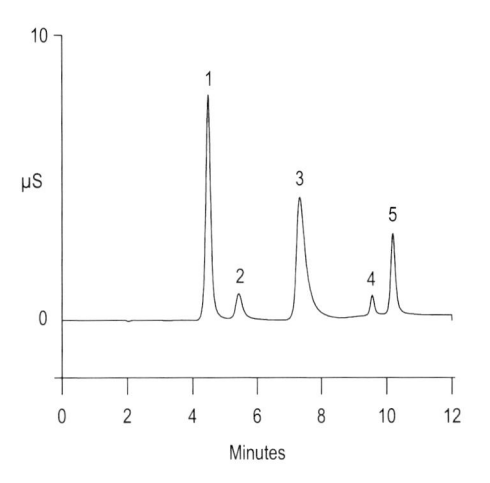

Fig. 9-152. Separation of inorganic cations in Diet Coke. – Separator column: IonPac CS12 with guard; eluant: methanesulfonic acid; gradient: 16 mmol/L isocratic for 5 min, then to 40 mmol/L for 5 min; flow rate: 1 mL/min; detection: suppressed conductivity; injection volume: 25 µL; sample: Diet Coke diluted 1:10 with de-ionized water; solute concentrations: 5.3 mg/L sodium (1), 2.3 mg/L ammonium (2), 9 mg/L potassium (3), 0.2 mg/L magnesium (4), and 1.7 mg/L calcium (5).

In many cases, a number of organic additives are added to refreshers. These additives include sweeteners such as saccharin or aspartame, preservatives such as benzoic acid, and flavours such as citric acid and caffeine.

Saccharin
(o-Benzenesulfimide)
pK_a = 2.3

Caffeine
pK_a = 4.2

Aspartame
(L-Aspartyl-L-phenylalanine-methyl ester)

All five compounds can be analyzed in one run using a multimode phase with anion exchange and reversed-phase properties and simultaneous conductivity and UV detection. A mixture of sodium carbonate and acetonitrile is used as an eluant. The chromatogram of a corresponding standard obtained under isocratic conditions is illustrated in Fig. 9-153. The top chromatogram represents the UV detection at 214 nm, the bottom chromatogram was obtained with suppressed conductivity detection. Saccharin can be detected very sensitively with either of these detection methods. As a typical example of a real sample, Fig. 9-154 shows the chromatogram of a Diet Coke. Inorganic constituents such as chloride, sulfate, and orthophosphate are completely separated from the organic components.

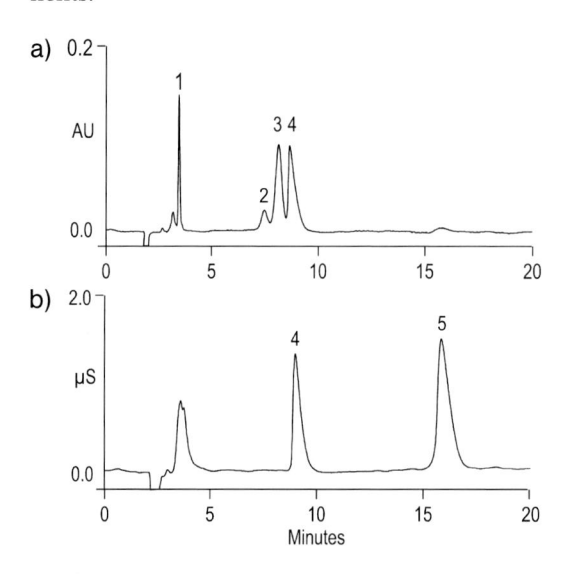

Fig. 9-153. Separation of typical organic additives used in refreshers. – Separator column: OmniPac PAX-500; eluant: 15 mmol/L Na_2CO_3 – acetonitrile (85:15 v/v); flow rate: 1 mL/min; detection: (a) UV (214 nm), (b) suppressed conductivity; injection volume: 25 µL; solute concentrations: 0.1 mmol/L benzoate (1), 0.2 mmol/L aspartame (2), 0.2 mmol/L caffeine (3), 0.1 mmol/L saccharin (4), and 0.2 mmol/L citric acid (5).

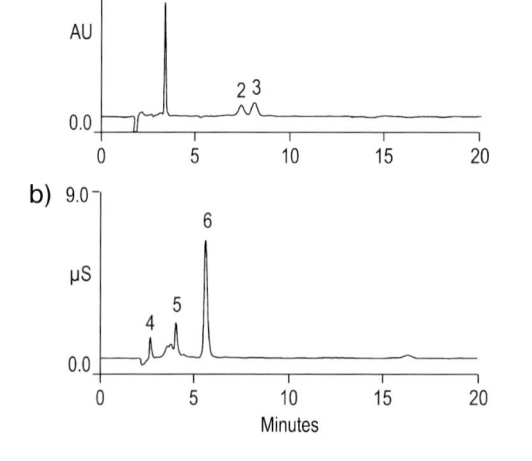

Fig. 9-154. Separation of organic additives in a Diet Coke. – Chromatographic conditions: see Fig. 9-153; peaks: (1) benzoate, (2) aspartame, (3) caffeine, (4) chloride, (5) sulfate, and (6) orthophosphate.

9.6.2
Dairy Products

For health purposes, the determination of inorganic anions such as nitrite, nitrate, and iodide in milk and milk products [225, 226] is of great importance. Together with secondary amines, nitrite, for example, can form carcinogenic nitrosamines; an excess of iodide can provoke thyroid malfunction. To prolong the lifetime of the stationary phase being used, it is recommended that milk proteins be removed from the sample. For extraction of nitrite and nitrate from milk or milk powder, 20 mL of de-ionized water are added to 1 g of sample and the resulting solution is filtered through a 1.2-µm filter. The filtrate is then centrifuged in a Centricon-3 filter at 5000 rpm for 30 minutes. After passing through a reversed-phase extraction cartridge, the sample can be injected into the ion chromatograph. The separation of nitrite and nitrate is performed on an IonPac AS7 latexed anion exchanger utilizing UV detection at 214 nm. A mixture of potassium chloride and Tris, adjusted to pH 7.5 with HCl, is used as an eluant. Figure 9-155 shows a chromatogram of low-fat milk. The nitrite content of 80 µg/L and the nitrate content of 1.12 mg/L in the extract correspond to a nitrite content of 1.6 µg/g and a nitrate content of 22 µg/g in the sample. Spiking experiments with 1 µg/g nitrite and 15 µg/g nitrate resulted in recovery rates of 100% and 93%, respectively. The detection limits for nitrite and nitrate in milk are 0.4 µg/g and 0.8 µg/g, respectively, and thus significantly below the commonly measured values. As an alternative to an extraction technique, a dialysis module can be used for sample preparation (e.g., Dialysis Module 754, Metrohm, Herisau, Switzerland). However, in the corresponding application note [227], only the separation of the major components chloride, orthophosphate, and sulfate using this sample preparation technique is described.

For the analysis of iodide in milk or whey, the milk proteins are precipitated with nitric acid. For this purpose, 50 mL of sample are put into a 100-mL measuring flask, 4 mL of a 3% acetic acid and 1 mL of concentrated nitric acid are added, and the flask is filled up to the meniscus with de-ionized water. This solution is then passed through a Whatman-2V filter, a Centriflo filter, and then through a reversed-phase extraction cartridge. The separation of iodide can be performed with both ion-exchange- and ion-pair chromatography [228]. Because anion exchangers are now available on which iodide can be rapidly eluted with purely aqueous eluants and without tailing, anion exchange chromatography is preferred. However, the official AOAC method [229] is still based on an ion-pair chromatographic separation. To detect the very low iodide concentrations in milk (100-300 µg/L) with sufficient precision, amperometric detection on a silver- or platinum working electrode is used. While silver electrodes require a low oxidation potential of 0.05 V and can be used directly, platinum electrodes have to be conditioned with a saturated potassium iodide solution prior to their use and require a significantly higher oxidation potential of 0.8 V that, in turn, leads to a higher susceptibility to interferences. As an example, Fig. 9-156 shows the separation of iodide in milk on an IonPac AS11 anion exchanger with a dilute nitric acid eluant. The detection limit for iodide with this method is approximately 3 µg/L.

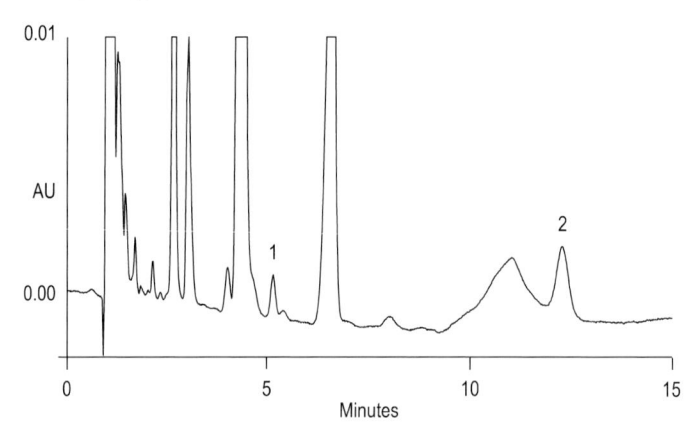

Fig. 9-155. Separation of nitrite and nitrate in a low-fat milk. — Separator column: IonPac AS7 with AG14 guard; eluant: 50 mmol/L KCl + 5 mmol/L Tris, pH 7.5 with HCl; flow rate: 1.5 mL/min; detection: UV (214 nm); injection volume: 90 µL; sample: low-fat milk diluted 1:10; sample preparation: see text; solute concentrations: 80 µg/L nitrite (1) and 1.1 mg/L nitrate (2).

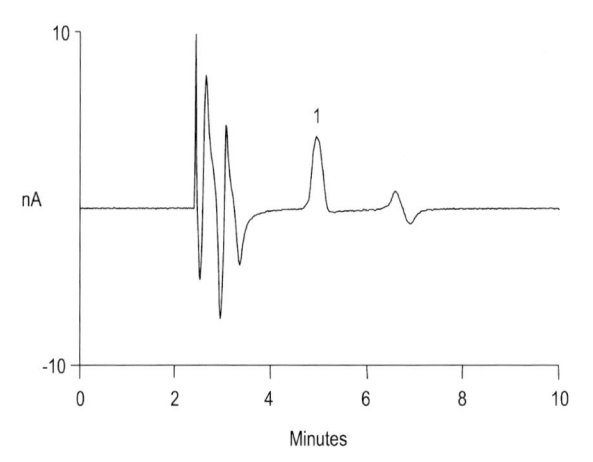

Fig. 9-156. Separation of iodide in low-fat milk. — Separator column: IonPac AS11; eluant: 5 mmol/L HNO_3; flow rate: 1.5 mL/min; detection: DC amperometry on a silver working electrode; oxidation potential: 0.05 V; injection volume: 100 µL; sample: low-fat milk; sample preparation: see text; solute concentration: 103 µg/L iodide (1).

In the field of cation analysis, Gaucheron et al. [230] developed an ion chromatographic method for determining ammonium in milk and dairy products. The origin and concentration of ammonium in dairy products depends on the nature of these products. In raw milk and in the absence of microbial contamination, the concentration of ammonium is less than 5 mg/kg [231]. However, after microbial contamination or different technological treatment (sterilization, acidification, ripening) the ammonium content in milk products can vary significantly. Ammonium can be produced by deamidation reactions of asparagine or glutamine residues present in the sequence of milk proteins, by degradation of urea, or by enzymatic degradation of proteins, peptides, and amino acids. The first two reactions can be induced by severe heat treatment of milk or by enzymes during the transformation of milk in yoghurt or cheese. To control the quality and the biochemical evolution of these products during these transformations, it is important to be able to determine ammonium, which can be considered as an indicator of food quality. Different methods to determine ammonium in milk products were used in the past including alkaline distillation followed by Kjeldahl titration [232], potentiometry with an ammonia-selective electrode [233], enzymatic methods [234], or flow injection analysis [235]. Each method has its own merits, but all of them are time-consuming. In contrast, cation exchange chromatography utilizing a crown ether modified IonPac CS15 cation exchanger and suppressed conductivity detection provides high resolution between ammonium and other mono- and di-valent cations in milk products within a reasonable amount of time. As an example, Fig. 9-157 illustrates the separation of cations in a commercial yoghurt sample. To prepare the sample, the aqueous phase of yoghurt was recovered by centrifugation (300 g for 15 min). The clear supernatant was filtered (0.42 µm) and diluted two-fold with de-ionized water prior to injection. The chromatogram in Fig. 9-157 is similar to that observed with milk treated at 120 °C for 30 minutes. Despite the high calcium and potassium content, the ammonium peak is clearly separated and, therefore, can be quantitated without any problem.

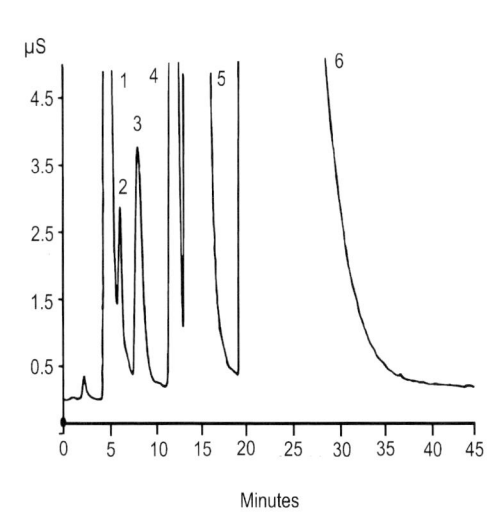

Fig. 9-157. Determination of ammonium in a commercial yoghurt sample. — Separator column: IonPac CS15 with CG15 guard; column temperature: 20 °C; eluant: 5 mmol/L H_2SO_4 — acetonitrile (91:9 v/v); flow rate: 1.2 mL/min; detection: suppressed conductivity; injection: 25 µL sample diluted two-fold with de-ionized water; sample preparation: see text; peaks: (1) sodium, (2) unknown, (3) 117 mg/kg ammonium, (4) magnesium, (5) calcium, and (6) potassium; (taken from [230]).

The possibility of detecting sulfur-containing antibiotics such as ampicillin and lincomycin via integrated amperometry utilizing a multicyclic pulse sequence was described in Section 7.1.2.2. Dasenbrock and LaCourse [236] developed this method to determine cephapirin and ampicillin in raw milk. At present, close attention is being paid to the detection of antibiotics in foods because the ubiquitous presence of antibiotics in foods can lead to severe health problems for the consumer. These problems include allergic reactions and, above all, the formation of bacteria that is resistent to antibiotics, such that currently existing drugs become ineffective in case of an emergency. Because antibiotics are routinely applied to milk cows to avoid mastitis, the U.S. FDA (Food and Drug Administration) has issued thresholds for antibiotics in milk. Trace analysis of antibiotics in milk is a difficult task, as many antibiotics do not have chromophores and so cannot be detected sensitively via photometry. However, many of the papers published so far [237-239] are based on photometric detection at low wavelength and, consequently, require very time-consuming and laborious sample preparation or post-column derivatization [240] to reach the required sensitivity. The advantages of integrated amperometry for detecting sulfur-containing antibiotics are based on sensitive and selective detection, which leads to less complex chromatograms with minimum sample preparation. Raw milk, however, has to be de-fatted and de-proteinated by centrifugation (5000 rpm for 30 minutes). For this purpose, Dasenbrock et al. utilized the precipitation method with acetonitrile and subsequently filtrated the supernatant through glass wool. The extract was then passed through a reversed-phase extraction cartridge, from which the analytes were eluted with 15% acetonitrile. After filtration (0.45 µm), this extract was injected directly. Separation was performed on a reversed-phase column (Luna C-8, Phenomenex, Torrance, CA, USA) under isocratic conditions using an acetate buffer in aqueous acetonitrile. Figure 9-158 shows the chromatogram of a milk extract spiked with 40 µg/L cephapirin and 20 µg/L ampicillin in comparison with a blank sample (bottom chromatogram). Both analytes could be eluted practically free of any interferences within 35 minutes, whereas in UV detection (254 nm) an interfering signal with a retention time typical for cephapirin was observed. The determination of ampicillin is virtually impossible because the peak height of this signal is about 40% that of cephapirin. The peak is also interfered by a much larger peak.

Sample preparation for determining carbohydrates in dairy products with the HPAEC-PAD method turns out to be much easier due to the high sensitivity of this detection method. For example, ice cream was dissolved by Baumgärtner [207] with de-ionized water, membrane filtered (0.45 µm), and directly injected. Figure 9-159 shows the corresponding chromatogram of a 0.16% solution in which major amounts of glucose, lactose, and sucrose and small amounts of maltose and its oligomers could be detected by using a gradient technique.

A number of years ago, the food industry began adding fructans to dairy products and other foods for nutritional and physiological reasons. These are carbohydrates such as inulin and fructooligosaccharides (FOS) that are not hydrolyzed by enzymes in the human body. Inulin is a linear polyfructan with $\beta(1\rightarrow2)$-linked D-fructofuranosyl units bound on the terminal end of a trisaccharide

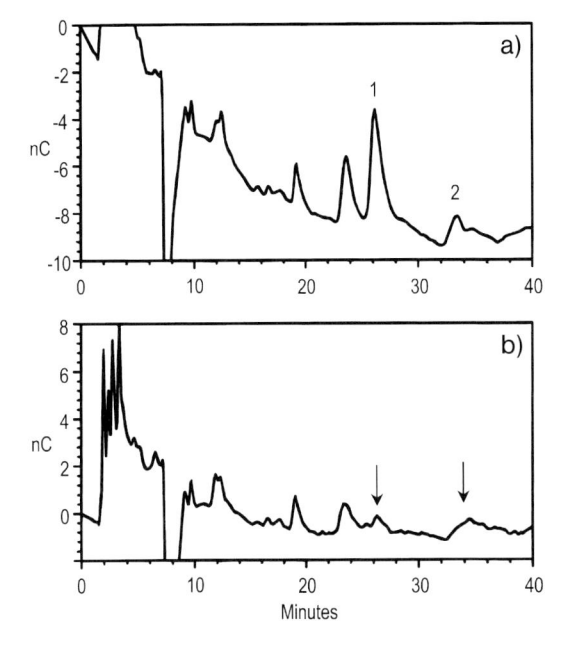

Fig. 9-158. Analysis of cephapirin and ampicillin in a milk extract utilizing integrated amperometry. — Separator column: 150 mm × 4.6 mm i.d. Luna C-8 (Phenomenex, USA); column temperature: 30 °C; eluant: 0.1 mol/L acetate buffer — acetonitrile (96:4 v/v), pH 3.75; detection: integrated amperometry on a gold working electrode; sample: (a) milk extract spiked with 40 µg/L cephapirin (1) and 20 µg/L ampicillin (2), (b) blank sample; sample preparation: see text; (taken from [237]).

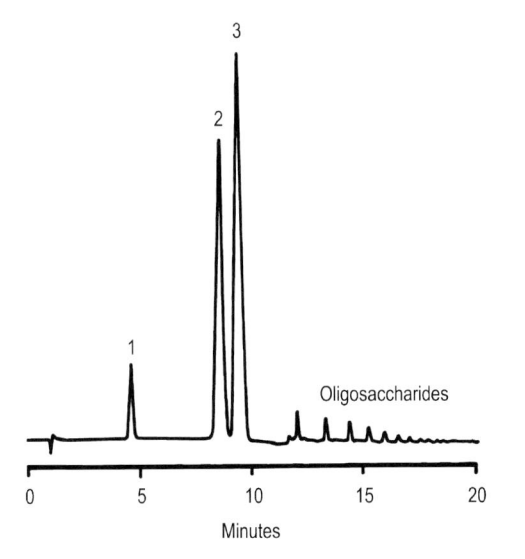

Fig. 9-159. Gradient elution of carbohydrates in ice cream. — Separator column: CarboPac PA1; eluant: (A) 0.1 mol/L NaOH, (B) 0.1 mol/L NaOH + 0.5 mol/L NaOAc; gradient: 100% A isocratic for 4 min, then linear to 100% B in 40 min; detection: pulsed amperometry on a gold working electrode; injection: 50 µL of a 0.16% solution of soft ice "Sundae"; peaks: (1) glucose, (2) lactose, and (3) sucrose; (taken from [207]).

(1-kestose or neo-kestose). Inulins exhibit a degree of polymerization from DP3 to DP60 with an average molecular weight of about 6000 Dalton. The lower molecular weight fraction (DP3 to DP_{20}) is termed fructooligosaccharides [241]. Due to their intensive commercialization in foods, fructooligosaccharides have to be declared and, thus, characterized and quantified. For this purpose, Durgnat et al. [242] developed a HPAEC-PAD technique, which is less cumbersome than gas chromatography due to its selectivity and sensitivity. GC can only be used in combination with a derivatization, and then only for the separation of DP2 to DP10 [243]. Because individual fructooligosaccharides are not available as reference compounds, Durgnat et al. recommend quantitating FOS sugars via commercial FOS mixtures such as Actilight-95®[1] with known contents of 1-kestose (38.6%), nystose (46.8%), and fructosylnystose (8.6%). Therefore, sugars such as 1-kestose and nystose can be used as tracer for quantification of Actilight-95. A typical example is the chromatogram of a strawberry yoghurt shown in Fig. 9-160. Sample preparation in this case was carried out in a traditional way by precipitating the proteins with Carrez I and II. The detection limit for this technique was determined by Durgnat et al. to be 80 ng for nystose; the recovery rates were between 95% and 108%.

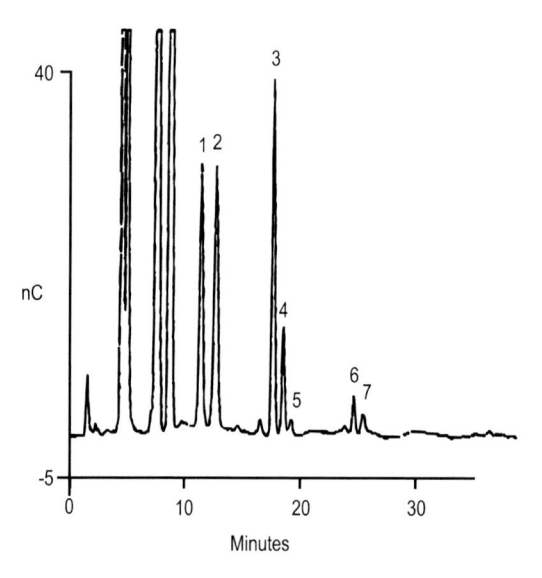

Fig. 9-160. HPAEC-PAD chromatogram of a strawberry yoghurt containing fructooligosaccharides. — Separator column: CarboPac PA1; eluant: NaOH/NaOAc gradient; flow rate: 1 mL/min; detection: pulsed amperometry on a gold working electrode; injection volume: 20 µL; sample: strawberry yoghurt no. 1; sample preparation: Carrez I and II; peaks: (1) 1-kestose, (2) maltose, (3) nystose, (4) maltotriose, (5) inulotriose, (6) fructosylnystose, and (7) maltotetraose; (taken from [242]).

Last but not least, milk caseins can also be separated on a suitable anion exchanger. The separation of caseins from bovine milk in Fig. 9-161 on ProPac SAX with a mixture of urea, 2-mercaptoethanol, imidazole, and NaCl as an eluant is clearly superior to a conventional separation on a wide-pore reversed-phase column, no matter whether an acetonitrile or an isopropanol gradient is used.

Another typical application example in the field of milk products is the determination of organic acids such as lactic acid, pyruvic acid, and citric acid in cheese products; such analysis is carried out by ion-exclusion chromatography after respective sample preparation. In addition, cheese contains polyphosphates, which serve as emulsion stabilizers. Commercial polyphosphates are mixtures of polyphosphates of different chain lengths. They are usually characterized by

[1] Actilight-95® is distributed by Béghin-Meiji Industries in France and manufactured enzymatically from sugar cane with fructosyl transferase [244].

end group titration, with which the average chain length can be determined. With ion chromatography, those chain length distributions can be determined much easier. Baluyot et al. [245] used a hydroxide-selective anion exchanger in the microbore format with an aqueous hydroxide gradient and subsequent conductivity detection. As can be seen from the chromatograms of two different hexamethaphosphate solutions in Fig. 9-162, fingerprint chromatograms are obtained in 40 minutes. In this case, both solutions were prepared from the same dry substance, however, the cheese products showed completely different characteristics. The comparison of the two resulting chromatograms in Fig. 9-162 indicates a different degree of hydrolysis during processing which, thereafter, could be optimized.

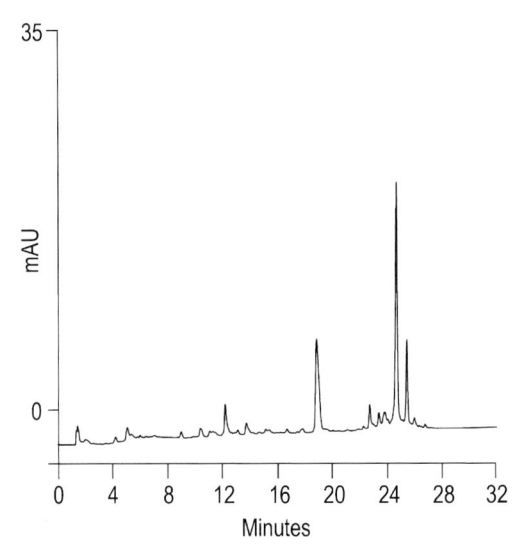

Fig. 9-161. Separation of caseins from bovine milk on an anion exchanger. — Separator column: ProPac SAX; eluant: (A) 4 mol/L urea + 0.1 mol/L 2-mercaptoethanol + 10 mmol/L imidazole, pH 7, (B) 1 mol/L NaCl in A; gradient: 100% A isocratic for 3 min, then from 5% B to 30% B in 27 min; flow rate: 1 mL/min; detection: UV (280 nm); Probe: 1 µg/mL caseins from bovine milk obtained by precipitation at pH 4.5 of whole bovine milk.

9.6.3
Meat Processing

A key problem in the investigation of meat and sausage products is the determination of nitrite and nitrate, whose tolerable concentrations are between 100 and 600 mg/kg depending on the product group. This regulation protects consumers and requires frequent control of the traded goods. In classical nitrite/nitrate analysis, nitrate is reduced to nitrite (cadmium reduction method) and then determined photometrically as total nitrite [246]. In 1990, the method "Determination of nitrite and nitrate in sausage products after enzymatic reduction" was accepted as part of the "Official collection of analytical methods according to § 35 LMBG". In comparison to the cadmium reduction method used so far, this new method is experimentally much simpler and the toxic cadmium is replaced

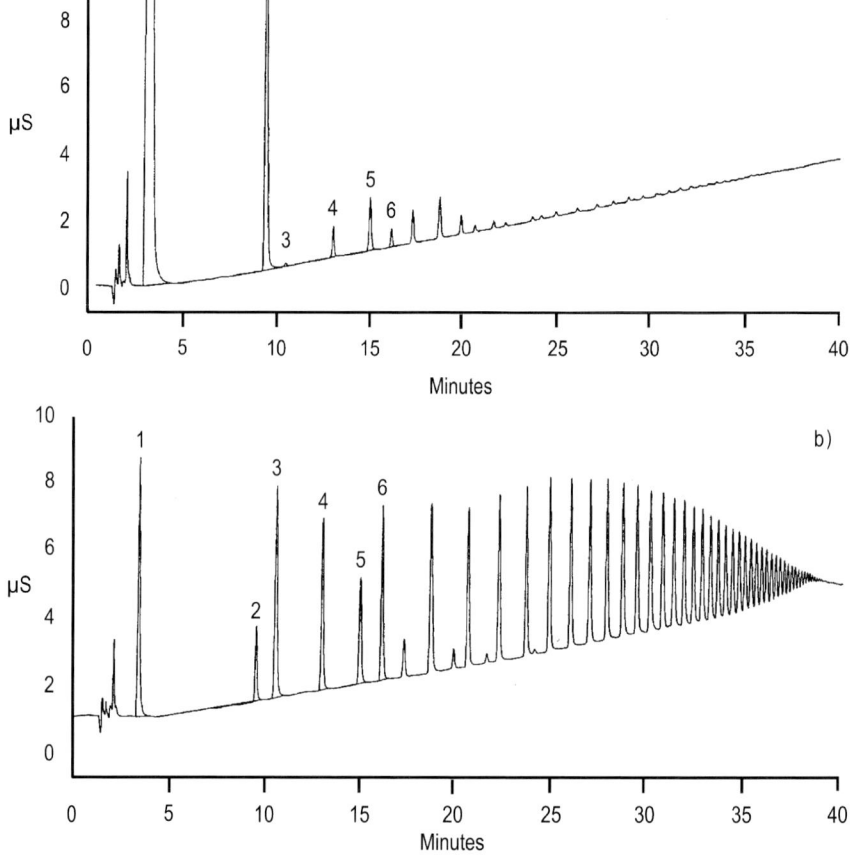

Fig. 9-162. "Fingerprint" analysis of polyphosphates in cheese products. − Separator column: IonPac AS11 (2-mm); eluant: NaOH gradient; flow rate: 0.3 mL/min; detection: suppressed conductivity; injection volume: 10 µL; Samples: (a) bad charge, (b) good charge; peaks: (1) orthophosphate, (2) pyrophosphate, (3) trimetaphosphate, (4) tripolyphosphate, (5) tetrametaphosphate and (6) tetrapolyphosphate; (taken from [245]).

by nitrate reductase. However, the costs of using enzymes in daily practice are significant. Therefore, ion chromatographic methods were developed at an early stage that significantly simplified nitrite/nitrate determinations [247, 248]. This does not include sample preparation, which at least in Germany is determined by § 35 LMBG, no matter which analytical method is applied. Sample preparation includes homogenization of the sample, its extraction with a 5% borax buffer in a hot water bath, and subsequent Carrez precipitation with 15% potassium hexacyanoferrate(II)- and 30% zinc sulfate solution. Depending on the concentration of the nitrogen compounds, the aqueous extracts thus obtained are further diluted with de-ionized water and membrane filtrated (0.22 µm) prior to injection.

In the ion chromatographic analysis of meat products a number of constituents elute close to the void volume of the anion exchanger used, so interferences cannot be excluded. Because chloride concentrations in meat products can easily exceed nitrite and nitrate concentrations by a factor of 5000, UV detection at low wavelengths (207-225 nm) is usually applied. At those wavelengths chloride absorption is usually very low. Recently, Siu and Henshall [249] described an alternative extraction method, which requires neither protein precipitation nor extraction with SPE cartridges. According to this method, 10 g of sample are mixed with de-ionized water to a total volume of 100 mL and homogenized in a blender for 1 minute. The homogenized sample is then heated to 70-80 °C for 15 minutes. After cooling down to room temperature, the sample is centrifuged at 6000 rpm for 10 minutes. The supernatant solution is then passed through a 1.2-µm Whatman No. 2, GF/A and a 0.2-µm Arodisc filter. The filtrate can be injected directly into the ion chromatograph. Figure 9-163 shows the chromatogram of a ham extract with a clear separation of nitrite and nitrate. To validate the signals, the sample was spiked with both components and chromatographed again. Nitrite and nitrate were clearly identified in this way with good recovery of the spiked amounts. The contents of sodium nitrite and sodium nitrate were 11.6 mg/kg and 5.37 mg/kg, respectively. Both values are clearly below the threshold of 100 mg/kg. Spiking experiments yielded recovery rates of 103 % for nitrite and 92 % for nitrate. The negative peak observed ahead of nitrite is chloride, which lowers eluant background absorption for a short time due to its high concentration. At the measuring wavelength of 225 nm the size of this negative peak in relation to the nitrite signal is the lowest.

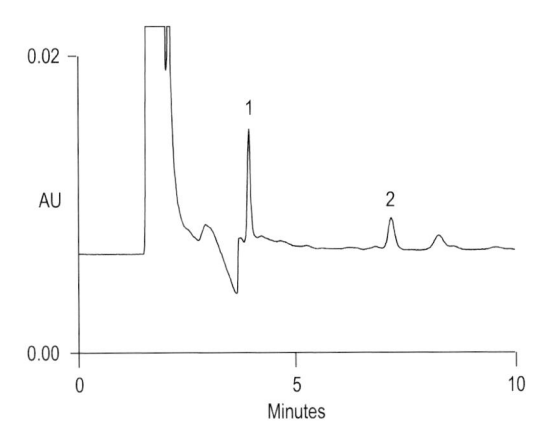

Fig. 9-163. Analysis of nitrite and nitrate in a ham extract. — Separator column: IonPac AS11; eluant: 5 mol/L NaOH; flow rate: 1 mL/min; detection: UV (225 nm); injection: 25 µL of a ham extract (weighted sample: 10 g ad 100 mL water); solute concentrations: 1.16 mg/L nitrite (1) and 0.54 mg/L nitrate (2); (taken from [249]).

Cutter and reddening agents used in meat processing can also be assayed ion chromatographically for their active constituents, which include inorganic anions such as chloride and diphosphate, ascorbic acid, and carbohydrates such as glucose, sucrose, and maltose. While the easiest way to analyze ascorbic acid is via ion-exclusion chromatography on a totally sulfonated cation exchanger, anion exchange chromatography with either suppressed conductivity or photometric detection after derivatization with iron(III) is used to analyze diphosphate. Selectivity and sensitivity of both methods are so high that an extensive sample preparation is not necessary. The analyte samples are only dissolved in de-ionized water and membrane-filtered (0.45 μm) prior to injection. Figure 9-164 displays the separation of diphosphate in a reddening agent.

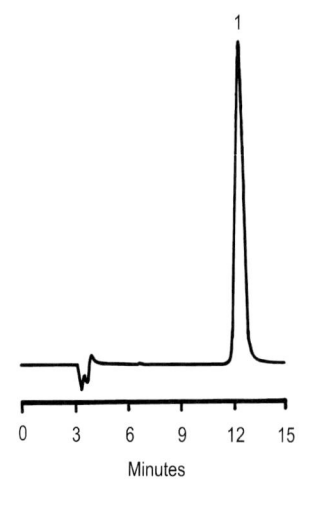

Fig. 9-164. Separation of diphosphate in a reddening agent. — Separator column: IonPac AS7; eluant: 30 mmol/L HNO_3; flow rate: 0.5 mL/min; detection: photometry at 330 nm after reaction with iron(III) nitrate; injection: 50 μL of a 1:5 diluted 0.1% solution; peak: diphosphate (1).

Sample preparation is different for investigating meat seasonings. Because of the high content of organic compounds in these seasonings, the aqueous solutions cannot be injected directly but must be treated with reversed-phase extraction cartridges. These cartridges contain either a non-polar polymer or a chemically bonded silica at which matrix components are retained. Ionic species are not affected by this sample preparation. The problem-free applicability of ion chromatography to such samples is illustrated in Fig. 9-165 with the alkaline-earth metal analysis in a meat seasoning.

9.6.4
Baby Food

The determination of choline, a biogenic amine in soy bean products which is added to baby food, is an important application. Usually, choline is separated by ion-pair chromatography on a chemically bonded reversed phase utilizing UV detection at 190 nm. This measuring wavelength is necessary as choline and its derivatives do not contain any chromophores. Because baby food does contain UV-absorbing matrix components, peak overlapping is more than likely. The separation of choline and acetylcholine on a cation exchanger utilizing suppressed conductivity as a detection method is much easier. For comparative purposes, Fig. 9-166 shows the separation of choline and acetylcholine with both

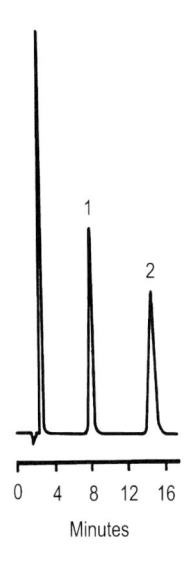

Fig. 9-165. Separation of alkaline-earth metals in a meat seasoning. — Separator column: IonPac CS3; eluant: 30 mmol/L HCl + 5 mmol/L 2,3-diaminopropionic acid; flow rate: 1 mL/min; detection: suppressed conductivity; injection: 50 µL of a 1:25 diluted 1% solution; solute concentrations: 134 mg/L magnesium (1) and 258 mg/L calcium (2).

chromatographic techniques. The different selectivity of the cation exchanger as compared with a reversed phase is also responsible for the different elution order of both compounds. When using conductivity detection, alkali metals such as sodium and potassium can be analyzed in the same run.

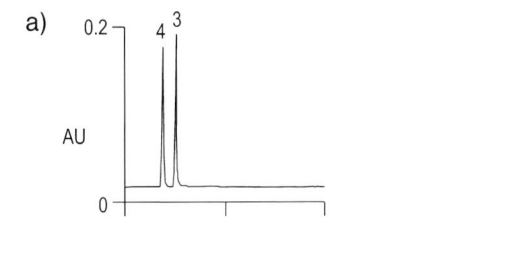

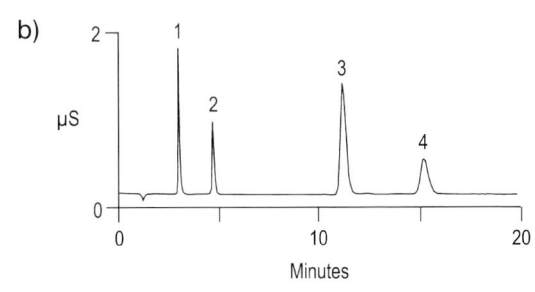

Fig. 9-166. Separation of choline and acetylcholine with different chromatographic techniques. — (a) reversed phase. Separator column: ODS; eluant: 5 mmol/L heptanesulfonic acid — acetonitrile (99:1 v/v), pH 4; flow rate: 1 mL/min; detection: UV (190 nm); solute concentrations: 10 µg each of acetylcholine (1) and choline (2), (b) cation exchange. Separator column: OmniPac PCX-100; eluant: 75 mmol/L HCl — methanol (99:1 v/v); flow rate: 1 mL/min; detection: suppressed conductivity; solute concentrations: 10 ng each of sodium (1) and potassium (2) as well as 100 ng each of choline (3) and acetylcholine (4).

Recently, the determination of phytic acid (*myo*-inositol hexaphosphate), which is believed to have an effect on the bioavailability of mineral substances, has become important. While phytic acid is not a food constituent in the United States, in other countries it is frequently added to foods as an antioxidant [250,

251]. To elucidate the nutritive scientific relevance of phytic acid, as early as 1983 Lee and Abendroth [252] developed an ion-pair chromatographic method for analyzing this compound. However, to be analyzed with ion exchangers, it required a pre-purification of the sample. In a modified form, ion-pair chromatography was also applied by Matthäus [253] for determining higher inositol phosphates in rape (the pomace of expressed grapes). Based on the post-column derivatization technique with iron(III) that was introduced by Fitchett and Woodruff [254] for determining polyphosphates in detergents, Philippy and Johnston [255] developed an analytical method for phytic acid that allows direct injection of food extracts. They employed an IonPac AS3 latexed anion exchanger with a dilute nitric acid eluant. The suitable measuring wavelength of 290 nm for the iron(III)-phytic acid complex is slightly lower than that of 330 nm used by Fitchett et al.

Figure 9-167 illustrates the application of this technique to the analysis of baby food. The baby food was extracted with 1.2% hydrochloric acid for 30 minutes; the extract is then centrifuged, diluted with de-ionized water to the suitable working range, and membrane filtered (0.45 µm) prior to injection. As can be seen in the chromatogram in Fig. 9-167, phytic acid can be separated from the matrix constituents without any problems. Phytic acid hydrolysis products such as inositol pentaphosphate and inositol tetraphosphate do not interfere with the analysis because they precipitate with iron(III) [256]. Meanwhile, the separator column used by Philippi is technologically outdated. Today, a modern anion exchanger with a comparable ion-exchange capacity and a much higher chromatographic efficiency would be used.

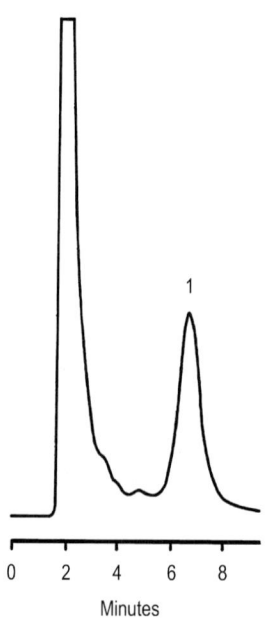

Fig. 9-167. Determination of phytic acid in a baby food sample. — Separator column: IonPac AS3; eluant: 0.11 mol/L HNO_3; flow rate: 1 mL/min; detection: photometry at 290 nm after reaction with iron(III) nitrate; injection: 100 µL of a diluted 1.2% HCl extract; peak: phytic acid (1).

Recently, Talamond et al. [257] published a procedure for determining phytic acid in foods that involves anion exchange chromatography and suppressed conductivity detection and, thus, avoids post-column derivatization. Separation is carried out with an OmniPac PAX-100 using a NaOH gradient in the presence of 1% (v/v) 2-propanol. The ASRS-1 suppressor is chemically regenerated with 25 mmol/L sulfuric acid.

In the area of inorganic constituents, the determination of nitrite/nitrate and iodide is very important. The sample preparation for instant milk is the same as that described above for milk products. As an example, Fig. 9-168 shows the separation of nitrite and nitrate in an instant baby food sample. As can be seen from this chromatogram, again both analytes are completely separated from matrix components. Precision, recovery rates, and detection limits are at a similar level as discussed for milk products.

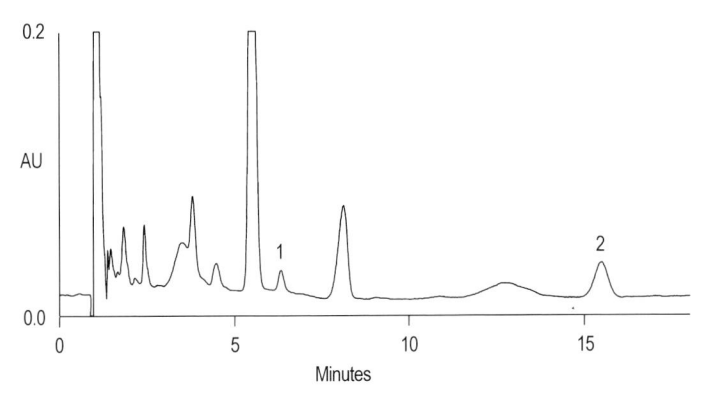

Fig. 9-168. Determination of nitrite and nitrate in an instant baby food sample. — Chromatographic conditions: see Fig. 9-155; injection volume: 25 µL; sample: instant milk; solute concentrations: 40 µg/L nitrite (1) and 0.6 mg/L nitrate (2), corresponding to 0.8 µg/g nitrite and 12 µg/g nitrate of the solid compound.

The chromatograms of iodide determination in baby food samples do not look much different from those of low-fat milk (see Fig. 9-156). However, iodide concentrations in baby food based on milk or soy are significantly lower than in milk. Taking cation analysis as an example, Fig. 9-169 illustrates the suitability of dialysis as an alternative sample preparation technique for such matrices [258].

Kaine et al. [260] successfully applied the method of HPAEC-PAD, which has already been cited several times in this book, to characterize various milk- and soy-based baby foods by determining the carbohydrate profiles. Although the carbohydrate portion itself does not vary much, the type of carbohydrates does vary depending on the product. Among the most important constituents are lactose, sucrose, maltodextrins, and starch, which are worked into the formulations in varying proportions. Because all constituents have to be declared, these

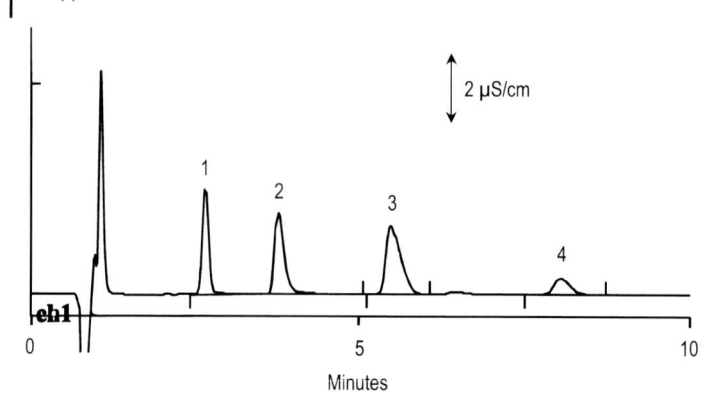

Fig. 9-169. Separation of alkali- and alkaline-earth metals in an instant baby food sample. – Separator column: Metrosep Cation 1-2; eluant: 4 mmol/L tartaric acid + 1 mmol/L pyridine-2,6-dicarboxylic acid; flow rate: 1 mL/min; detection: non-suppressed conductivity; injection volume: 20 µL after dialysis; sample: 0.5 g instant milk dissolved in 250 mL 2.5 mmol/L HNO_3; solute concentrations: 2.46 mg/g sodium (1), 6.18 mg/g potassium (2), 4.89 mg/g calcium (3), and 0.63 mg/g magnesium (4); (taken from [259]).

products can be monitored by ion chromatography in a fast and efficient way. For this purpose, 0.2 g powder is dissolved in de-ionized water by shaking it for 2 minutes. After filtration (0.45 µm) and treatment with a reversed-phase extraction cartridge, the samples can be injected directly. Using the gradient technique developed by Kaine et al. all relevant carbohydrates can be analyzed in the same run. The key parameter is clearly lactose, which is not contained in soy formulations. Thus, adulterations and mistakes can be identified, for example, by the presence of lactose. This is extremely important, because a mix up of milk- and soy products can have severe consequences for health.

9.6.5
Groceries and Luxuries

An analytical problem of general interest for the food industry is the detection of preservatives. The necessity of analytical control results from the fact that the preservatives allowed in the European Union are physiologically safe, but can significantly affect the taste of foods. Moreover, allergic reactions towards benzoic acid or sorbic acid with a constantly growing number of people being sensitive to it cannot be excluded. Detection and quantification of the preservatives benzoic acid and sorbic acid in foods can be performed via liquid chromatography within a short time. In most cases, separation is carried out on chemically modified silica utilizing an ion-suppression technique, i.e., the dissociation of both compounds is suppressed by choosing the respective pH value in the mobile phase. Because both preservatives contain chromophores, detection is carried out via measuring the light absorption. However, the absorption maxima of benzoic acid and sorbic acid differ significantly, so that a multi-wavelength UV detector or a photodiode array detector should be used. Figure 9-170 exemplifies this with the analysis of a potato salad sample, which has been extracted with hexane to remove the lipids [201]. To achieve maximal sensitivities for benzoic acid and sorbic acid, the sample was measured at two different wavelengths (230 nm and 254 nm). Information on the peak purity is obtained by evaluation of the signals via external calibration and standard addition at both wavelengths which, in the present case, led to the same result.

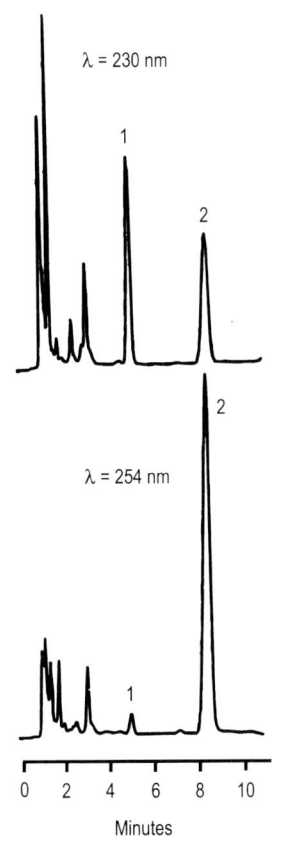

Fig. 9-170. Analysis of benzoic acid and sorbic acid in a potato salad sample. — Separator column: 124 mm × 4 mm i.d. RP18 (Macherey & Nagel); column temperature: 30°C; eluant: 50 mmol/L phosphate buffer — methanol; detection: UV (230 nm and 254 nm); solute concentrations: 0.11 g/kg benzoic acid (1) and 0.09 g/kg sorbic acid (2); (taken from [201]).

A typical example of the high performance of modern ion chromatography is the chromatogram of biogenic amines in canned herring shown in Fig. 9-171. Biogenic amines are regarded as an indicator for the quality and freshness of fish. A number of HPLC procedures were developed for the determination of biogenic amines. However, all of them are based on derivatization techniques, because compounds such as putrescine, cadaverine, spermidine and others do not contain any chromophores. Draisci et al. [261] succeeded to determine biogenic amines directly using integrated amperometry. For separating these compounds, they used a strong acid latexed cation exchanger, eluting the biogenic amines with a perchlorate gradient under acidic conditions. The top chromatogram in Fig. 9-171 represents a rotten canned herring sample; the bottom chromatogram is from the same sample spiked with 300 µg/g each of putrescine, histidine, cadaverine, histamine, and spermidine.

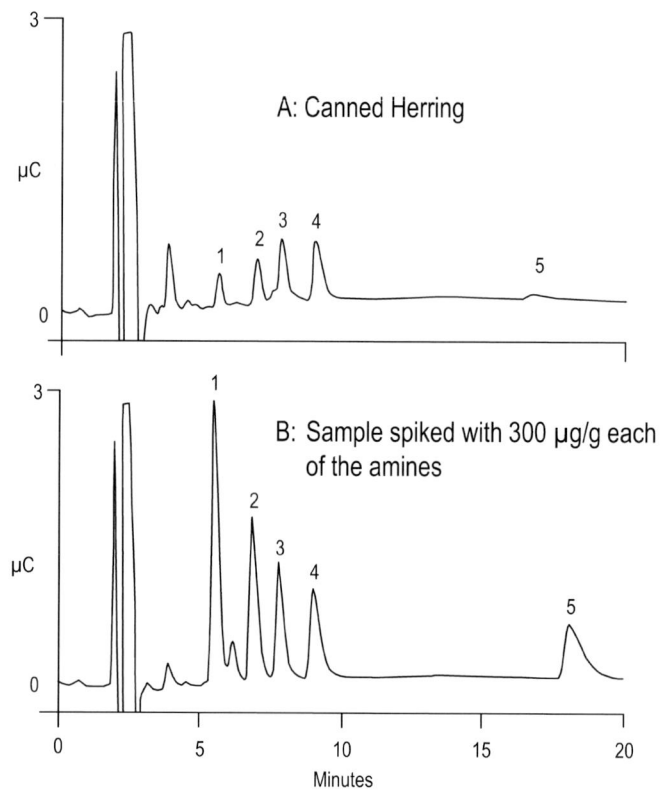

Fig. 9-171. Analysis of biogenic amines in a canned herring sample with direct detection via integrated amperometry. — Separator column: IonPac CS10 with guard; eluant: (A) 50 mmol/L $HClO_4$ + 0.6 mol/L $NaClO_4$ — acetonitrile (95:5 v/v), (B) 50 mmol/L $HClO_4$ + 0.85 mol/L $NaClO_4$ — acetonitrile (95:5 v/v); gradient: 100% A isocratic for 1 min, then linearly to 100% B in 9 min; detection: integrated amperometry on a gold working electrode; solute concentrations: (sample) 16 µg/g putrescine (1), 103 mg/g histidine (2), 187 µg/g cadaverine (3), 172 µg/g histamine (4), and 294 µg/g spermidine (5), (spiked sample) 300 µg/g each of the five amines; (taken from [261]).

Food samples usually contain a number of organic acids that elute within or close to the system void of an anion exchanger under isocratic conditions. If these components are to be separated and analyzed together with inorganic anions in the same run, the application of a gradient technique is inevitable. This is demonstrated here with the analysis of an aqueous tobacco extract. When analyzing this extract under isocratic conditions with a carbonate/bicarbonate eluant mixture, the chromatogram illustrated in Fig. 9-172 is obtained. The group of peaks close to the system void points to the presence of short-chain organic acids. In addition to chloride, nitrate and orthophosphate can be identified. The peak appearing at t_{ms} = 5.4 min, which normally should be assigned to sulfate, is unusually broad. This implies that at least one more component (possibly a short-chain dicarboxylic acid) elutes at that time.

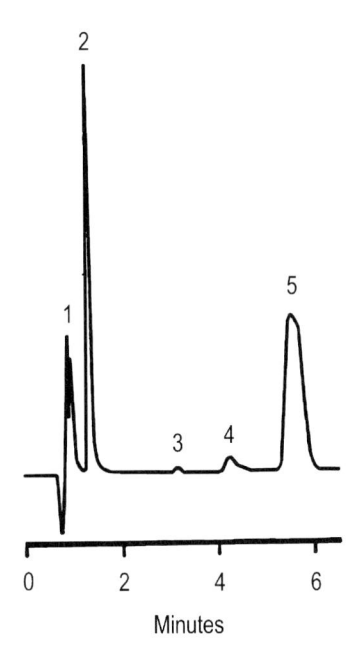

Fig. 9-172. Analysis of inorganic and organic anions in a tobacco extract. — Separator column: IonPac AS4A; eluant: 1.7 mmol/L NaHCO$_3$ + 1.8 mmol/L Na$_2$CO$_3$; flow rate: 2 mL/min; detection: suppressed conductivity; injection: 50 µL of a 1:25 diluted 1% extract; sample constituents: organic acids (1), chloride (2), nitrate (3), orthophosphate (4), and sulfate (5).

To avoid co-elutions of early eluting species and those in the retention range of sulfate, a gradient technique has to be applied. As can be seen from the chromatogram in Fig. 9-173, malic acid elutes just ahead of sulfate, thus interfering with the sulfate signal under isocratic conditions. Another advantage of applying a gradient elution technique is the fact that early eluting monocarboxylic acids are almost completely resolved, and more strongly retained anions such as citric acid can be analyzed in the same run. However, such high chromatographic resolution is only obtained when two separator columns are used in series. Moreover, it is important that the sample is injected into the ion chromatograph as dilute as possible to avoid column overloading. In the present case, an Eluant Generator™ was used for preparing hydroxide eluant, without which the required reproducibility of the retention times, especially in the front part of the chromatogram, cannot be achieved.

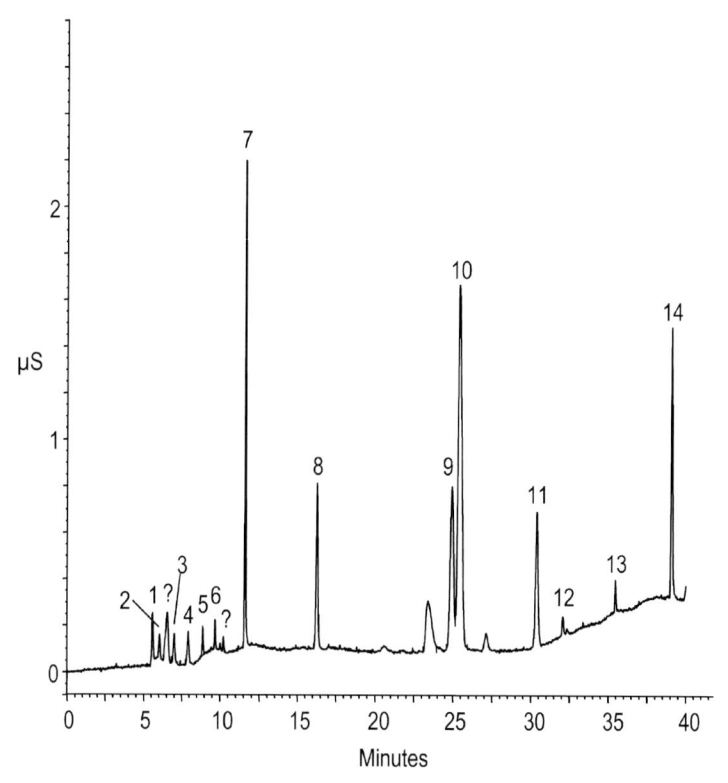

Fig. 9-173. Gradient elution of inorganic and organic anions in a tobacco extract. – Separator columns: 2 IonPac AS17 (2-mm); eluant: NaOH gradient (EG40); flow rate: 0.35 mL/min; detection: suppressed conductivity; injection volume: 5 µL; peaks: quinate (1), fluoride (2), acetate (3), propionate (4), formate (5), pyruvate (6), chloride (7), nitrate (8), succinate (int. standard) (9), malate (10), sulfate (11), oxalate (12), orthophosphate (13), and citrate (14).

The use of an Eluant Generator™ is also essential for analyzing carbohydrates in instant coffee (ISO 11292, AOAC 995.13) [262-264] by means of the HPAEC-PAD method. Sugars relevant for coffee include mannitol, fucose, arabinose, rhamnose, galactose, glucose, sucrose, xylose, mannose, fructose, and ribose, which in the past could only be separated on CarboPac PA1 by using pure deionized water as an eluant. Because the optimal eluant pH for pulsed amperometry is pH 13, a concentrated NaOH solution had to be added to the column effluent prior to entering the detector cell. Under these eluant conditions, carbonate is rapidly accumulated at the stationary phase, leading to a steady decrease of retention times. Thus, stable retention times are only obtained if the separator column is rinsed with 0.2 mol/L NaOH after every injection, which means a total analysis time of approximately 80 minutes. With the introduction of CarboPac PA10, which exhibits a much higher selectivity for monosaccharides, a significant improvement of this AOAC method was achieved regarding separation. The problem of continuously decreasing sensitivity was solved by applying a new pulse sequence with four different potentials and by preparing a very dilute carbonate-free hydroxide eluant with an Eluant Generator™, so that total analysis time is half of what it used to be. The chromatogram of a corresponding standard with eleven sugars relevant for coffee obtained with an eluant of 2.3 mmol/L KOH is illustrated in Fig. 9-174. After the elution of ribose, eluant concentration is raised to 100 mmol/L for five minutes as a rinsing step before the column is equilibrated again with the initial KOH concentration. The robustness of this modified AOAC method is remarkable; the relative standard deviations for the retention times of the individual sugars are between 0.5% and 0.8% based on non-stop operation over more than 100 hours. Figure 9-175 shows the chromatograms of a coffee sample and a corresponding standard. Sample preparation for measuring soluble carbohydrates only requires treatment with a RP extraction cartridge and subsequent membrane filtration (0.2 µm). For determining the total amount of carbohydrates, the sample has to be hydrolyzed first. For this purpose, about 300 mg of sample are dissolved in 1 mol/L HCl and heated to 100 °C for four hours. After cooling down, the sample is passed through a fluted filter, neutralized by treating it with a cation exchanger in the silver form, and finally passed through a membrane filter (0.2 µm). The sample can then be injected into the chromatograph.

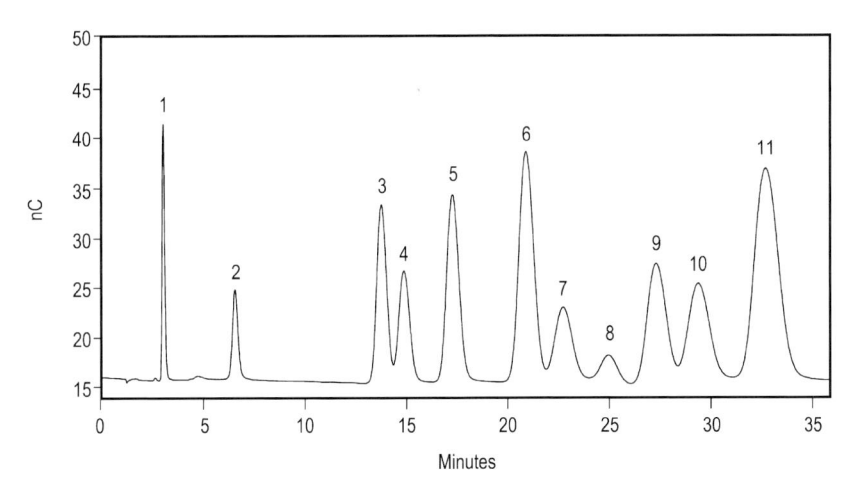

Fig. 9-174. Separation of carbohydrates relevant for coffee. – Separator column: CarboPac PA10; eluant: 2.3 mmol/L KOH (EG40), rinsing step with 0.1 mol/L for 5 min; flow rate: 1 mL/min; detection: integrated amperometry on a gold working electrode; peaks: (1) mannitol, (2) fucose, (3) arabinose, (4) rhamnose, (5) galactose, (6) glucose, (7) sucrose, (8) xylose, (9) mannose, (10) fructose, and (11) ribose.

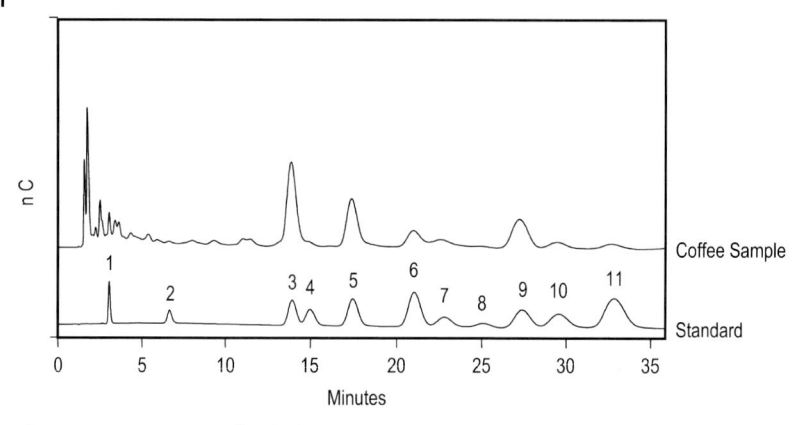

Fig. 9-175. Separation of carbohydrates in an instant coffee sample. – Chromatographic conditions and peak identification: see Fig. 9-174.

Moreover, green coffee beans contain a number of chlorogenic acids such as derivatives of caffeoylquinic acid, dicaffeoylquinic acid, and feruloylquinic acid [265], whose amount and distribution in the coffee beans depend on the bean type and degree of ripeness. During the roasting process, a significant portion of the chlorogenic acids decomposes to form volatile and non-volatile compounds [266]. The remaining portion of the chlorogenic acids and their degradation products strongly affect the sensoric properties of coffee due to their solubility in water. For this reason, the analysis of chlorogenic acids in various coffee products is of interest. While the total content of chlorogenic acids can be determined by derivative spectrophotometry [267], whose results are in good agreement with traditional UV absorption, liquid chromatography allows the quantification of individual isomers [268].

9.6.6
Sweeteners

More and more frequently, food is sweetened with sugar substitutes instead of sugar or glucose syrup. Because of the diversity of products containing artificial sweeteners, the matrices that need to be investigated are very complex. In addition, sweeteners may be present in these products individually or in combination.

The best-known sugar substitutes include saccharin, sodium cyclamate, and acesulfam-K.

Sodium cyclamate

Acesulfam-K

While RPLC and RPIPC techniques that employ UV detection have already been elaborated for analyzing saccharin and acesulfam-K [269, 270], the liquid chromatographic analysis of sodium cyclamate is barely substantiated in the literature because of the non-chromophoric structure of this compound. The only exception is the HPLC method introduced by Hermann et al. with indirect photometric detection [271].

As the above-mentioned compounds exist as anions in an alkaline medium, anion exchange chromatography with subsequent conductivity detection provides a welcome alternative to RPLC with UV detection. This is especially true because the flavors and dyes contained in food do not represent any interference upon application of conductivity detection. Based on these discoveries, Biemer [272] developed an ion chromatographic technique for determining these three sugar substitutes and applied it to the analysis of chewing gum. The analyte samples were extracted with acetic acid/chloroform and the aqueous phase was injected directly after appropriate dilution. Figure 9-176 shows the chromatogram of a chewing gum sample spiked with cyclamate. A pure sodium bicarbonate

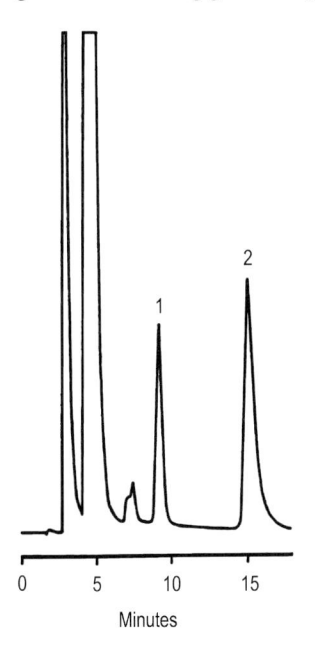

Fig. 9-176. Analysis of cyclamate in a chewing gum sample. — Separator column: IonPac AS4A; eluant: 1.7 mmol/L $NaHCO_3$; flow rate: 2 mL/min; detection: suppressed conductivity; injection: 50 µL of a spiked chewing gum sample; peaks: (1) cyclamate and (2) bromide; (taken from [272]).

solution served as an eluant. Bromide, eluting after cyclamate, was added as an internal standard. Acesulfam-K and saccharin are much more strongly retained and, thus, are eluted with pure sodium carbonate solution. The corresponding chromatogram is displayed in Fig. 9-177. In this case, fumaric acid served as an internal standard. Inorganic anions such as chloride, nitrate, orthophosphate, and sulfate do not interfere with the analysis. When applying a gradient elution technique, cyclamate and acesulfam-K can be analyzed in the same run.

Another sugar substitute is palatinitol. The preparation of palatinitol uses sucrose as a starting material. After enzymatic rearrangement into palatinose (isomaltulose) and hydration, palatinitol is formed, representing an equimolar mixture of the isomers (α-D-glucopyranosido-1,6-mannitol and (α-D-glucopyranosido-1,6-sorbitol:

Sucrose   Isomaltulose

Isomaltulose   α-D-Glucopyranosido-1,6-mannitol   α-D-Glucopyranosido-1,6-sorbitol

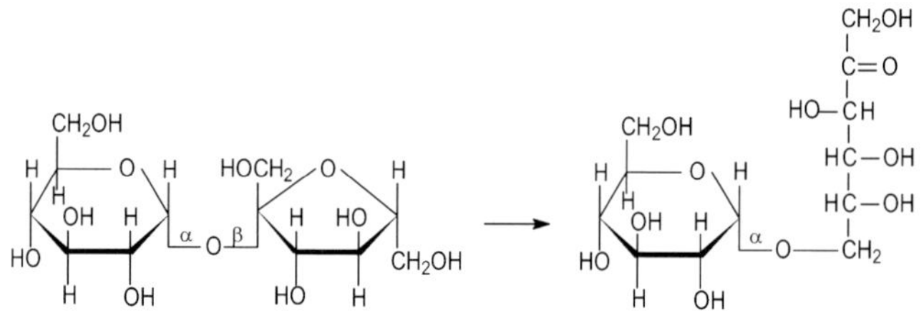

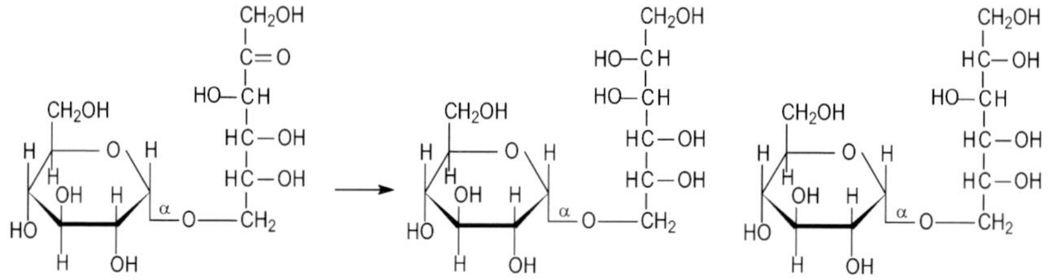

Like all carbohydrates, palatinitol may be separated in an alkaline medium on a strong basic anion exchanger in the hydroxide form and may be detected via pulsed amperometry. Figure 9-178 shows a chromatogram with the separation of both isomers. Sorbitol, mannitol, and isomaltose can be detected as impurities in the same run.

In principle, honey can also be used to sweeten food and beverages; it represents an exceptionally complex product. Nectar collected by bees contains, in addition to sucrose, varying amounts of invert sugar. In the digestive tract it is hydrolyzed to cane sugar, so that the sugar delivered by the bees contains only about 10% unchanged sucrose. In addition to these main components, various di- and tri-saccharides are found in honey in different concentrations [273]. In

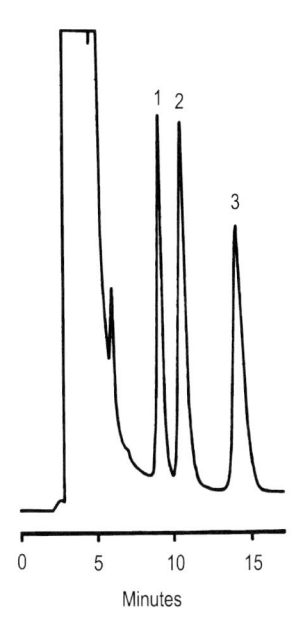

Fig. 9-177. Analysis of acesulfam-K and saccharin in a chewing gum sample. – Separator column: IonPac AS4A; eluant: 2.8 mmol/L Na$_2$CO$_3$; flow rate: 2 mL/min; detection and injection: see Fig. 9-176; peaks: (1) fumarate, (2) acesulfam-K, and (3) saccharin; (taken from [272]).

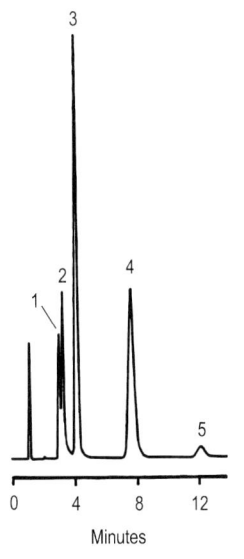

Fig. 9-178. Separation of the palatinitol isomers. – Separator column: CarboPac PA1; eluant: 0.1 mol/L NaOH; flow rate: 1 mL/min; detection: pulsed amperometry on a gold working electrode; injection volume: 50 µL; peaks: (1) sorbitol, (2) mannitol, (3) 10 mg/L α-D-glucopyranosido-1,6-sorbitol (GPS), (4) 10 mg/L α-D-glucopyranosido-1,6-mannitol (GPM), and (5) isomaltose.

1988, Swallow and Low [274] were the first to report on the analysis and quantification of carbohydrates in honey by HPAEC-PAD. At that time, it was already known that oligosaccharide profiles can be correlated to a certain extent to the origin of the honey. A special feature of forest honey and fir tree honey is the presence of melizitose, a trisaccharide that is contained in honey dew. These excretes found on conifers are picked up by the bees and, consequently, are only detectable in such types of honey. Thus, the detection of melizitose in fir tree honey represents a point of quality. As seen in the chromatogram of such a forest fir honey in Fig. 9-179, the product investigated by Baumgärtner [207] meets this quality requirement. In addition to the main components glucose and fructose, trehalose and maltose and small concentrations of maltose oligomers can also be detected. The structures of the other carbohydrates present in trace amounts could not be elucidated.

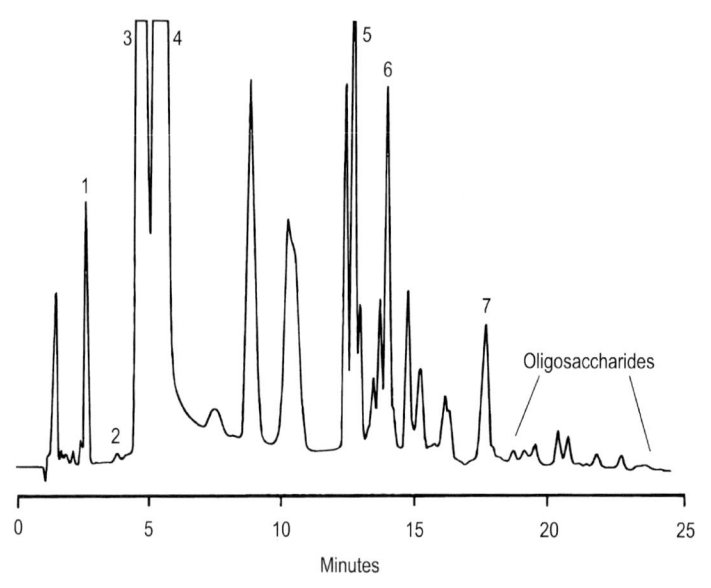

Fig. 9-179. Analysis of carbohydrates in a forest honey sample. – Separator column: CarboPac PA1; eluant: (A) 0.1 mol/L NaOH, (B) 0.1 mol/L NaOH + 0.5 mol/L NaOAc; gradient: 100 % A isocratic for 2 min, then linearly to 25 % B in 30 min; flow rate: 1 mL/min; detection: pulsed amperometry on a gold working electrode; injection: 50 µL of a 0.3 % solution; peaks: (1) trehalose, (2) arabinose, (3) glucose, (4) fructose, (5) melizitose, (6) maltose, and (7) maltotriose; (taken from [207]).

9.7
Ion Chromatography in the Pharmaceutical Industry

Ion chromatography also finds a large number of applications in the field of pharmaceutical analysis. Here, only selected examples will be explained.

Ion chromatography plays a prominent role in the characterization of pharmaceutically relevant compounds, predominantly in the early stages of research [275]. This includes the trace analysis of impurities and metabolites, the elemental analysis, and the structural elucidation of counter ions. The latter are often inorganic anions such as chloride and bromide or simple organic acids such as acetate. Somewhat rare counter ions include methyl sulfate, methanesulfonate, and trifluoroacetate.

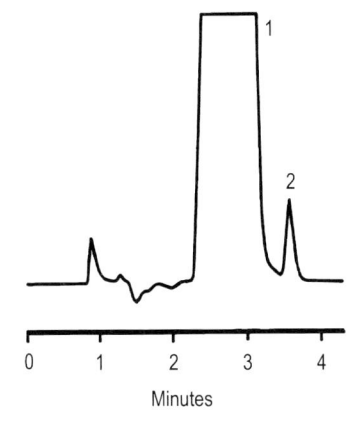

Fig. 9-180. Analysis of methyl sulfate in a pharmaceutical sample. – Separator column: IonPac AS4A(-SC); eluant: 1 mmol/L TBAOH; flow rate: 2 mL/min; detection: suppressed conductivity; injection: 50 µL sample; peaks: (1) chloride and (2) methyl sulfate; (taken from [276]).

Methyl sulfate is not only of analytical interest as a counter ion in pharmaceutically relevant compounds. Rychtman [276] describes the trace analysis of this compound in pharmaceutical products that have been produced via alkylation with dimethyl sulfate. Tetraalkylammonium compounds, for example, may be synthesized by quaternation of tertiary amines with dimethyl sulfate:

$$R_3N + (CH_3O)_2SO_2 \rightarrow R_3N^+CH_3 + CH_3OSO_3^- \tag{322}$$

Because methyl sulfate must not be present in pharmaceutical products, it is later replaced by chloride via anion exchange. The exchange efficiency can be monitored by determining the remaining methyl sulfate content. The analytical procedure developed by Rychtman uses a conventional latexed anion exchanger with tetrabutylammonium hydroxide as a mobile phase. Methyl sulfate detection is performed via suppressed conductivity utilizing a membrane-based suppressor system. Figure 9-180 displays the chromatogram of a research sample. Although chloride is present in 50-fold excess, the separation of methyl sulfate is baseline-resolved. The methyl sulfate peak corresponds to a sulfur content of 2.8 mg/g.

For the separation of trifluoroacetate, a modern grafted polymer such as IonPac AS14 is the most suitable stationary phase. Trifluoroacetate may be detected, in addition to chloride and sulfate, in very low concentrations in peptides and stems from mobile phases used for the HPLC separation of peptides. With a mixture of carbonate and bicarbonate as an eluant, all these anions can be determined in one run. The chromatogram shown in Fig. 9-181 was obtained by injecting 10 µL of a peptide sample (40 µg/mL) purified by gel permeation chromatography, in which 6.5 mg/L TFA could be detected.

A less common counter ion for the formation of a stable salt of S-adenosylmethionine is 1,4-butanedisulfonic acid. Because the pharmacokinetic properties of this compound in animals were not known, Moro et al. [277] developed an analytical method for this acid in physiological fluids. The authors used an IonPac

AS9-SC with acrylate-based latex particles as a stationary phase, from which 1,4-butanedisulfonic acid can be eluted together with an internal standard (1,2-ethanesulfonic acid) using a carbonate/bicarbonate eluant mixture. Detection was carried out via electrical conductivity. Because an extraction of this compound from biological matrices with organic solvents is not feasible due to its high polarity, Moro et al. deproteinated the samples by precipitating with perchlorate. The excess of perchlorate is then precipitated with potassium carbonate solution. However, 10% of the perchlorate remains in solution, which limits sensitivity because the sample has to be diluted a little bit to obtain complete separation between perchlorate and the internal standard under the given chromatographic conditions. Figure 9-182 shows the chromatogram of a plasma sample of a rat, to which the salt was orally administered with a dose of 225 mg/kg over a long period of time in the course of a toxicity study. Moro et al. determined the detection limit of this method for 1,4-butanedisulfonic acid to be 2.5 μg/mL in plasma samples.

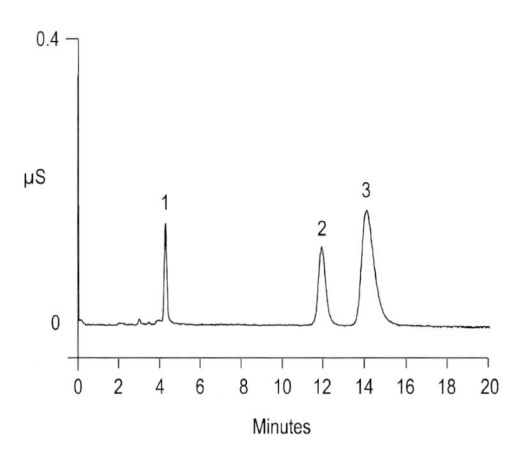

Fig. 9-181. Analysis of trifluoroacetate in a peptide sample. – Separator column: IonPac AS14; eluant: 0.8 mmol/L $NaHCO_3$ + 3.5 mmol/L Na_2CO_3; flow rate: 1.2 mL/min; detection: suppressed conductivity; injection: 10 μL of a peptide sample (40 μg/mL); solute concentrations: 0.3 mg/L chloride (1), 0.9 mg/L sulfate (2), and 6.5 mg/L trifluoroacetate (3).

In the field of cation analysis, ion chromatography has become a welcomed alternative to atomic absorption spectrometry. Waterworth et al. [278], for example, recently reported on the analysis of calcium in mupirocin, a broad-range antibiotic, with which predominantly bacterial skin infections such as furuncles are treated. While the purity of this compound is determined by reversed-phase chromatography via the analysis of the free acid, the calcium content has until now been determined by AAS using a nitrous oxide/acetylene flame. Waterworth et al. saw the advantage of ion chromatography in its ability to determine calcium together with other cations in the same run. For the separation of calcium they used an IonPac CS12 weak acid cation exchanger with a methanesulfonic acid eluant; detection was performed via suppressed conductivity. While the data obtained by ion chromatography were in good agreement with those obtained by ICP, the AAS values were 25% lower. The robustness of the ion chromatographic method allowed its validation without any difficulties.

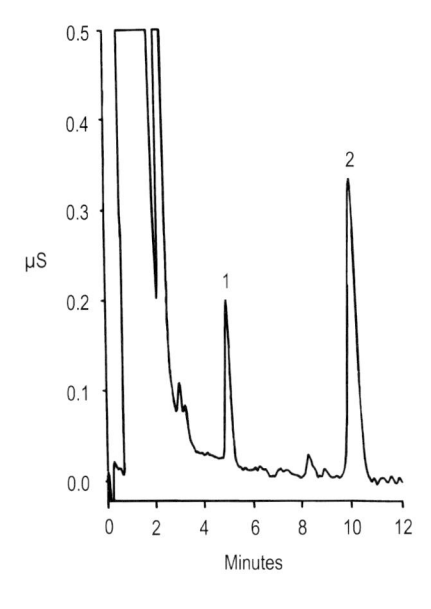

Fig. 9-182. Analysis of 1,4-butanedisulfonic acid in the plasma of a rat. — Separator column: IonPac AS9-SC; eluent: 1.8 mmol/L $NaHCO_3$ + 5 mmol/L Na_2CO_3; flow rate: 2 mL/min; detection: suppressed conductivity; injection: 100 µL of a plasma sample (40 µg/mL); peaks: (1) 1,4-butanedisulfonic acid and (2) 1,2-ethanedisulfonic acid (int. standard); (taken from [277]).

A major problem in pharmaceutical analysis is the analysis of aliphatic amines. Because they are non-chromophoric compounds, UV detection is only possible at very low wavelengths (<200 nm) which are prone to interferences by matrix and eluant components. Moreover, amines undergo interactions with free silanol groups if chemically bonded silica is used as a stationary phase. Thus, conventional HPLC methods are not very well suited for the analysis of amines. When applying gas chromatography for analyzing underivatized primary amines, the resulting peaks exhibit a strong tailing which can be attributed to adsorption phenomena. In both cases, the only way out is a derivatization. Apart from the time required, this is especially problematic for complex matrices, which pharmaceutical formulations sometimes are. In contrast, primary, secondary, and tertiary amines can easily be separated by cation chromatography on suitable stationary phases and detected via universal conductivity detection or integrated amperometry [279] on a gold working electrode (see Section 7.1.2.2). A typical example is the analysis of amylamine and *tert.*-butylamine on an IonPac CS14 cation exchanger developed by Jagota et al. [280]. While amylamine is used in the synthesis of certain drug components and, consequently, can appear as an impurity in the finished product, *tert.*-butylamine serves as a counter ion of a drug component. Both amines can be eluted with methanesulfonic acid, but different concentrations are used to account for the different affinities of the two amines towards the stationary phase. Moreover, amine retention can be controlled by a small amount of acetonitrile in the mobile phase. In both cases, it is fortunate that the respective drug components are soluble in water and, thus, can be injected directly into the ion chromatograph. Figure 9-183 shows a typical chromatogram of amylamine at the detection limit in the presence of the drug component. The detection limit was determined to be 0.02% (w/w). Interferences by other impurities resulting from the synthesis were not observed. Figure 9-184 shows the respective chromatogram of *tert.*-butylamine in the presence of the drug component.

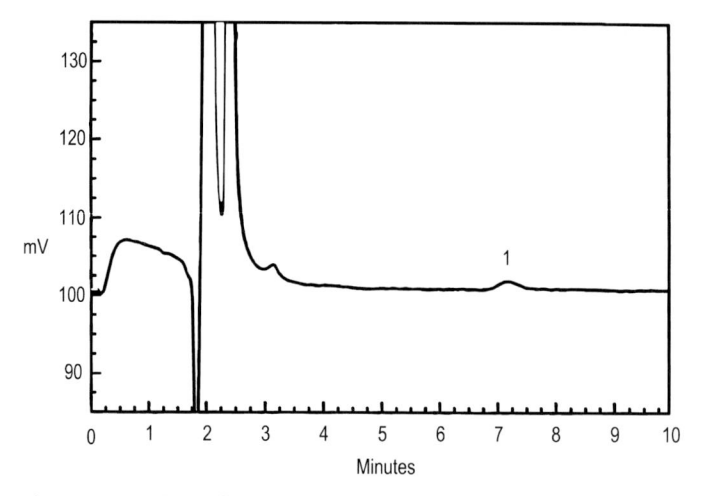

Fig. 9-183. Analysis of amylamine in a pharmaceutical drug. – Separator column: IonPac CS14; eluant: 50 mmol/L methanesulfonic acid – acetonitrile (95:5 v/v); flow rate: 1 mL/min; detection: suppressed conductivity; injection: 20 µL drug solution (5 mg/mL BMS-181 866-02); solute concentration: 0.5 µg/mL amylamine (1); (taken from [280]).

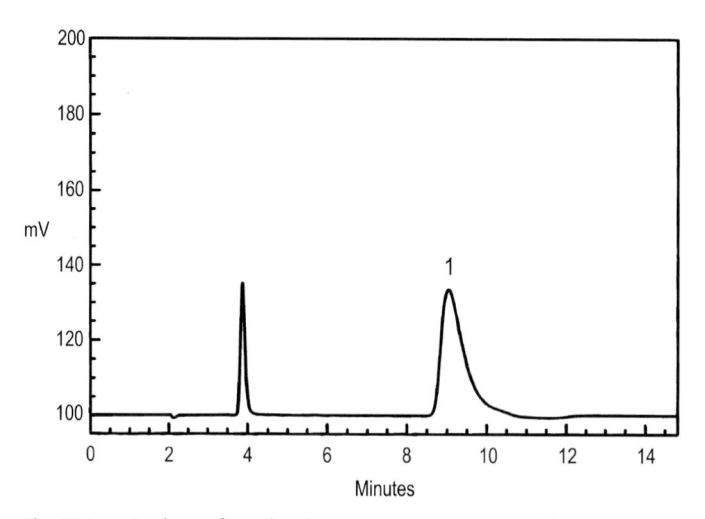

Fig. 9-184. Analysis of *tert.*-butylamine as a counter ion of a pharmaceutical drug. – Separator column: IonPac CS14; eluant: 10 mmol/L methanesulfonic acid – acetonitrile (99:1 v/v); flow rate: 1 mL/min; detection: suppressed conductivity; injection: 20 µL drug solution (1 mg/mL BMS-188 494-04); solute concentration: 10.6% (w/w) *tert.*-butylamine (1); (taken from [280]).

Sodium sulfite, which is added to some pharmaceuticals as an antioxidant, can be determined ion chromatographically with high precision [281, 282]. However, to avoid autoxidation of sulfite to sulfate, the samples must be stabilized with formaldehyde solution. For that, 0.5 mL of a 25% formaldehyde solution is added per liter sample. Alternatively, dilution is carried out with appropriately pre-treated water. The resulting formaldehyde-sulfite complex exhibits the same retention time as the sulfite ion itself. Figure 9-185 shows the chromatogram of a local anesthetic diluted 1:10. Its therapeutic agent is present as hydrochloride, thus explaining the high chloride peak at the beginning of the chromatogram.

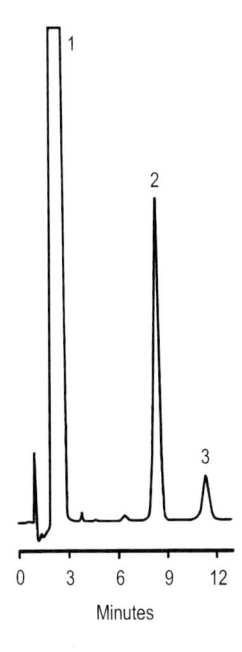

Fig. 9-185. Analysis of sulfite in a local anesthetic. − Separator column: IonPac AS9 (-SC); eluant: 0.75 mmol/L $NaHCO_3$ +2 mmol/L Na_2CO_3; flow rate: 2 mL/min; detection: suppressed conductivity; injection: 50 µL sample (diluted 1:10); peaks: (1) chloride, (2) sulfite, and (3) sulfate.

Another example of this kind is the chromatographic separation of benzalkonium chloride in nose drops, as illustrated in Fig. 9-186.

Benzalkonium chloride

$R = C_8$ to C_{18}

Benzalkonium chloride is a quaternary ammonium base that is added to some pharmaceuticals as a disinfectant. Usually, it is a mixture with alkyl chains R that vary in length and have even-numbered members between C_{12} and C_{16}. Separation is performed with ion-pair chromatography using hydrochloric acid as an ion-pair reagent. Because benzalkonium chloride is an aromatic compound, it can be detected very sensitively with UV detection at 215 nm. As can be seen in the chromatogram shown in Fig. 9-186, three signals are obtained for this compound, which may be attributed to the various chain lengths of the alkyl residue. The signal marked by an asterisk (*) was used for the calibration, as the distribution of the signals is identical both in the raw material and in the sample.

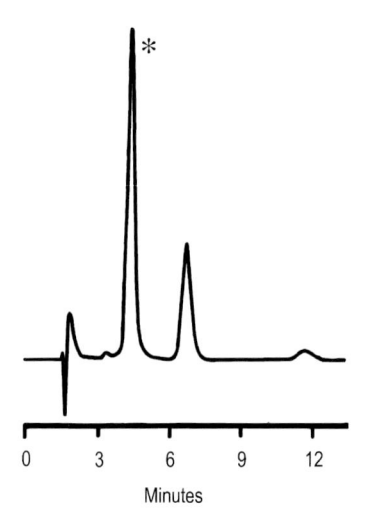

Fig. 9-186. Analysis of benzalkonium chloride in nose drops. — Separator column: IonPac NS1 (10-µm); eluant: 20 mmol/L HCl — acetonitrile (37:63 v/v); flow rate: 1 mL/min; detection: UV (215 nm); injection: 50 µL sample (diluted 1:10).

Often, pharmaceutical formulations contain anionic components, which can be determined much more easily by ion chromatography than by classical HPLC. Typical examples are cough suppressants containing the drug component dextromethorphan as a hydrobromide salt, and other anionic components such as citric acid, sodium benzoate, and saccharin. All these components can be analyzed in one run by using a hydroxide-selective IonPac AS11 anion exchanger and applying a gradient elution technique with subsequent suppressed conductivity detection. The resulting chromatogram is illustrated in Fig. 9-187 with an enlarged segment. It clearly shows that minor components such as chloride and sulfate as well as major components such as bromide, benzoate, citrate, and saccharin can be determined in the same run. Sample preparation involves only a dilution of the sample with de-ionized water and subsequent filtration.

The cough suppressant also contains the sugar alcohol sorbitol, which can be determined together with secondary and tertiary alcohols such as propylene glycol and glycerol via the HPAEC-PAD method. To achieve complete resolution between the alcohols the CarboPac MA1, which has the highest ion-exchange capacity of all CarboPac columns, is used as a stationary phase. The three components mentioned above can be separated under standard conditions without any difficulties. The high sensitivity of this method allows the detection of small amounts of mannitol and maltitol as well as a 100-fold dilution of the sample, so that interferences by matrix components are not observed. The respective chromatogram — again with an enlarged segment — is shown in Fig. 9-188.

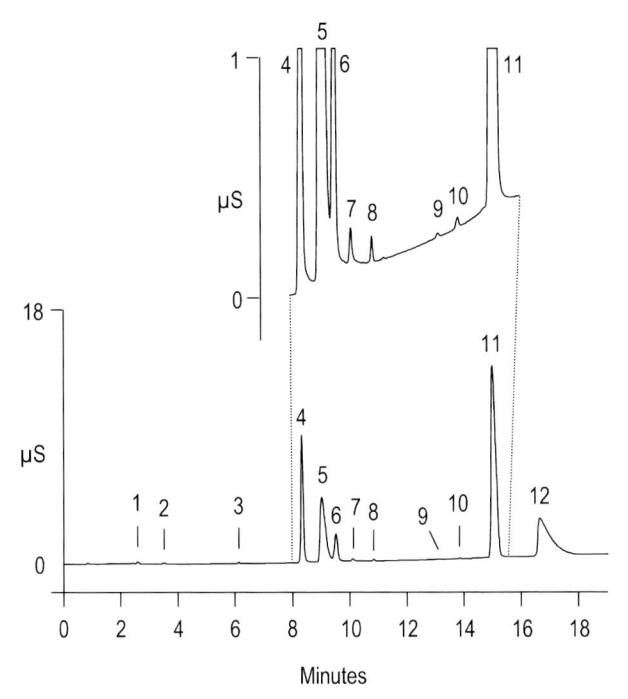

Fig. 9-187. Gradient elution of anionic components in a cough suppressant. — Separator column: IonPac AS11; eluant: NaOH; gradient: 0.5 mmol/L isocratic for 2.5 min, then linearly to 5 mmol/L in 3.5 min, then to 38 mmol/L in 12 min; flow rate: 2 mL/min; detection: suppressed conductivity; injection: 10 µL of a 1:10 (w/w) diluted sample; solute concentrations: (1) and (2) unknown, 0.08 mg/L chloride (3), 30 mg/L bromide (4), 80 mg/L benzoate (5), (6) and (7) unknown, 0.2 mg/L sulfate (8), 0.09 mg/L orthophosphate (9), (10) unknown, 130 mg/L citrate (11), and 120 mg/L saccharin (12).

Among the many pharmaceutical drugs there are some which can be analyzed directly by ion chromatography. A known and important example is clodronate, the disodium salt of dichloromethylenebisphosphonic acid, which is registered for use in the effective management of hypercalcaemia and bone pain associated with skeletal metastases in patients with multiple myeloma or carcinoma of the breast.

Clodronate

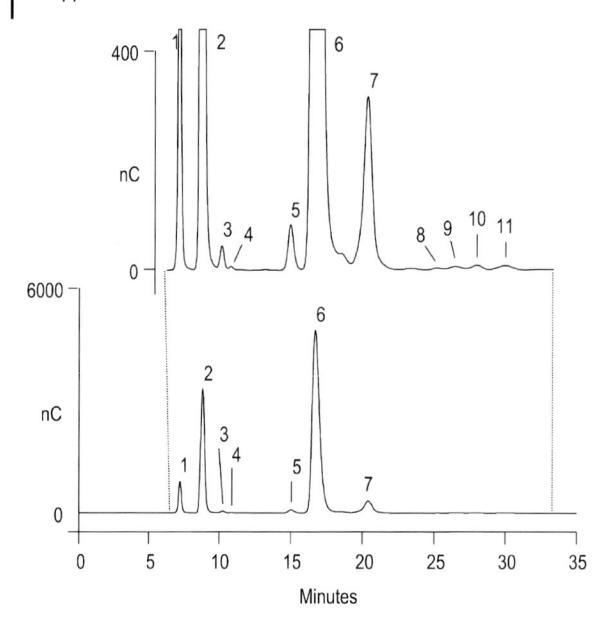

Fig. 9-188. Analysis of polyvalent alcohols and sugar alcohols in a cough suppressant. — Separator column: CarboPac MA1; eluant: 0.48 mol/L NaOH; flow rate: 0.4 mL/min; detection: pulsed amperometry on a gold working electrode; injection: 10 µL of a 1:100 (w/w) diluted sample; solute concentrations: 1.1 g/L propylene glycol (1), 0.6 g/L glycerol (2), (3) to (5) unknown, 1.5 g/L sorbitol (6), 0.1 g/L mannitol (7), (8) to (10) unknown, 5 mg/L maltitol (11).

It has been in clinical use for more than ten years [283]. Moreover, clodronate is an attractive candidate in the fight against osteoporosis [284]. Clodronate is synthesized in a two-step reaction from tetraisopropylmethylenebisphosphonic acid, which is first reacted with hypochlorite to the chloro-substituted ester followed by de-esterification with refluxing hydrochloric acid [285]. Possible impurities from this synthesis include, among others, chloride, chloro-substituted partial esters, methylenebisphosphonic acid, and monochloromethylenebisphosphonic acid. Under alkaline conditions orthophosphate and carbonylphosphonate can also occur as degradation products. With a few exceptions, all liquid- and gas chromatographic methods described in literature for analyzing clodronate are based on pre- [286] or post-column derivatizations [287], which require either special instrumentation or time-consuming sample preparation.

Because regulatory authorities specify the identification and quantification of impurities in drug components down to the level of 0.1% (w/w) before they are cleared, Taylor [288] developed an ion chromatographic procedure with which the drug component itself and possible impurities can be separated on an anion exchanger by applying a hydroxide gradient and suppressed conductivity detection. This procedure was based on the idea that the components to be analyzed are multivalent anions that elute under alkaline conditions in the order of increasing valency. While some phosphonic acids can also be eluted from an anion exchanger after partial protonation in an acidic environment (but detected only after derivatization), it is almost hopeless to separate clodronate from structurally very similar impurities under isocratic conditions at low pH. However, under alkaline conditions, all compounds are totally dissociated, so that interactions with the anion exchanger are maximal. Considering that the separation of bisphosphonates not only depends on their charge but also on their hydrophobic interactions with the stationary phase, Taylor tested a series of hydroxide-selective anion exchangers for their suitability for this separation problem; IonPac AS5 delivered the best results. As can be seen from the chromatogram of a model solution in Fig. 9-189, all components of interest can be separated to baseline using a purely aqueous hydroxide gradient. Elevation of column temperature to 45 °C is critical for the resolution between orthophosphate and the tribasic acid; it is not possible at ambient temperature because of the lower retention of the polyvalent anions. The model solution comprised the following components: 830 µg/mL monotributylammonium clodronate containing 2.7% monochloromethylenebisphosphonic acid, 6 µg/mL methylenebisphosphonic acid, 167 µg/mL monotributylammonium clodronate degraded under alkaline conditions, and 167 µg/mL of a mixture of partially esterified components. The concentrations of inorganic anions were 5 mg/L each.

The similar structured pamidronate (3-amino-1-hydroxypropylidene-1,1-bisphosphonate), which is also used for treating the above-mentioned diseases, can be analyzed via anion exchange chromatography.

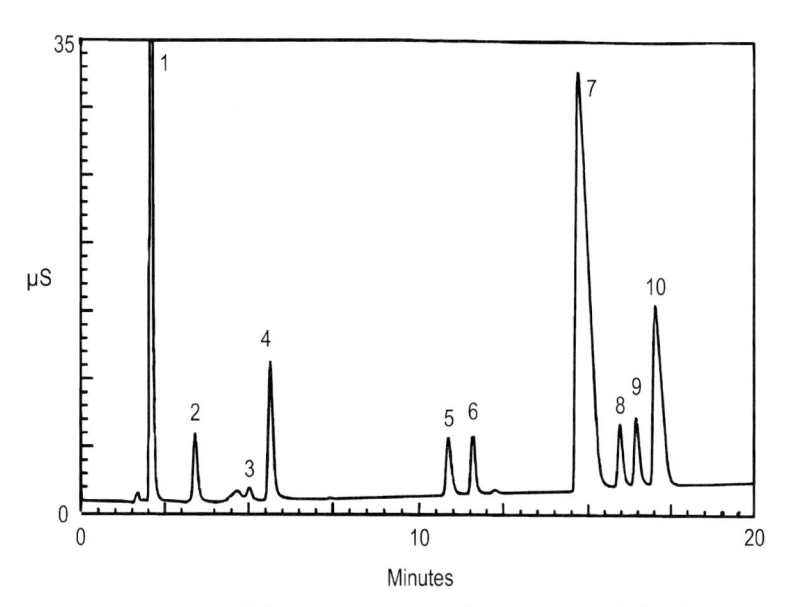

Fig. 9-189. Separation of clodronate and potential impurities on a hydroxide-selective anion exchanger. − Separator column: IonPac AS5; column temperature: 45 °C; eluant: NaOH; gradient: linear, 20 mmol/L to 100 mmol/L in 20 min; flow rate: 1 mL/min; detection: suppressed conductivity; solute concentrations: (1) chloride, (2) nitrate, (3) diester of dichloromethylenebisphosphonic acid, (4) sulfate, (5) orthophosphate, (6) monoester of dichloromethylenebisphosphonic acid, (7) clodronate, (8) monochloromethylenebisphosphonic acid, (9) methylenebisphosphonic acid, and (10) carbonylbisphosphonic acid; (taken from [288]).

$$\text{Pamidronate: } HO-\underset{\underset{-O}{|}}{\overset{\overset{O}{||}}{P}}-\underset{\underset{OH}{|}}{C}(NH_2)-\underset{\underset{O^-}{|}}{\overset{\overset{O}{||}}{P}}-OH \cdot 5H_2O$$

Pamidronate

Quitasol et al. [289] used a polyhydroxyethylmethacrylate (Universal Anion, Alltech, USA) as a stationary phase; 5 mmol/L potassium nitrate adjusted to pH 3.5 with nitric acid served as an eluant. Under these chromatographic conditions the organic phosphate can be separated to baseline from orthophosphate and orthophosphite within ten minutes. Because Quitasol et al. determined only the drug component in various pharmaceutical formulations, they used a refractive index detector for the sake of simplicity. This procedure can be validated in terms of precision, accuracy, and linearity without any difficulties.

A typical example in the field of aliphatic quaternary ammonium compounds is bethanechol chloride (2-[(aminocarbonyl)oxy]-N, N, N-trimethyl-1-propanaminium chloride).

$$H_2N-\overset{\overset{O}{||}}{C}-O-\underset{\underset{}{|}}{\overset{\overset{CH_3}{|}}{C}H}-CH_2-\underset{\underset{CH_3}{|}}{\overset{\overset{CH_3}{|}}{N^+}}-CH_3 \quad Cl^-$$

Bethanechol chloride

It is a cholinergic that is administered by either an injection or a tablet. It was recently proposed that the gravimetric assay for bethanechol chloride injection and tablet described in the *USP 24 NF 19* (page 230) will be replaced with a more specific ion chromatography assay. This assay separates bethanechol from other cations present in either preparation by using an IonPac CS14 weak acid cation exchanger and a methanesulfonic acid eluant. Bethanechol is detected directly by suppressed conductivity. Figure 9-190 shows the separation of bethanechol and its degradation product, 2-hydroxypropyltrimethyl ammonium (2-HPTA). The large sodium peak is from NaOH used to prepare 2-HPTA via hydrolysis of bethanechol. The minimum detection limits calculated from seven replicate injections of 50 µg/L 2-HPTA and 100 µg/L bethanechol were 30 µg/L and 36 µg/L, respectively.

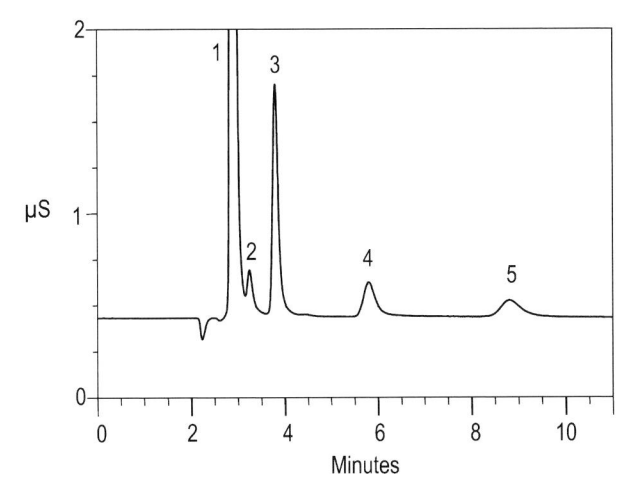

Fig. 9-190. Separation of bethanechol and 2-hydroxypropyltrimethyl ammonium. — Separator column: IonPac CS14; eluant: 20 mmol/L methanesulfonic acid (EG40); flow rate: 1 mL/min; detection: suppressed conductivity; injection volume: 25 µL; solute concentrations: (1) sodium, (2) unknown, (3) magnesium, (4) 2 mg/L 2-hydroxypropyltrimethyl ammonium, and (5) 2 mg/L bethanechol.

Succinylcholine is another example of a cationic drug component.

$$H_3C-\overset{CH_3}{\underset{CH_3}{\overset{|}{N^+}}}-CH_2-CH_2-O-\overset{O}{\overset{\|}{C}}-CH_2-CH_2-\overset{O}{\overset{\|}{C}}-O-CH_2-CH_2-\overset{CH_3}{\underset{CH_3}{\overset{|}{N^+}}}-CH_3$$

Succinylcholine

It is a neuromuscular blocking agent used as an adjunct to anesthesia to induce skeletal muscle relaxation. Succinylcholine slowly degrades in aqueous solution due to the hydrolysis of the ester bonds. Choline is the major degradation product. The determination of low concentration choline in succinylcholine chloride bulk drugs and formulations is important and challenging, because choline has no detectable chromophores. The detection method for analyzing choline and acetylcholine that was used until recently employed a post-chromatographic enzymatic reaction with choline oxidase and acetylcholinesterase to form hydrogen peroxide, which can be detected electrochemically [290, 291]. In order to simplify the instrumentation the enzyme was immobilized in an enzyme reactor. Chemiluminescence detection [292] and derivatization techniques [293] are also described, with which sensitivities down to 30 fmol can be achieved. Alternatively, Chen et al. [294] described a method using reversed-phase separation and suppressed conductivity detection for determining choline. Because choline is an ionic compound, an ion-pair reagent has to be added to the mobile phase. As outlined in Section 4.4, choline can also be separated on a cation exchanger, for example on OmniPac PCX-100. However, the eluant used in this case (0.1 mol/L HCl – 10% (v/v) acetonitrile) is not strong enough to elute succinylcholine. In contrast, by means of ion-pair chromatography, the drug component itself, its degradation products (succinylmonocholine, succinic acid), and other components of the formulation such as methyl paraben and propyl paraben can be eluted from the column by applying a solvent gradient. For the separation of choline in an aged formulation (Fig. 9-191), hexanesulfonic acid proved to be suitable as an ion-pair reagent. After the elution of choline the acetonitrile content in the mobile phase is increased step-wise to elute all the other components as one peak. Chen et al. determined the minimum detection limit for choline based on suppressed conductivity detection to be 10 pmol. With a drug component concentration of 2 mg/mL in the sample this corresponds to a choline content of 0.1%.

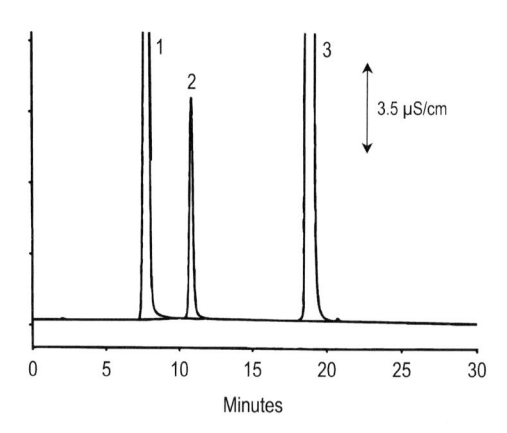

Fig. 9-191. Analysis of choline in an aged succinylcholine formulation. – Separator column: 250 mm × 4.6 mm i.d. Alltima C18 (5-μm); eluant: 5 mmol/L hexanesulfonic acid – acetonitrile (95:5 v/v), after 15 min switch to 5 mmol/L hexanesulfonic acid – acetonitrile (50:50 v/v); flow rate: 1 mL/min; detection: suppressed conductivity; injection: 50 μL drug solution with 2 mg/mL succinylcholine; peaks: (1) chloride, (2) choline, and (3) sum peak of the other components; (taken from [294]).

Apart from the necessary purity analysis of water [295] used for injectable pharmaceutical formulations, a classical application area for ion chromatography in the pharmaceutical industry is the micro-analytical determination of hetero atoms such as phosphorus, sulfur, and halides in organic compounds. Synthetic and clinical chemists use this application to check chemical structures. These methods are based on the Schöninger combustion, with which the elements are mineralized and transferred into a consistent form after chemical treatment. In the past, the subsequent quantification was carried out via titration [296] which is prone to chemical and physical interferences. It was recognized very early that sulfate and chloride can be simultaneously determined by ion chromatography free of interferences and with higher precision [297]. Due to the initial success with these two anions, the method was extended to the elements bromine, fluorine, and phosphorus. As illustrated in Fig. 9-192, the Schöninger combustion is carried out in a 1-L flask filled with oxygen and 20 mL of an absorber solution [298]. The weighed sample is wrapped in ash-free paper. After combustion, the flask is shaken for at least 20 minutes. The absorber solution can then be injected directly into the ion chromatograph. For the determination of sulfur, chlorine, and bromine, a 0.6% hydrogen peroxide solution is used for absorbing the combustion gases. All anion exchangers with sufficient resolution between fluoride and the system void can be used as stationary phases. Because samples combusted according to Schöninger contain nitrate, a baseline-resolved separation between bromide and nitrate is also required. Due to poor recovery rates the determination of fluorine has not been listed as a routine method, although the separation of fluoride no longer represents a problem. With phosphorus-containing compounds the Schöninger combustion leads to the formation of different phosphates. In addition to orthophosphate, significant amounts of pyrophosphate and tripolyphosphate are formed. Although the determination of higher phosphates is possible by ion chromatography, it is advantageous to convert all phosphates to orthophosphate, which can be eluted under isocratic conditions together with all other mineral acids. Using an enzymatic method to avoid another acid digest should be the basis for a future micro-analysis of phosphorus.

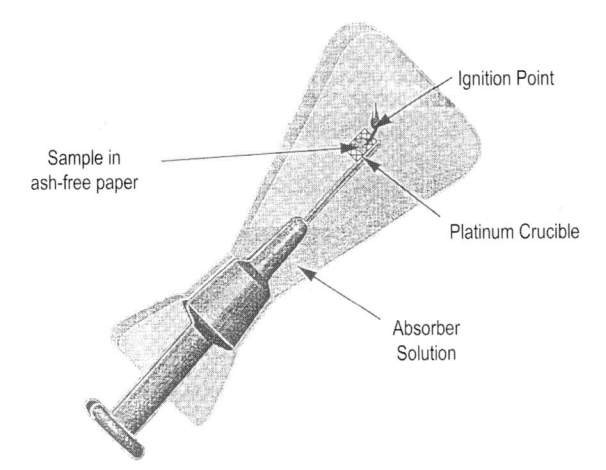

Fig. 9-192.
The Schöninger flask.

In the area of vitamins, Fig. 9-193 shows the gradient elution of water-soluble vitamins using Spherisorb ODS 2, at which the most important compounds of this kind can be analyzed in less than 15 minutes. Under these conditions, ascorbic acid elutes close to the system void, which hampers its determination in real-world samples. Among the numerous HPLC methods published in literature for determining ascorbic acid, those that operate under acidic conditions are preferable because ascorbic acid oxidizes under alkaline conditions. Such methods include ion-exclusion chromatography [299] and ion-suppression techniques [300]. An optimal separation of ascorbic acid and isoascorbic acid (which interferes with the determination of ascorbic acid in plasma) is obtained on macroporous polymeric stationary phases with a purely aqueous sodium dihydrogenphosphate eluant of high concentration [301]. Maximal selectivity between the two compounds is obtained at pH 2.14. To avoid ascorbic acid oxidation,

1 mmol/L DL-homocysteine is added to the standard, with which dehydroascorbic acid, the first intermediate product in the oxidative degradation of ascorbic acid, is reduced. Figure 9-194 shows a typical chromatogram of the separation of ascorbic acid and isoascorbic acid under these conditions.

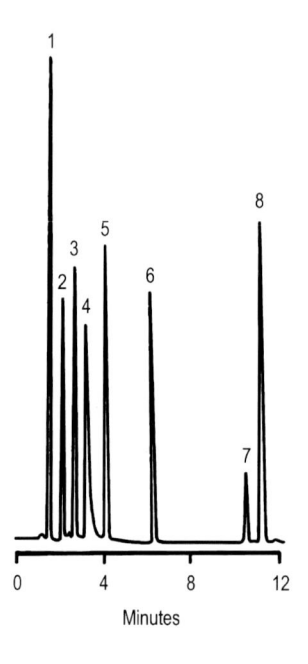

Fig. 9-193. Analysis of water-soluble vitamins. – Separator column: Spherisorb ODS 2 (5-µm); eluant: (A) 0.1 mol/L KOAc (pH 4.2 with HOAc), (B) water/methanol/acetonitrile (50:10:40 v/v/v); gradient: linear, 6% B in 30 min to 100% B; flow rate: 2 mL/min; detection: UV (254 nm); injection volume: 50 µL; solute concentrations: 5 nmol each of ascorbic acid (1), nicotinic acid (2), thiamine (3), pyridoxine (4), nicotinic acid amide (5), p-aminobenzoic acid (6), cyanocobalamine (7), and riboflavine (8).

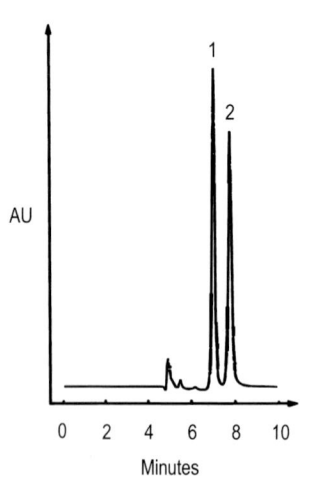

Fig. 9-194. Separation of ascorbic acid and isoascorbic acid. – Separator column: PLRP-S (5-µm), (Polymer Laboratories, Amherst, MA, USA); eluant: 0.2 mol/L NaH$_2$PO$_4$ (pH 2.14 with HCl); flow rate: 0.5 mL/min; detection: UV (220 nm); injection volume: 20 µL; peaks: (1) ascorbic acid and (2) isoascorbic acid; (taken from [301]).

One of the lesser known vitamins is lipoic acid, also called thioctic acid, which consists of eight carbon atoms and may exist in two forms: as lipoic acid – a cyclic disulfide – and in an open-chain form as dihydrolipoic acid, which carries a sulfhydryl group at the positions 6 and 8.

9.7 Ion Chromatography in the Pharmaceutical Industry

Lipoic acid

Dihydrolipoic acid

Lipoic acid acts as one of the coenzymes in the oxidative decarboxylation of pyruvate and other α-keto acids. It can be separated in an alkaline environment on a strong basic anion exchanger in the hydroxide form, and can be detected like carbohydrates via pulsed amperometry on a gold working electrode. The corresponding chromatogram of a lipoic acid standard is shown in Fig. 9-195. This method enables the accurate detection of 0.1 nmol lipoic acid.

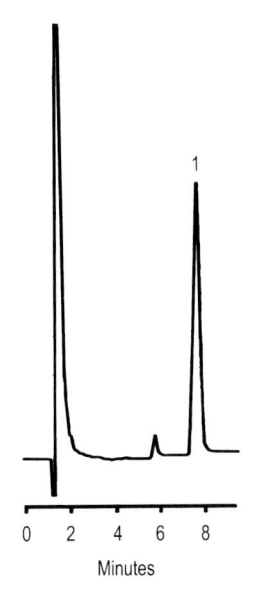

Fig. 9-195. Analysis of lipoic acid. — Separator column: CarboPac PA1; eluant: 0.1 mol/L NaOH + 0.5 mol/L NaOAc; flow rate: 1 mL/min; detection: pulsed amperometry on a gold working electrode; injection volume: 50 µL: solute concentration: 40 mg/L lipoic acid (1).

An impressive example of the high efficiency of anion exchange chromatography comes from the process of determining sulfate in penicillin V (α-phenoxyethyl penicillin). As revealed in the corresponding chromatogram in Fig. 9-196, in addition to the sulfate signal, three other peaks are obtained. The first two could be attributed to acetic acid and phenoxyacetic acid. Before the introduction of ion-exchange chromatography they were determined enzymatically in order to monitor the manufacturing process. When looking at the penicillin V structure, it is not surprising that this substance can also be eluted from an anion exchanger, because the carboxylate group located at the thiazolidine ring is fully dissociated under alkaline conditions.

The fact that it can be determined together with acetic acid, phenoxyacetic acid, and sulfate in the same run is unexpected. Thus, all three methods for characterizing penicillin V and its precursors may be replaced by one ion chromatographic procedure, thus drastically reducing the time expenditure.

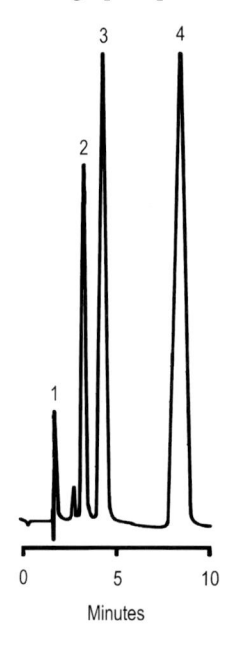

Fig. 9-196. Analysis of penicillin V. — Separator column: IonPac AS4A; eluant: 3 mmol/L NaHCO$_3$ + 2.4 mmol/L Na$_2$CO$_3$; flow rate: 1 mL/min; detection: suppressed conductivity; injection volume: 50 µL; solute concentrations: 20 mg/L acetate (1), 30 mg/L phenoxyacetate (2), 500 mg/L penicillin V (3), and 30 mg/L sulfate (4).

9.7.1
Fermentation

Biotechnological techniques in the production of food and feed stuff have been known for ages. Dairy products, sour vegetables, and rising agents such as yeast and sourdough for bakery products represent biotechnological products that are known to everybody. In addition to traditional fermentation techniques, gene-technological methods for producing a number of active agents and substances have become more prevalent. A well-known example is the microbial production of antibiotics such as penicillin. Other pharmaceuticals, enzymes, amino acids, and organic acids such as citric acid are also produced. To optimize the yield of such products, a number of investigations are necessary during the fermentation. These investigations concern the nutrient solution, the analysis of components that control or induce the metabolic activity, the reprocessing of products, and the analysis of minor products and precursors.

In many cases, process control comprises only the online monitoring of temperature and pH value and an accompanying enzymatic analysis of glucose and lactic acid. Because this represents an indirect process control, product yields may be only marginally improved as a result of these analytical results. In contrast, a number of instrumental methods are now available that allow the analysis of components essential to the fermentation process within a short time, so that a timely intervention in the process is possible and the product yield may be optimized.

Usually, glucose, lactose, sucrose, maltose, etc. are used as a carbon source for cell growth and product synthesis. Sugars and starch-containing raw materials are the most important fermentation substrates for microorganisms in the production of antibiotics, baking yeast, and enzymes. To optimize the fermentation process it is necessary to carry out time-dependent substrate analyses. The method of anion exchange chromatography with pulsed amperometric detection (HPAEC-PAD) is especially suitable; due to the high sensitivity, the matrix can largely be eliminated by diluting the samples. Another advantage of this method is the ability to analyze metabolic by-products such as alcohols (ethanol, methanol, and sugar alcohols) and glycols (glycerol) in the same run. The only separator column with which such a complex mixture can be separated is CarboPac MA1. This is a macroporous grafted polymer with an extremely high ion-exchange capacity of 1450 µequiv/column (see Section 3.10). As can be seen from Fig. 9-197, all alcohols, glycols, and saccharides relevant for fermentation monitoring can be separated at this stationary phase using a NaOH eluant. Despite the high NaOH concentration (0.48 mol/L) disaccharides elute very late. Shorter analysis times are obtained when using CarboPac PA1. However, this involves sacrificing the resolution between the individual alcohols, glycols, and sugar alcohols. Even though ethanol, methanol, and 2,3-butanediol are separated on CarboPac MA1, one has to consider that the response factor for ethanol is about 570 times smaller than that for glucose; with methanol the ratio is even less favourable. However, these ratios become an advantage when large amounts of ethanol or methanol are formed during the fermentation process, as is the case with alcoholic beverages. For quantification of glycerol and ethanol at trace level, ion-exclusion chromatography on a totally sulfonated cation exchanger is more suitable [302]. Ion-exclusion chromatography can also be combined with pulsed amperometry, using a platinum working electrode with a respectively modified pulse sequence.

In order to illustrate carbohydrate analysis in fermentation broths [303], a yeast and a bacterial fermentation broth were used: *Saccharomyces cerevisiae* in a YPD broth (Yeast Extract-Peptone-Dextrose) and *Escherichia coli* (*E. coli*) in a LB broth (Luria-Bertani). These are common eucaryotic and procaryotic fermentation systems, respectively. They represent a great challenge for most separation

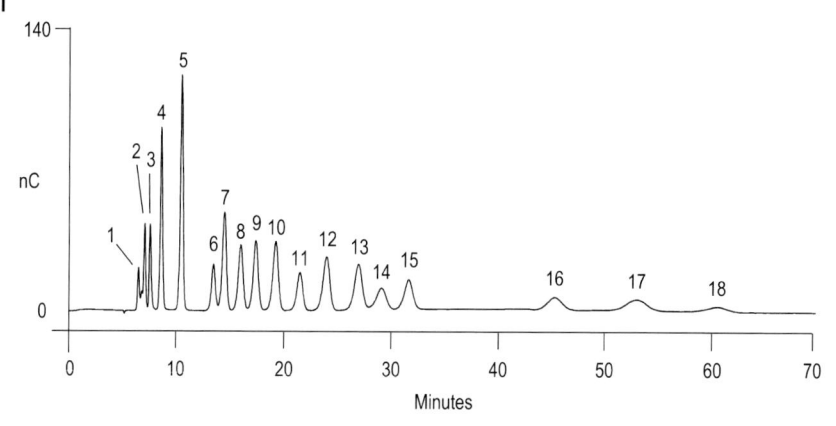

Fig. 9-197. Separation of alcohols, glycols, and saccharides relevant for fermentation monitoring. – Separator column: CarboPac MA1; eluant: 0.48 mol/L NaOH; flow rate: 0.4 mL/min; detection: pulsed amperometry on a gold working electrode; peaks: (1) 2,3-butanediol, (2) ethanol, (3) methanol, (4) glycerol, (5) erythritol, (6) rhamnose, (7) arabitol, (8) sorbitol, (9) galactitol, (10) mannitol, (11) arabinose, (12) glucose, (13) galactose, (14) lactose, (15) ribose, (16) sucrose, (17) raffinose, and (18) maltose.

and detection systems due to their complex composition. For comparison, the two chromatograms in Fig. 9-198 show the separation of alcohols and saccharides in a YPD fermentation broth before and after incubation with *Saccharomyces cerevisiae* over 24 h at 37 °C. Before incubation, glucose was the dominant component. During the first three hours, the glucose level decreased, and after three hours glucose was not detected at all. At the beginning, ethanol was also present at relatively high concentration, along with trace levels of glycerol, erythritol, rhamnose, trehalose, arabinose, and cellobiose. In the same time period in which glucose is depleted, an increase in the glycerol concentration is observed that remains constant after three hours. Ethanol concentration decreases after seven hours, presumably due to evaporative losses. While erythritol and rhamnose concentrations do not change, cellobiose concentration decreases by 50%, and trehalose and arabinose are depleted between 7 and 24 hours. When the cell culture broth was modified to contain ten different saccharides and alditols, at the same total carbohydrate concentration as the standard Bacto YPD broth, it was apparent that yeast prefers using certain carbohydrates over others, and that some saccharides or alditols could not be used as a carbon source during the incubation period. Glucose and raffinose were metabolized within the first hour. After one hour, the yeast began to consume maltose and galactose. Sorbitol, arabinose, lactose, and ribose were either unchanged or decreased slightly.

In a completely analogous way, Fig. 9-199 shows the comparison based on a LB fermentation broth before and after incubation with *E. coli* over 24 h at 37 °C. Ethanol, glucose, and trehalose are completely metabolized during the incubation, while the glycerol concentration significantly increases.

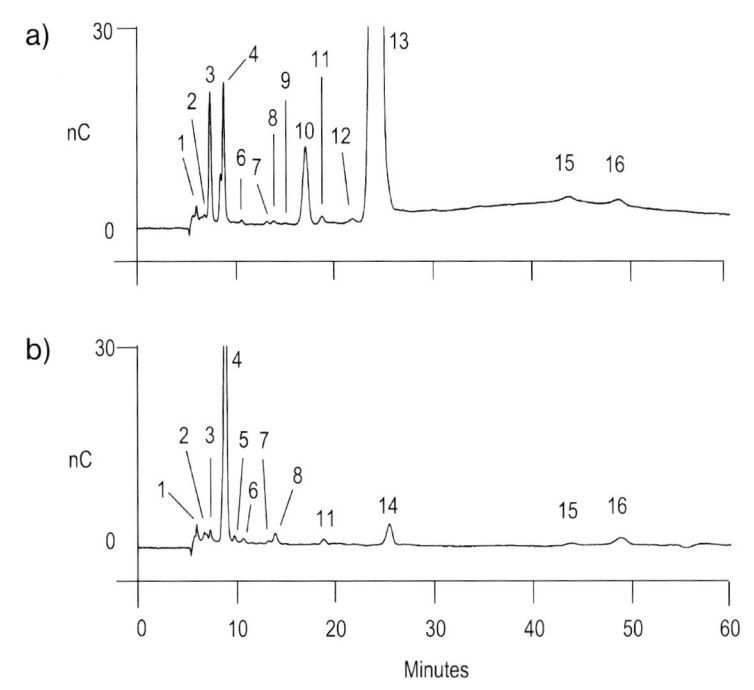

Fig. 9-198. Separation of alcohols, glycols, and saccharides in a YPD fermentation broth before (a) and after incubation (b) with Saccharomyces cerevisiae. – Separator column: CarboPac MA1; eluant: 0.48 mol/L NaOH; flow rate: 0.4 mL/min; detection: pulsed amperometry on a gold working electrode; sample preparation: centrifuged sample diluted 1:100 with de-ionized water; peaks: (1) unknown, (2) unknown, (3) ethanol, (4) glycerol, (5) unknown, (6) erythritol, (7) unknown, (8) rhamnose, (9) arabitol, (10) trehalose, (11) unknown, (12) arabinose, (13) glucose, (14) unknown, (15) cellobiose, and (16) unknown.

One of the technical fermentation processes used for some time is the isolation of gluconic acid, which is produced by several fungi – especially *Aspergilles niger*. Glucose is used as a substrate, from which gluconolacton is produced by glucose dehydrogenase, and finally gluconic acid forms. Organic acids that are directly derived from carbohydrates may be separated chromatographically together with the carbohydrates on an anion exchanger under strongly alkaline conditions. Organic acids, because of their carboxyl group, exhibit a much higher affinity towards the stationary phase than carbohydrates, so a mixture of sodium hydroxide and sodium acetate is used as an eluant. Detection is performed via pulsed amperometry on a gold working electrode.

Regarding fermentation control, organic acids [304, 305] and inorganic anions and cations [306, 307], in addition to alcohols and carbohydrates, also play a significant role. Inorganic anions can be separated together with organic acids on an IonPac AS11 hydroxide-selective, latexed anion exchanger and detected via suppressed conductivity detection. Such a chromatogram is illustrated in Fig.

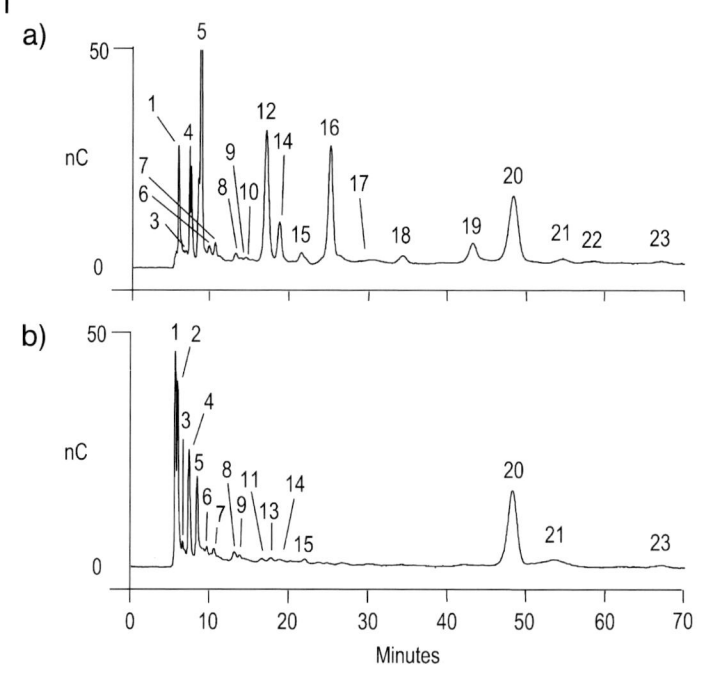

Fig. 9-199. Separation of alcohols, glycols, and saccharides in a LB fermentation broth before (a) and after incubation (b) with *E. coli*. — Chromatographic conditions: see Fig. 9-198; sample preparation: centrifuged sample diluted 1:10 with de-ionized water; peaks: (1) unknown, (2) unknown, (3) 2,3-butanediol, (4) ethanol, (5) glycerol, (6) unknown, (7) erythritol, (8) unknown, (9) rhamnose, (10) arabitol, (11) unknown, (12) unknown, (13) galactitol, (14) unknown, (15) arabinose, (16) unknown, (17) lactose, (18) to (21) unknown, (22) maltose, and (23) unknown.

9-200. Due to the high number of possible constituents, a separation under isocratic conditions is not possible. Again, Fig. 9-200 demonstrates this with the separation of inorganic and organic anions in the above-mentioned YPD fermentation broth before (A) and after (B) incubation with *Saccharomyces cerevisiae* over 24 hours at 37 °C. As expected, after 24 hours of incubation, higher concentrations are observed for lactate, acetate, and formate as a result of metabolic processes. In contrast, orthophosphate concentration decreases, presumably due to its incorporation into the bio mass (DNA, RNA, membrane phospholipids, etc.). At least ten unidentified peaks were observed. Eight of these peaks changed their size over the course of the incubation period. To examine anions at lower concentrations, injections of a more concentrated culture are needed. The high-capacity version of IonPac AS11 is used as a stationary phase in this case, as it exhibits a slightly different selectivity. Phenylacetate, bromide, and nitrate, for example, elute on the AS11 column a few minutes ahead of malate, whereas on the AS11-HC column they elute immediately ahead of malate. Figure 9-201 illustrates the advantages of the high-capacity AS11-HC column as compared with the AS11 column for the separation of the early eluting monocarboxylic acids, which are much better resolved on IonPac AS11-HC, even though the analysis time is significantly longer. Moreover, some of the trace components cannot be detected when using IonPac AS11 because of the strong dilution. This, in turn, is necessary to avoid column overloading. For comparison, Fig. 9-202 shows the two resulting chromatograms of a fermentation broth before (A) and after (B) incubation with E. coli over 24 hours at 37 °C.

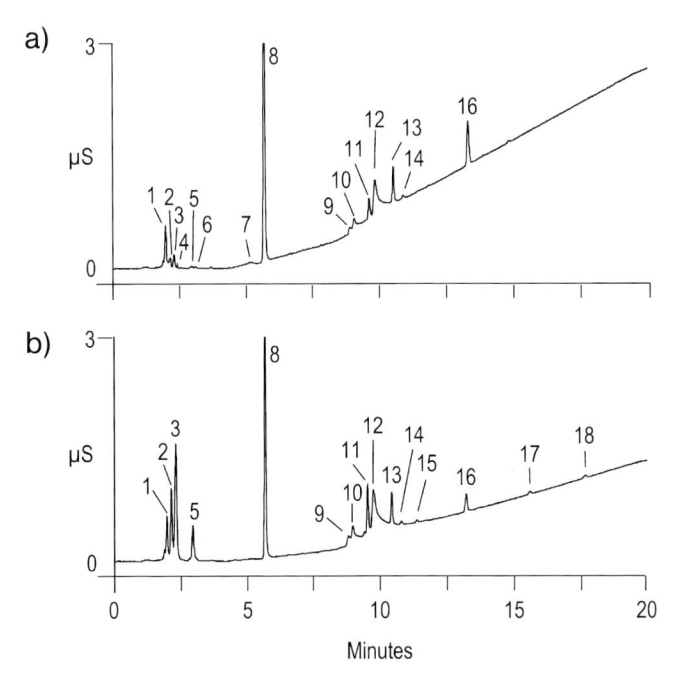

Fig. 9-200. Separation of inorganic and organic anions in a YPD fermentation broth before (a) and after incubation (b) with Saccharomyces cerevisiae. – Separator column: IonPac AS11; eluant: NaOH; gradient: 0.5 mmol/L isocratic for 2.5 min, then linearly to 5 mmol/L in 3.5 min, then to 38 mmol/L in 12 min; flow rate: 2 mL/min; detection: suppressed conductivity; injection volume: 10 µL; sample preparation: centrifuged sample diluted 1:100 with de-ionized water; peaks: (1) unknown, (2) lactate, (3) acetate, (4) unknown, (5) formate, (6) unknown, (7) unknown, (8) chloride, (9) unknown, (10) unknown, (11) malate/succinate, (12) malonate/carbonate, (13) sulfate, (14) fumarate/oxalate, (15) unknown, (16) orthophosphate, (17) and (18) unknown.

The fact that inorganic cations also play a role for the cell growth of microorganisms and for certain metabolic processes is often ignored. For example, alkali- and alkaline-earth metals as well as ammonium are needed at mmol/L concentrations to boost cell growth. Also, the concentration ratio between H^+, Na^+, and K^+ is of great importance for membrane physiology. However, the optimal concentration very much depends on the type of species and the cell density. Halophilic bacteria, for example, require sodium concentrations of more than

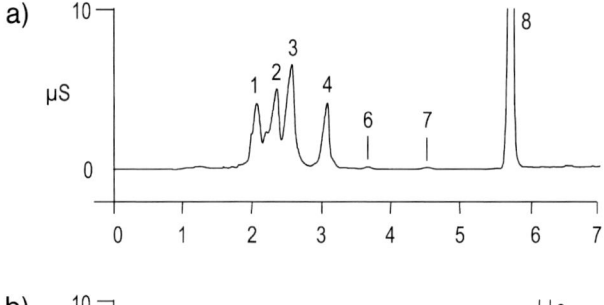

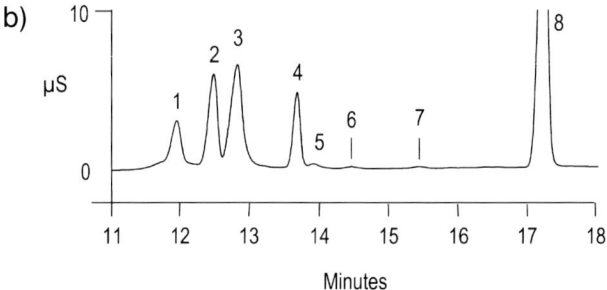

Fig. 9-201. Comparison of the separation of early eluting anions between IonPac AS11 (a) and AS11-HC (b) in a YPD fermentation broth after incubation with *Saccharomyces cerevisiae*. — (a) Separator column: IonPac AS11; eluant: NaOH; gradient: see Fig. 9-200; flow rate: 2 mL/min; (b) Separator column: IonPac AS11-HC; eluant: NaOH; gradient: 1 mmol/L isocratic for 8 min, then linearly to 15 mmol/L in 10 min, then to 30 mmol/L in 10 min, then to 60 mmol/L in 10 min; detection: suppressed conductivity; injection volume: 10 µL; sample preparation: (a) centrifuged sample diluted 1:100 with de-ionized water, (b) centrifuged sample diluted 1:10 with de-ionized water; peaks: (1) unknown, (2) lactate, (3) acetate/glycolate, (4) formate, (5) butyrate, (6) pyruvate/isovalerate, (7) valerate, and (8) chloride.

1 mol/L for survival. Such concentrations are inhibiting for other microorganisms [308]. In contrast, transition metals are only present at trace level, but are essential as co-factors for an optimal metabolization of certain compounds. Thus, production of bacitracin requires a manganese concentration of 0.7 µmol, while concentrations above 40 µmol are inhibiting [309]. *Streptomyces griseus* needs five times more iron for producing streptomycin than for growth, and *Corynebacterium diphteriae* only produces the respective toxin if the iron content is below 0.8 ppm [310]. These examples prove how critical the observance of the respective concentration ranges is for certain fermentations. In recent times, more and more HPLC methods are being employed for the analysis of inorganic cations.

Bell [306] analyzed transition metals together with magnesium and calcium utilizing ion-pair chromatography on a chemically bonded reversed phase, which can be combined with both indirect conductivity detection and photometric detection after derivatization with PAR. Because Bell used predominantly ashed

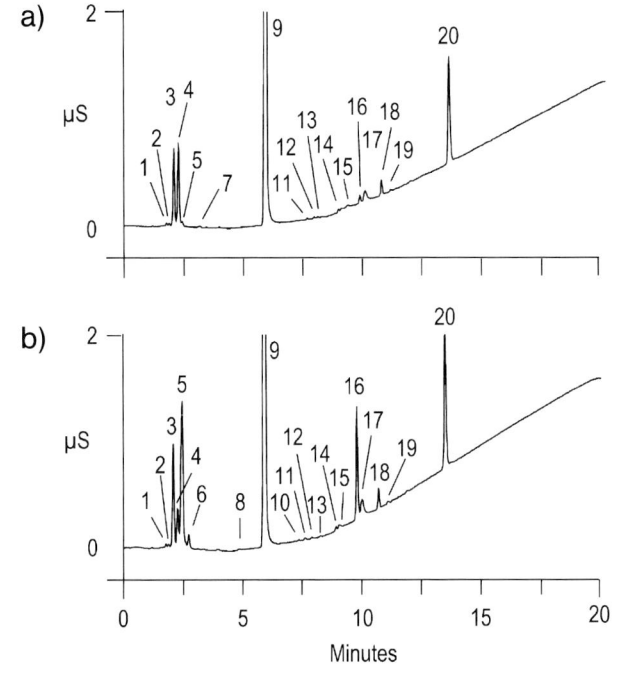

Fig. 9-202. Separation of inorganic and organic anions in a LB fermentation broth before (a) and after incubation (b) with E. coli. – Separator column: IonPac AS11; chromatographic conditions: see Fig. 9-200; sample preparation: centrifuged sample diluted 1:100 with de-ionized water; peaks: (1) to (3) unknown, (4) lactate, (5) acetate, (6) propionate, (7) formate, (8) valerate, (9) chloride, (10) unknown, (11) phenylacetate, (12) bromide, (13) nitrate/5-keto-D-gluconate, (14) unknown, (15) unknown, (16) malate/succinate, (17) malonate/carbonate, (18) sulfate, (19) fumarate/oxalate, and (20) orthophosphate.

fermentation samples, his chromatograms are free of interferences. Simultaneous separations of alkali- and alkaline-earth metals in methanotropic bacterial cultures and other media on surface-sulfonated latexed cation exchangers were first reported by Joergensen et al. [304] and Robinett et al. [302]. Based on these three publications it is apparent that the treatment of samples with reversed-phase cartridges is more than sufficient as sample preparation for the analysis of alkali- and alkaline-earth metals; an acid digest does not affect the measuring values. In contrast, a dependence of the measuring values on the type of sample preparation is observed for transition metals. This can be attributed to the fact that transition metals form stable complexes with a number of inorganic and organic complexing agents such as EDTA, citric acid, oxalic acid, etc. and, conesquently, disturb the complexation equilibrium that is required for the chromatographic separation. Therefore, compounds of this kind have to be removed from the sample prior to the analysis.

A rather topical work on the analysis of alkali- and alkaline-earth metals on a weak acid cation exchanger with very little sample preparation was published by Robinett et al. [307]. Because the concentrations of inorganic cations in fermentation broths are several orders of magnitude above the detection limits, samples can be strongly diluted. Typical dilution factors are between 100 and 2,000. The selectivity of this method is even higher when using conductivity detection in combination with a suppressor system. Thus, other sample constituents such as inorganic and organic anions, alcohols, and carbohydrates do not interfere at all. Amines such as monoethanolamine and choline are potential interferences, but they can be separated from inorganic cations on weak acid cation exchangers. Robinett et al. checked the applicability of this method with an *E. coli* fermentation broth. The time course of fermentation is illustrated in Fig. 9-203 through an overlay of representative chromatograms showing isocratic cation analysis. These chromatograms reveal that the original concentrations of ammonium and potassium decreased by 16-20% and magnesium decreased by 60-65% over the course of the fermentation. In contrast, the calcium level increased by 10% relative to the starting concentration. Levels of sodium remained virtually unchanged throughout the fermentation.

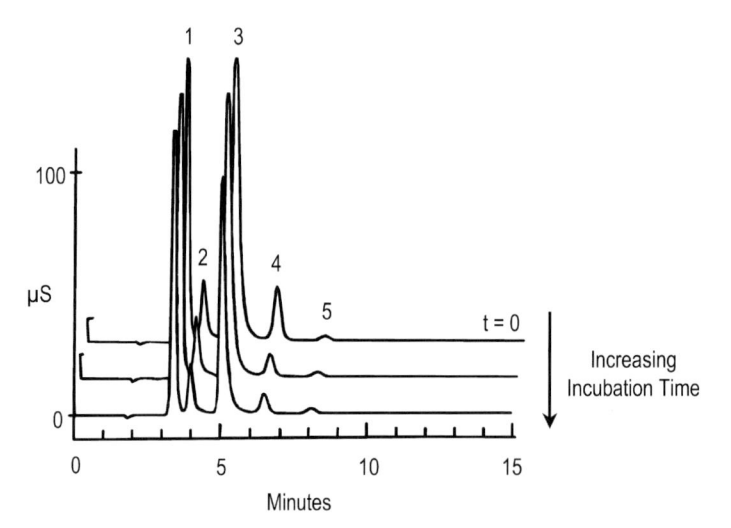

Fig. 9-203. Time dependence of the analysis of alkali- and alkaline-earth metals in a fermentation broth incubated with *E. coli* with glucose as a primary carbon source over the course of fermentation. – Separator column: IonPac CS12; eluant: 20 mmol/L methanesulfonic acid; flow rate: 1 mL/min; detection: suppressed conductivity; injection volume: 10 µL; peaks: (1) sodium, (2) ammonium, (3) potassium, (4) magnesium, and (5) calcium; (taken from [307]).

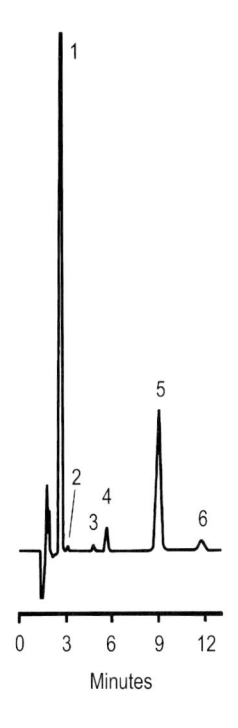

Fig. 9-204. Separation of inorganic anions in human saliva. — Separator column: IonPac AS4A; eluant: 1.7 mmol/L $NaHCO_3$ + 1.8 mmol/L Na_2CO_3; flow rate: 1.5 mL/min; detection: suppressed conductivity; injection: 50 µL of a solution of 390 mg saliva in 25 mL water; peaks: (1) chloride, (2) nitrite, (3) bromide, (4) nitrate, (5) orthophosphate, and (6) sulfate.

Due to the robustness and the high specificity of ion chromatography for fermentation monitoring as described above, its application in online mode is conceivable, as in many cases sample preparation is restricted to dilution and filtration.

9.8
Ion Chromatography in Clinical Chemistry

One of the most important applications of ion chromatography in clinical chemistry is the investigation of body fluids such as saliva, urine, and serum for inorganic and organic anions and cations.

In this field, saliva [311] and teardrops [312] represent rather simple matrices and may be directly injected without any sample preparation after appropriate dilution with de-ionized water. As seen in the anion chromatogram of a human saliva sample (390 mg ad 25 mL H_2O) in Fig. 9-204, the constituents may be separated without any interference. The detection of small quantities of bromide is remarkable. Its occurrence was unknown until now and the source cannot be explained at present.

Bromide, the almost forgotten first antiepileptic, is again successfully used for epilepsy treatment, especially with patients for whom other antiepileptics do not show any effect [313]. Therefore, frequent control of the bromide level in the patient's serum and urine is required. Because it has been observed that an above-average intake of sodium chloride leads to a rapid decrease of the bromide level in adolescent patients, Juergens [314] investigated the possibility of a correlation between bromide and chloride concentrations in serum and urine. For this type of analysis, he chose anion exchange chromatography with non-suppressed conductivity detection, because both halide anions can be simultaneously detected. To avoid having to collect urine samples over a longer period of time (24 h), the creatinine contents were measured in all samples. Because the excretion of creatinine is a constant process, the creatinine concentration in urine can be used as a measure for its degree of dilution [315]. However, halide ion excretion in urine is subject to large variations during the day, which cannot be compensated by converting to creatinine as an internal scale basis. Nevertheless, Juergens could show that an accelerated detoxication after bromide intoxication can be achieved by controlled administration of sodium chloride.

Ion chromatography is a decisive analytical tool for solving analytical problems in the field of nephrology. Today, for example, the analysis of oxalate in urine is one of the most important determinations in the examination of patients with kidney stones [316]. Many analysis methods used in the past, such as manganometry after oxidation of oxalate to carbon dioxide, or spectrophotometry after reduction of oxalate to glyoxalic acid or glycolic acid, are very labor-intensive and time-consuming, as a preceding separation of oxalate with a corresponding precipitation reaction is necessary. The direct oxalate determination by enzymatic analysis is also difficult, as the enzymes are inhibited by matrix constituents such as sulfate and orthophosphate, resulting in a systematic underestimation. In the past, therefore, one had to be content with measuring the calcium content in urine to obtain evidence of irregularities in patients with kidney stones. Not until the introduction of chromatographic techniques such as GC [317, 318] and HPLC [319], was it possible to determine oxalate ions directly in urine. Based on the work of Menon et al. [320], Robertson et al. [321] developed an ion chromatographic method that allows the analysis of oxalate in less than 20 minutes with very little sample preparation. They employed a conventional anion exchanger as a stationary phase with a mixture of sodium carbonate and sodium bicarbonate as an eluant. To achieve the highest possible sensitivity, they passed the separator column effluent through a membrane-based suppressor system prior to entering the detector cell. With this setup, Robertson et al. obtained a detection sensitivity of 0.5 µmol/L oxalate in acidified, diluted urine. The sample acidification with 1 mol/L hydrochloric acid is necessary to re-dissolve calcium oxalate crystals that may have precipitated in urine. Figure 9-205 shows a corresponding chromatogram; the peak attributed to oxalate corresponds to a concentration of 10 µmol/L.

Under similar chromatographic conditions it is also possible to determine p-aminohippuric acid, which represents one of the control substances for diagnosing kidney function. On the basis of certain analytical values, it is possible

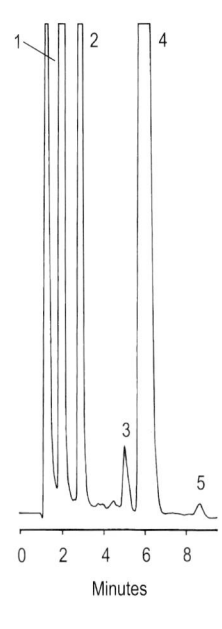

Fig. 9-205. Separation of oxalate in urine. — Separator column: IonPac AS4; eluant: 2.8 mmol/L NaHCO$_3$ + 2.2 mmol/L Na$_2$CO$_3$; flow rate: 2 mL/min; detection: suppressed conductivity; injection: 50 µL sample (diluted 1:10); peaks: (1) chloride, (2) orthophosphate, (3) nitrate, (4) sulfate, and (5) oxalate.

to determine the total blood circulation through the kidneys and, at higher plasma concentrations of p-aminohippuric acid, the maximal secretion performance of the tubular system. In contrast to Baranowski and Westenfelder [322] who used chemically bonded silica as a stationary phase for separating p-aminohippuric acid, separation on a conventional anion exchanger is much easier. In both cases, detection is carried out by measuring light absorption at 254 nm. The chromatogram of a urine sample in which a p-aminohippuric acid content of 730 mg/L was detected, is represented in Fig. 9-206. Further urine constituents such as mandelic acid and phenylglyoxylic acid may also be determined very sensitively via UV detection. Their low affinity towards the stationary phase of an anion exchanger suggests that a 15 mmol/L sodium hydroxide solution would be a suitable eluant.

The sample preparation for ion chromatographic analyses of sera is much more elaborate. Proteins present in sera in high concentration must be removed before the sample is injected, because they negatively affect the separation efficiency of the ion exchangers under the chromatographic conditions suited for determining inorganic ions. The originally used precipitation method based on acetonitrile [323] has not succeeded because proteins are not quantitatively precipitated by this method, which limits the lifetime of the separator column. In contrast, quantitative protein precipitation is obtained with perchloric acid [324], a procedure that was successfully applied by Reiter et al. [325] to determine sulfate in human sera. Because high perchlorate concentrations interfere with the ion chromatographic determination of mineral acids, perchlorate must also be removed after protein precipitation. For this purpose, the following procedure was developed by Reiter et al.:

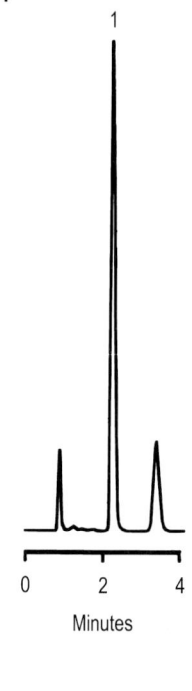

Fig. 9-206. Separation of p-aminohippuric acid in urine. — Separator column: IonPac AS4A; eluant: 1.7 mmol/L NaHCO$_3$ + 1.8 mmol/L Na$_2$CO$_3$; flow rate: 2 mL/min; detection: UV (254 nm); injection: 50 µL sample (diluted 1:100); solute concentration: 730 mg/L p-aminohippuric acid (1).

1 mL of the serum sample is added to 1 mL of 0.665 mol/L cold perchloric acid. After shaking, the mixture is kept at 4 °C for 10 minutes to wait for complete precipitation. After centrifugation at 3100 g for 15 minutes, 1 mL of the supernatant is added to 1 mL of 0.7 mol/L cold potassium carbonate solution to precipitate the perchlorate. This mixture is centrifuged at 3100 g for 15 minutes. The supernatant is diluted with de-ionized water according to the concentration of the analyte ions.

A chromatogram of a 1:10 diluted serum sample is displayed in Fig. 9-207. Apart from orthophosphate, sulfate, and small amounts of oxalate, bromide and nitrate could also be detected. This corroborates the observations of de Jong [326] who, together with other authors, attributed the presence of bromide in serum to nutrition habits [327], environmental influences [328], and anesthetics [329]; the occurence of nitrate in serum was attributed to fertilizers as indirect sources [330]. In contrast, sulfate plays a significant role in the metabolism of many endogenous compounds. So there is hope that a better understanding of these processes can be obtained by knowing the sulfate turnover as reflected by the sulfate concentrations in serum and urine.

As can be seen from Fig. 9-207, nitrite cannot be quantified under these chromatographic conditions because of the high chloride excess (usually about 100 mmol/L). However, nitrite and nitrate are metabolites of nitric oxide radicals which are involved in a number of physiological and pathophysiological processes [331], for example, in intercellular communication and neurotransmission

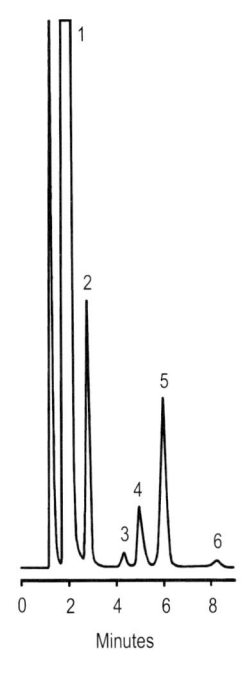

Fig. 9-207. Separation of inorganic and organic anions in serum. — Separator column: IonPac AS4A; eluant: 1.7 mmol/L NaHCO$_3$ + 1.8 mmol/L Na$_2$CO$_3$; flow rate: 1.5 mL/min; detection: suppressed conductivity; injection: 50 µL sample (diluted 1:10); peaks: (1) chloride, (2) orthophosphate, (3) bromide, (4) nitrate, (5) sulfate, and (6) oxalate.

in the peripheral and central nerve system. Analytical methods for detecting NO$^{\bullet}$-radicals are based, among other things, on chemiluminescence assays through the reaction of NO$^{\bullet}$ with ozone [332] or Luminol-H$_2$O$_2$ [333], but cannot be used for *in vivo* clinical, pharmacokinetic studies due to the short half life of NO$^{\bullet}$-radicals (<5 s). In an oxygen-containing aqueous solution, NO$^{\bullet}$-radicals are stoichiometrically oxidized to nitrite ions [334]:

$$4NO^{\bullet} + O_2 + 2H_2O \rightarrow 4NO_2^- + 4H^+ \tag{323}$$

Endogenous nitrite cannot be determined in plasma, as nitrite is almost completely oxidized to nitrate by, for example, oxyhaemoglobin [334]:

$$4HbO_2 + 4NO_2^- + 4H^+ \rightarrow 4Hb^+ + 4NO_3^- + O_2 + 2H_2O \tag{324}$$

Therefore, reduction of nitrate to nitrite [335] is required in order to apply the classical Griess method [336]. According to the Griess method, nitrite is diazotized with sulfanilamide and then reacted with N-1-naphthylethylenediamine to form a coloured product. However, this method is not very reliable when applied to complex matrices such as serum, as the reduction of nitrate to nitrite is chemically difficult and enzymatically very expensive. Stains and clouding in the sample also cause interferences, and the presence of copper in the sample catalyzes the degradation of the diazo salt which, in turn, leads to smaller measuring values.

Various methods were developed for the ion chromatographic solution of this problem [337, 338], of which anion exchange chromatography with UV detection at 214 nm [339] is the most suitable one. In this case, the choice of stationary phase is of decisive importance. Stratford et al. [337] obtained the best separation under isocratic conditions utilizing IonPac AS9-SC with a phosphate buffer as an eluant. Because the authors observed severe problems of contamination by nitrate when applying ultrafiltration, they preferred the acetonitrile precipitation as a sample preparation technique. For this purpose, 50 µL of serum is mixed with 50 µL of acetonitrile in a 300-µL glass tube (Sci-Vi, Chromacol, Welwyn, UK) and centrifuged at 2000 g for 2 minutes. The clear supernatant can then be injected directly into the chromatograph. Stratford et al. regard the removal of chloride with a cation exchanger in the silver form to be difficult to handle when dealing with small sample volumes (<0.5 mL), and justly point to the increased contamination risk when doing so. Moreover, in the long run the silver ions released from the resin material will lead to adverse effects with the anion exchanger.

Excellent separations were obtained by Crowther et al. [339] on CarboPac PA-100 when applying a gradient elution technique with a chloride eluant. The use of a chloride eluant for determining nitrite and nitrate in the presence of high amounts of chloride had been published by Pastore et al. [340], and Haddad et al. [341] successfully applied this technique for the trace analysis of nitrite in sea water. The advantage of this technique is due to the fact that prior to injection, the ion-exchange groups are in the chloride form, so the column equilibrium is fairly undisturbed by large amounts of chloride in the sample. In contrast to the commonly used carbonate/bicarbonate eluant mixture, a chloride eluant does not have buffer capacity, which can be compensated by adding Tris. An example chromatogram is illustrated in Fig.9-208. By choosing CarboPac PA-100 with its high ion-exchange capacity of 90 µequiv/column, chloride can be used at relatively high concentration. Thus, an accumulation of buffer ions at the stationary phase at low eluant strength is avoided, which would lead to artefacts when increasing the eluant strength during the gradient run. Under the given chromatographic conditions in Fig. 9-208, nitrite and nitrate are separated in less than ten minutes. Ten more minutes are required for a short column rinse with subsequent equilibration. In contrast to Stratford et al., Crowther et al. chose ultrafiltration of the 1:4 diluted serum samples as sample preparation. For ultrafiltration, they centrifuged the samples at 7500 g for 120 minutes through a Centricon filter (Amicon, Stonehouse, UK). Each filter was intensively rinsed with de-ionized water before use to remove residual nitrate.

With a modified gradient program, starting with a lower initial chloride concentration and a more shallow gradient, some organic acids such as lactic acid, pyruvic acid, formic acid, acetic acid, malonic acid, hippuric acid, fumaric acid, citric acid, and maleic acid can also be determined in serum. The respective example in Fig. 9-209 is a first attempt, and shows that more time has to be invested to optimize the separation.

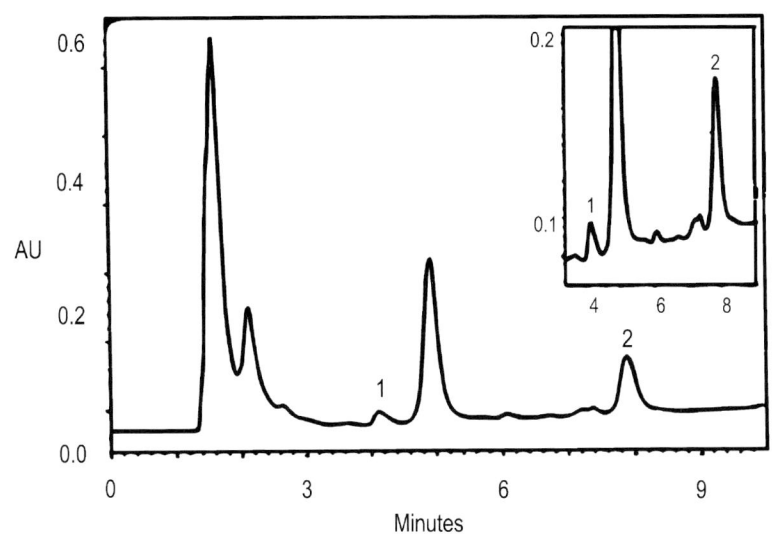

Fig. 9-208. Separation of nitrite and nitrate in human serum. — Separator column: CarboPac PA-100; eluants: NaCl — Tris (5 mmol/L, pH 7.5); gradient: 0.12 mol/L linearly to 0.3 mol/L in 10 min, then rinsing step with 1 mol/L for 4 min; flow rate: 1 mL/min; detection: UV (214 nm); injection volume: 30 µL; sample preparation: see text; peaks: (1) nitrite and (2) nitrate; (taken from [339]).

Protein samples are usually provided with sodium azide as a fungicide. Due to the toxicity of azide its concentration has to be determined in protein samples dedicated for pharmaceutical research. Electrochemical, spectrophotometric, gas chromatographic, volumetric, liquid chromatographic, and ion chromatographic methods are described in literature for the analysis of azide. Of all these methods, only the liquid chromatographic and ion chromatographic ones offer the ability to determine azide in a simple way that is free of interference in the protein matrix. With a pK_a value of 4.8, azide is an ideal candidate for anion exchange chromatography using alkaline eluants, although it is difficult to separate azide from other equally sized anions such as bromide and nitrate. As early as 1982, Yang et al. [342] reported a separation on a Vydac 302 (15 µm, 250 mm × 4.6 mm i.d.) anion exchanger utilizing non-suppressed conductivity detection with o-phthalic acid (pH 3.5) as an eluant. Although this technique allowed the separation of azide from other mineral acids as potential constituents, the reported minimum detection limit of 2 mg/L, based on an injection volume of 100 µL, is not sufficient. A similar separation with a nicotinic acid eluant on Wescan 269-031 was shown in Fig. 3-103 (Section 3.7.3). In contrast, the separation on an IonPac AS4A(-SC) latexed anion exchanger illustrated in Fig. 3-81 (Section 3.7.2) was carried out with a carbonate/bicarbonate eluant mixture utilizing suppressed conductivity detection. Though the sensitivity is significantly higher

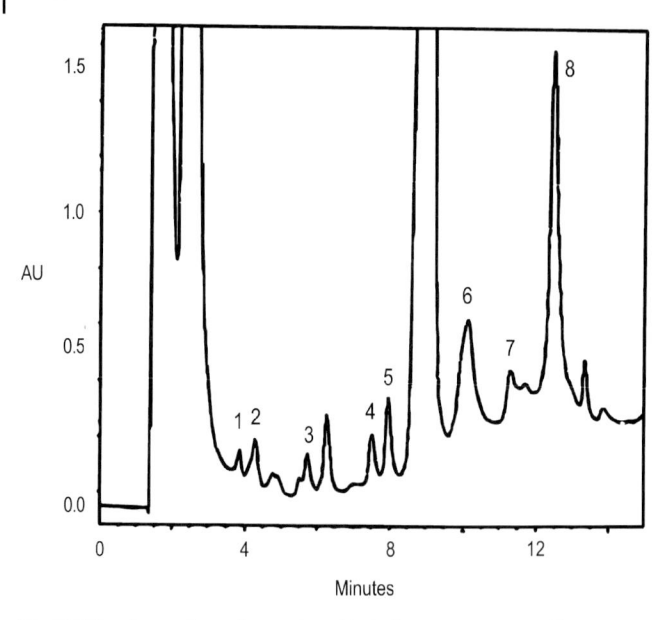

Fig. 9-209. Separation of organic acids in human serum. — Separator column: CarboPac PA-100; eluant: NaCl — Tris (5 mmol/L, pH 7.5); gradient: 0.03 mol/L isocratic for 1 min, then linearly to 0.3 mol/L in 15 min, then rinsing step with 1 mol/L for 4 min; flow rate: 1 mL/min; detection: UV (214 nm); injection volume: 30 µL; sample preparation: see text; peaks: (1) lactate, (2) acetate, (3) pyruvate, (4) malonate, (5) nitrite, (6) fumarate, (7) citrate, and (8) nitrate; (taken from [339]).

when using a suppressor system, this separation lacks complete resolution between azide and nitrate. Annable and Sly [343] obtained this complete resolution with a tetraborate eluant, even though sulfate is strongly retained under these conditions. When preparing tetraborate eluants, attention has to be paid that the de-ionized water being used is thoroughly degassed with helium and subsequently kept under inert gas to avoid retention time shifts caused by CO_2 from the atmosphere. Figure 9-210 shows a typical chromatogram of azide in BSA (Bovine Serum Albumin). Because BSA is soluble in the tetraborate eluant, a filtration was the only sample preparation required. The authors report a minimum detection limit for azide of approximately 30 µg/L.

Ion chromatographic methods were also developed for determining alkali- and alkaline-earth metals in human serum [344-347]; they serve as reference methods for flame AAS. As early as 1994, Thienpont et al. [344] utilized cation exchange chromatography with chemically regenerated suppressed conductivity detection for determining magnesium and calcium in human serum. Method validation was carried out with certified reference compounds from NIST (Gaithersburg, MI, USA) and BCR (Brussels, Belgium). The lyophilized sera were dissolved under defined conditions, acidified with HCl, and filtrated through

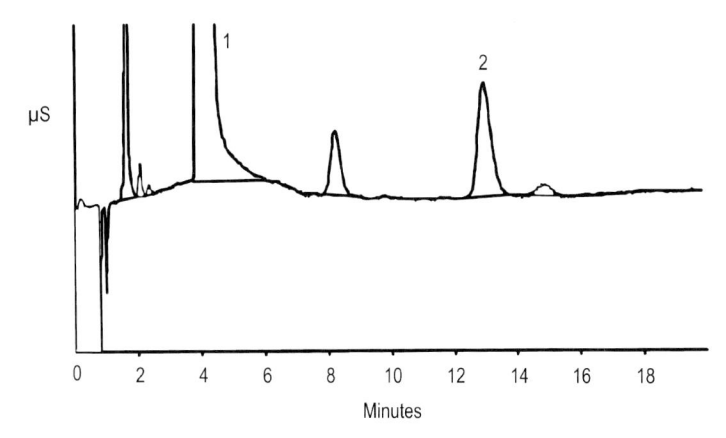

Fig. 9-210. Determination of azide in BSA. – Separator column: IonPac AS4A(-SC); eluant: 15 mmol/L $Na_2B_4O_7$; flow rate: 2 mL/min; detection: suppressed conductivity; sample: 1000 mg/L BSA (Sigma); solute concentrations: 51 mg/L chloride (1) and 3.2 mg/L azide (2); (taken from [343]).

Arodisc LC13 PVDF filters (Gelman Sciences, Ann Arbor, MI, USA). Thienpont et al. used two IonPac CG10 guard columns in series as a stationary phase and a mixture of 2,3-diaminopropionic acid (DAP) and HCl as an eluant. However, different DAP concentrations were used for the analysis of magnesium and calcium to achieve an optimal separation of the corresponding metal ion from matrix components. One year later, the same authors [345] published a complementary paper on the determination of sodium and potassium in human serum, for which they used an IonPac CS12 weak acid cation exchanger. On this stationary phase, alkali- and alkaline-earth metals can be analyzed in the same run using a methanesulfonic acid eluant. A representative chromatogram is depicted in Fig. 9-211. To account for the strongly different concentrations of sodium and the other cations, the detector range was switched from 100 µS/cm to 3 µS/cm after the elution of sodium, which is indicated in the chromatogram by a small arrow. As a result of the two extensive works, Thienpont et al. confirmed that ion chromatography can indeed be used as a reference method in terms of accuracy and precision.

In addition to alkali- and alkaline-earth metals, living organisms also contain metals such as copper, iron, zinc, and aluminum. Copper and iron, especially, play an important role as essential elements in maintaining biological processes such as metabolism and transport [348]. Zinc is an essential trace element for the synthesis of nucleic acids and proteins and, moreover, is responsible for growth. For some time, it has been suspected that aluminum is involved in various neurological diseases, for example Alzheimer's disease. As a consequence, all these metals are determined in serum in the course of clinical analysis. Because these metals are traditionally analyzed with flame- or graphite-furnace AAS, there are very few literature citations for ion chromatographic meth-

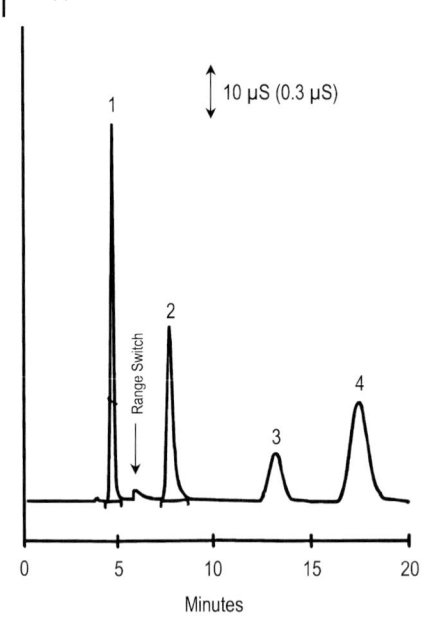

Fig. 9-211. Analysis of alkali- and alkaline-earth metals in human serum. – Separator column: IonPac CS12; eluant: 14 mmol/L methanesulfonic acid; flow rate: 1 mL/min; detection: suppressed conductivity; range switch from 100 µS/cm to 3 µS/cm after elution of sodium; injection volume: 25 µL; sample preparation: see text; peaks: (1) sodium, (2) potassium, (3) magnesium, and (4) calcium; (taken from [345]).

ods. This is indeed surprising, as the samples to be analyzed using flame AAS can be only slightly diluted to obtain adequate sensitivity [349]. In addition, absorption abnormalities can occur in flame- and graphite-furnace AAS that are caused by the high protein content in biological samples [350]. In 1988, Ong et al. [351] published a paper on the determination of copper and zinc in serum and full blood; it introduced the classical ion chromatographic method of anion exchange chromatography in combination with photometric detection after derivatization with PAR as a welcomed alternative to atomic spectrometric methods. The authors used an IonPac CS5 with defined anion- and cation exchange capacities as a stationary phase; copper and zinc were eluted with an oxalic acid eluant. Ong et al. precipitated the proteins with trichloroacetic acid by mixing 400 µL of serum with 200 µL of trichloroacetic acid in an Eppendorf vial and centrifugating the mixture at 9000 rpm for three minutes. In extensive investigations, Ong et al. compared the ion chromatography results with those of graphite-furnace AAS and discovered they were in good agreement. Only at very low copper concentrations are the AAS results somewhat higher than the IC results. However, this difference is smaller than the variation in the copper content between the individual probands, which is influenced by their health situation and the daily intake of this element via food.

Another application of ion chromatography in clinical chemistry is the analysis of carbohydrates [352] in physiological samples, which is more and more often carried out with the HPAEC-PAD method. In light of this, the analysis of mannitol, 3-O-methylglucose, and lactulose described by Fleming et al. [353] is of great interest. Via urinary excretion, the concentration ratio of these three sugars after oral administration allows statements on the permeability of the intestinal tract and the integrity of the intestinal mucous membrane. The clinical application of such a method exists in the monitoring of intestinal malfunction during HIV infection. Difficulties with the quantification of these sugars have severely limited the distribution of this test. Using HPAEC-PAD, Fleming et al. could simultaneously determine all three sugars in plasma and urine. The urine samples were collected over a period of six hours and preserved by adding 1 mL of 20% (w/w) chlorohexidine. After dilution with de-ionized water and addition of melibiose as an internal standard, the samples were desalted with an anion exchanger in the H^+ form (Amberlite IR-120 H^+). Melibiose was also used as an internal standard for plasma samples. Deproteination was carried out with ice-cold 5-sulfosalicylic acid solution (c = 35 g/L). After 20 minutes of ice cooling the samples were centrifuged at 9000 g for five minutes and subsequently de-salted as described above. Fleming et al. used CarboPac PA1 as a stationary phase and NaOH as an eluant. To ensure a baseline-resolved separation between lactulose and lactose, zinc acetate was added to the eluant. Plasma and urine samples were analyzed with different concentrations of NaOH and zinc acetate. In both cases, the separator column was rinsed with 1 mol/L NaOH after the elution of the analyte sugars to ensure stable retention times for the subsequent chromatographic run. Figure 9-212 shows representative chromatograms of plasma and urine samples from a healthy patient. The plasma sample was obtained 90 minutes after oral administration of the sugars, while the urine sample was collected over a period of six hours after administration.

Particular attention is also given to the biochemistry of inositol phosphates, because previous investigations suggest that, for example, inositol-1,4,5-tri-phosphate (IP_3) serves as an intracellular second messenger for a multitude of hormones, growth-stimulating factors, and neurotransmitters. In the past, such compounds were determined by costly radiochemical methods. Smith et al. [354, 355] developed an ion chromatographic method that involves a conventional anion exchanger and chemically suppressed conductivity detection. Its advantage is the ability to determine a diversity of physiologically relevant anions in one run, enabling the biochemist to follow several reactions or metabolic processes simultaneously. Radioactive labeling of the compounds can be omitted, because the detection method exhibits high sensitivity. Thus, detection limits for the investigated samples are in the range between 20 pmol and 100 pmol. Other sample constituents such as phosphocreatine, amino acids, and nucleosides do not interfere with the inositol phosphate determination, as they are converted into the cationic form and exchanged for hydronium ions in the suppressor. Figure 9-213 shows the separation of IP_3 in a rat brain extract which was passed through an extraction cartridge prior to injection. This cartridge contained a cation exchange resin in the silver form that served to precipitate the chloride present in the sample. p-Cyanophenol was used as an eluant. Its elution power is high enough to elute the hexavalent IP_3 in a comparatively short time.

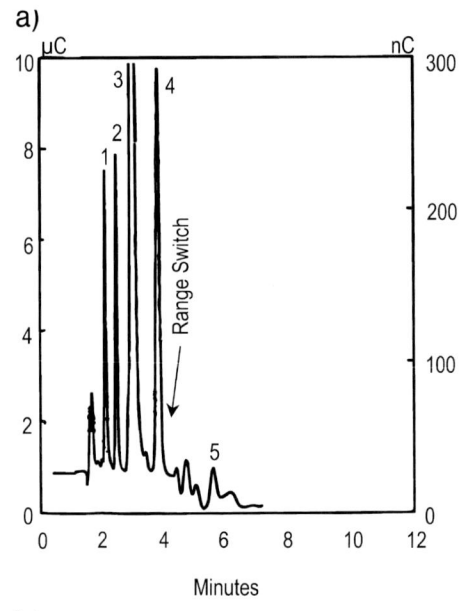

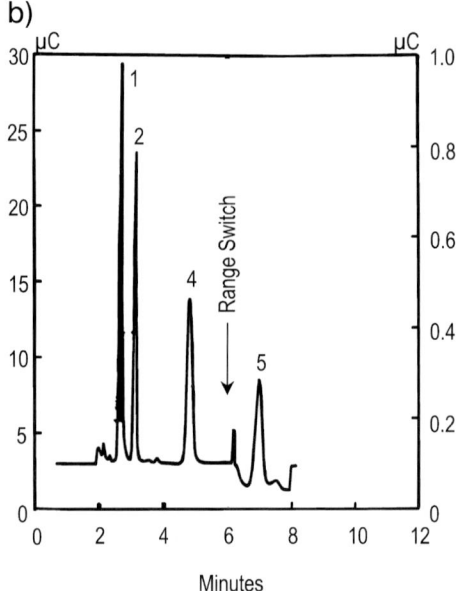

Fig. 9-212. Analysis of mannitol, 3-O-methylglucose, and lactulose in plasma and urine of a healthy patient. — Separator column: CarboPac PA1; (a) eluant: 160 mmol/L NaOH + 0.675 mmol/L zinc acetate, (b) 120 mmol/L NaOH + 0.5 mmol/L zinc acetate; flow rate: 1 mL/min; detection: pulsed amperometry on a gold working electrode; injection volume: 25 µL; sample: (a) plasma 90 minutes after oral administration, range switch from 10 µC to 300 nC, (b) 6-h collective urine sample after oral administration, range switch from 30 µC to 1 µC; peaks: (1) mannitol, (2) 3-O-methylglucose, (3) glucose, (4) melibiose, and (5) lactulose; (taken from [353]).

The most important isomers of the mono-, di-, and tri-phosphates of inositol can be separated in one run using a gradient elution technique. The separation represented in Fig. 9-214 was obtained on a hydroxide-selective IonPac AS11 anion exchanger with a NaOH concentration gradient. If radiochemical detection is used instead of conductivity detection, even higher NaOH concentrations can be used, thus enabling the elution of higher inositol phosphates.

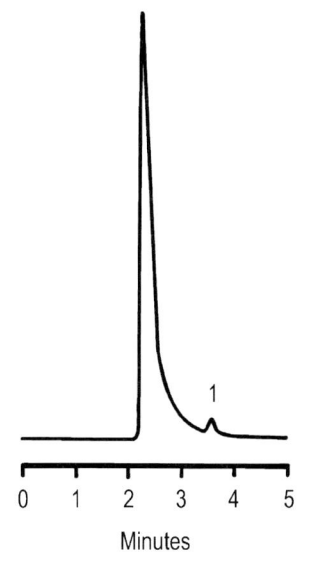

Fig. 9-213. Separation of inositol-1,4,5-triphosphate in a rat brain extract. — Separator column: IonPac AS4A; eluant: 26.4 mmol/L p-cyanophenol; flow rate: 2 mL/min; detection: suppressed conductivity; injection: 50 µL of a chloroform/methanol extract (2:1 v/v) with IP$_3$ (1); (taken from [354]).

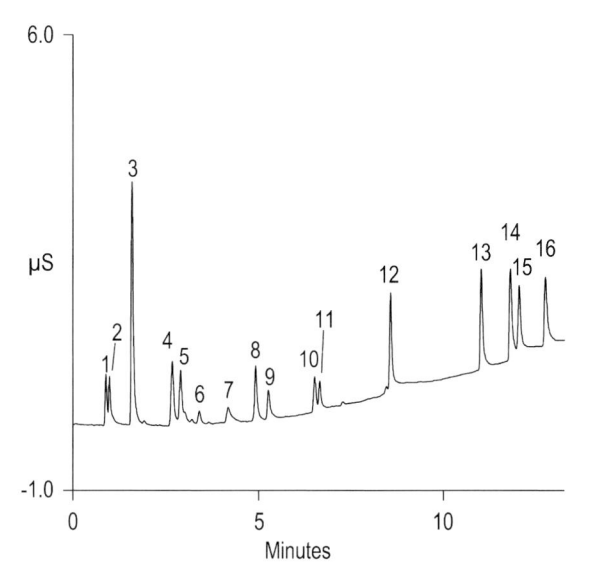

Fig. 9-214. Gradient elution of inositol phosphate isomers. — Separator column: IonPac AS11 (2-mm); eluant: NaOH; gradient: linear, 5 mmol/L to 10 mmol/L in 2 min, then to 90 mmol/L in 10 min; flow rate: 0.5 mL/min; detection: suppressed conductivity; peaks: (1) and (2) unknown, (3) chloride, (4) inositol-2-phosphate, (5) inositol-1-phosphate, (6) unknown, (7) carbonate, (8) sulfate, (9) unknown, (10) inositol-1,4-diphosphate, (11) orthophosphate, (12) inositol-4,5-diphosphate, (13) inositol-1,5,6-triphosphate, (14) inositol-1,2,5,6-tetraphosphate, (15) inositol-1,4,5-triphosphate, and (16) inositol-1,3,4,5-tetraphosphate.

Vogt et al. [356] used the same stationary phase and a very similar gradient program for analyzing metabolites of glycolysis, glycogenolysis, and the Krebs cycle, which was not possible with the HPLC methods published in literature. Sugar phosphates as intermediate products of the carbohydrate metabolism, as well as the nucleotide phosphates AMP, ADP, and ATP involved in energy transmission, can be analyzed in the same run. Utilizing conductivity detection in combination with a membrane-based suppressor system, Voigt et al. obtained minimum detection limits in the pmol range, so this method can be applied to very small sample amounts. Despite the high separation power of the IonPac AS11, Voigt et al. observed a few co-elutions, for example between fructose-6-phosphate and fumaric acid. If the conductivity cell effluent is passed through a UV detector cell (206 nm), this problem can be solved because only fumaric acid absorbs at this wavelength. Thus, the fructose-6-phosphate content can be calculated by differential measurement.

Recently, integrated pulsed amperometry has also been applied for the simultaneous analysis of homocysteine and methionine in plasma [357]. Homocysteine has been established as an independent risk factor for cardiovascular disease [358]. More than any other amino thiol, it has been a subject of intensive analytical method development. A number of different HPLC-based techniques are presently in use for the analysis of plasma and urine samples. Two of them, for example, utilize fluorescence detection after derivatization with appropriate reagents [359, 360]. An application of cation exchange chromatography in combination with integrated pulsed amperometry was first reported in 1995 by Evrovski et al. [361]. It allowed the simultaneous analysis of other amino thiols, such as methionine together with homocysteine in reduced plasma; such analysis is increasingly important for the better understanding of metabolic pathways and related genetic issues. Evrovski et al. utilized a 250-mm standard bore PCX-500 column in combination with a 50-mm PCX-500 Guard column as stationary phases and a mixture of 0.1 mol/L perchloric acid, 0.15 mol/L sodium perchlorate, and 5% (v/v) acetonitrile as an eluant. This eluant composition was optimized for an ideal resolution of homocysteine from other plasma components eluting close to this compound. However, some plasma components exhibit very long retention under these conditions if allowed to enter the longer of the two multimode columns. This problem was solved by the application of a column switching technique, in such a way that only a narrow fraction eluting from the shorter of the two columns was allowed to continue through the main column. The sample components eluting immediately before and after homocysteine were sent directly to waste.

Another improvement implemented over the past five years deals with sample preparation. In plasma and blood, only 5-15% of the total homocysteine is free. The rest is bound to disulfide bridges, either to other thiols or to various proteins. A reduction step is thus necessary for a meaningful monitoring of the homocysteine concentration. By switching from sodium borohydride to tris(2-carboxyethyl)phosphine (TCEP), the disulfide reducing agent most widely used today [362], sample preparation was simplified.

The original cation exchange/IPAD procedure required two pumps with two separate eluant containers and a second column switching valve. Further improvements of the original procedure became possible with the introduction of a narrow-bore version of the OmniPac PCX-500 column which makes column switching for the analysis of homocysteine and methionine obsolete. Figure 9-215 shows a standard chromatogram of eight amino thiols on the narrow-bore version of this cation exchange column. The eluant composition is the same as in the original report [361], except that the concentration of perchloric acid was lowered from 100 mmol/L to 20 mmol/L. With the original eluant composition, oxidized thiols and cystathionine could not be measured because injections of these compounds never produced a peak in the usual time window of 0-30 min. Avdalovic et al. [357] calculated the retention times of cystathionine, cysteine, and oxidized glutathione at the original perchloric acid concentration of 100 mmol/L to be 92, 89, and 35 min, respectively. Another aspect of the separation depicted in Fig. 9-215 is the close co-elution of cysteinylglycine (CysGly) and glutathione (GSH) at low perchloric acid concentrations. For an optimum separation of the pair, the perchloric acid concentration must be approximately

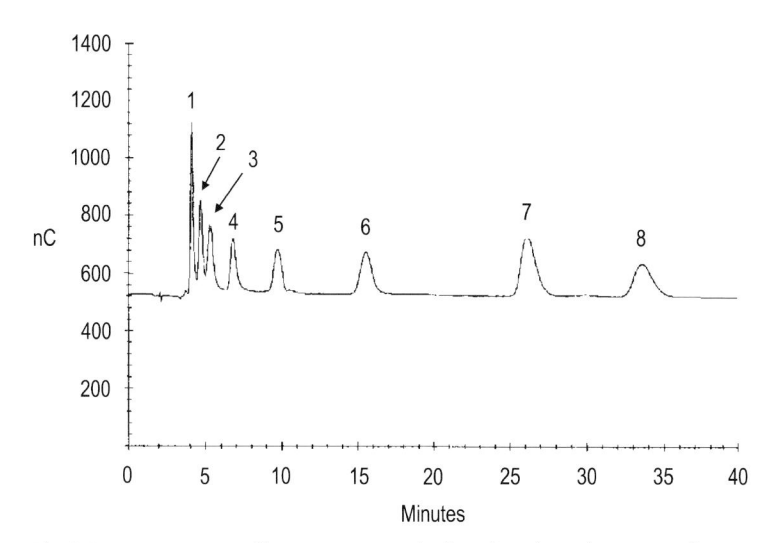

Fig. 9-215. Separation of homocysteine and other physiological amino acids. – Separator column: OmniPac PCX-500 (250 mm × 2 mm i.d.); column temperature: 30 °C; eluant: 0.15 mol/L NaClO$_4$ – acetonitrile (95:5 v/v) + 19.7 mmol/L HClO$_4$; flow rate: 0.25 mL/min; detection: integrated pulsed amperometry on a gold working electrode; Injection volume: 20 µL; solute concentrations: 3.8 µmol/L each of cysteine (1), cysteinylglycine (2), glutathione, red. (3), homocysteine (4), methionine (5), glutathione, ox. (6), cysteine (7), and cystathionine (8); (taken from [357]).

20 mmol/L. The resolution of the same peak pair is also affected by column temperature. At an elevated column temperature of 30 °C the resolution of the CysGly/GSH is only slightly affected, but the retention time of the late eluting peaks is decreased by 5-10 min relative to room temperature.

Finally, the determination of catecholamines and their metabolites shall be mentioned. Their separation is usually performed by ion-pair chromatography utilizing chemically modified ODS as stationary phase and gradient elution. The high sensitivity required for investigating physiological samples is only accomplished by means of DC amperometric detection. Catecholamines can be oxidized very easily on a glassy carbon electrode at a potential of +0.8 V. However, urine contains a number of electroactive compounds which are retained on a reversed-phase column and, therefore, interfere with the separation. For this reason, strong acid cation exchangers are preferred for the separation of catecholamines in urine [363, 364], because less interfering compounds are retained on a cation exchanger. Nevertheless, a clean-up of urine samples is indispensable. Of all the procedures in use, adsorption on aluminum oxide in combination with a cation exchanger is the most common one [365]. Figure 9-216 shows a standard chromatogram of the most important catecholamines and metabolites on a solvent-compatible OmniPac PCX-100 strong acid cation exchanger in the microbore format. All seven components can be separated to baseline in less than 15 minutes with a mixture of sodium perchlorate and perchloric acid in the presence of a small amount of acetonitrile. The sensitivity of this procedure can be enhanced 30-fold by increasing the injection volume to 100 µL and the oxidation potential to +0.85 V. In this way, the minimum detection limits are in the range of 0.6 pmol/mL.

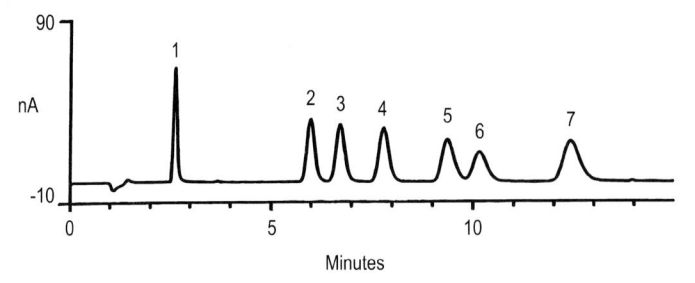

Fig. 9-216. Analysis of catecholamines and their metabolites on a strong acid cation exchanger. − Separator column: OmniPac PCX-100; eluant: 0.11 mol/L HClO$_4$ + 37.5 mmol/L NaClO$_4$ − acetonitrile (93.6:6.4 v/v); flow rate: 0.3 mL/min; detection: DC amperometry on a glassy carbon working electrode; oxidation potential: +0.6 V; injection volume: 30 µL; solute concentrations: 2 µmol each of 5-hydroxy-3-indolylacetic acid (HIAA) (1), norepinephrine (2), epinephrine (3), 3,4-dihydroxybenzylamine (DHBA) (4), normetanephrine (5), metanephrine (6), and dopamine (7).

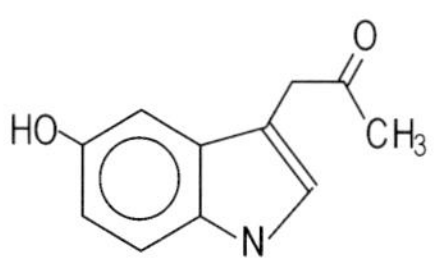

5-Hydroxyindolyl-3-acetic acid

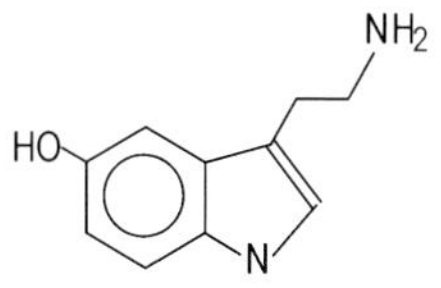

Serotonin

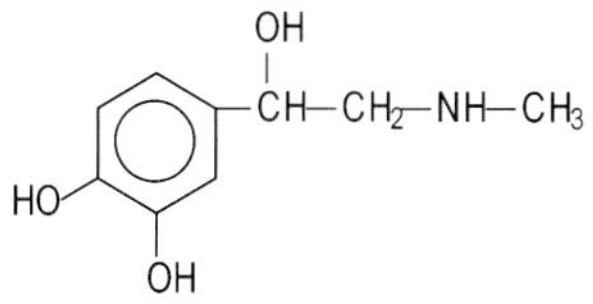

Epinephrine

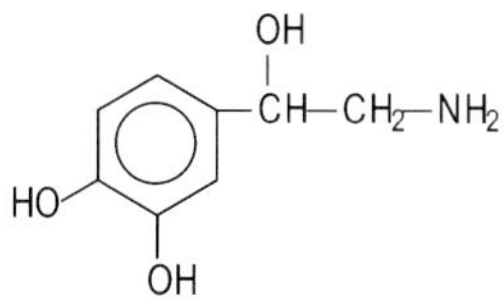

Norepinephrine

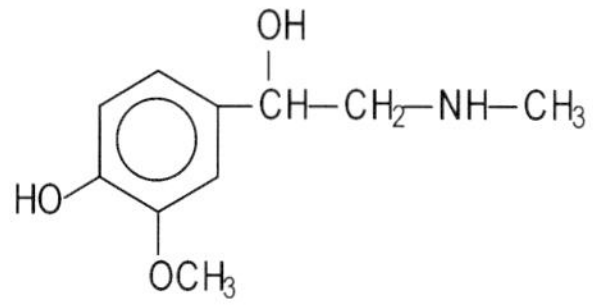

Metanephrine

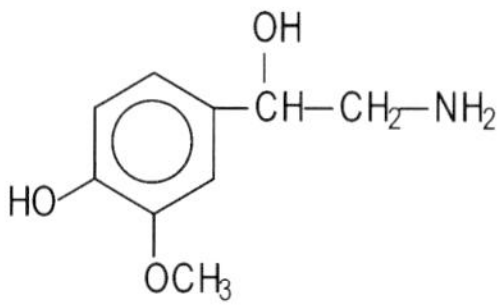

Normetanephrine

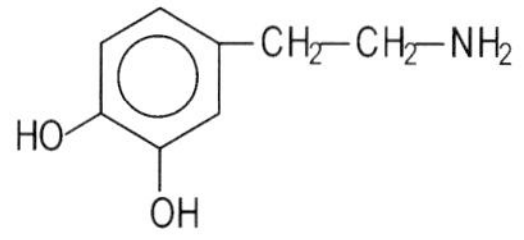

Dopamine

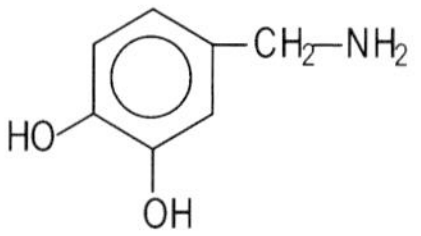

3,4-Dihydroxybenzylamine

Because typical catecholamine concentrations in plasma are between 1 and 10 pmol/mL, sample pre-concentration is recommended. For this purpose, 50 mL of urine are mixed with 10 µL DHBA (c = 1 mmol/L). To 5 mL of this mixture 15 mL EDTA solution (c = 2.7 mmol/L) are added and the pH is adjusted to 6.5. The sample thus prepared is passed through an extraction cartridge containing a cation exchanger in the H^+ form (Biorex 70, 200 mesh). After rinsing the cartridge twice with 8 mL of water, the catecholamines are eluted with 0.65 mol/L boric acid. This eluate can be injected directly into the chromatograph. A respective chromatogram is shown in Fig. 9-217.

9.9 Oligosaccharide Analysis of Membrane Coupled Glycoproteins

In this section, the determination of glyco moieties of a special class of glycoproteins, the class G immunoglobulins (IgG), is described. Immunoglobulins G consist of two heavy and two light polypeptide chains and carry their carbohydrates mainly in the conservative region of the heavy chains (F_C). Additionally,

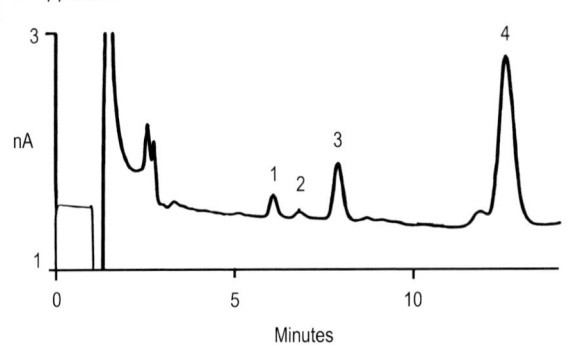

Fig. 9-217. Analysis of catecholamines and their metabolites in a urine sample. – Separator column: OmniPac PCX-100; eluant: 0.75 mmol/L HClO$_4$ + 25 mmol/L NaClO$_4$ – acetonitrile (93.6:6.4 v/v); flow rate: 0.3 mL/min; detection: DC amperometry on a glassy carbon working electrode; oxidation potential: +0.85 V; injection volume: 100 µL; peaks: norepinephrine (1), epinephrine (2), 20 nmol 3,4-dihydroxybenzylamine (DHBA) (3), and dopamine (4).

oligosaccharide structures are in the variable region of the light (V$_L$) and heavy chains (V$_H$). The schematic structure of the proteins and the positions of the carbohydrate binding sites were already shown in Fig. 3-213 (Section 3.11). A major characteristic of the IgG carbohydrates is their high heterogeneity with relatively small amounts of N-acetylneuraminic acids [366]. They seem to influence not only the higher half time of the proteins in blood but also their immune reactivity [366-368]. Because immunoglobulins are often used as monoclonal antibodies (Mab) for therapeutic purposes, the analysis of their carbohydrate structures becomes more and more important.

The experimental strategy for analyzing glycans of IgG's and Mab's is sketched in Fig. 9-218:

- Separation and purification of the protein via SDS-PAGE and electro-transfer of immunoglobulins onto an inert PVDF membrane
- Determination of monosaccharides after acid hydrolysis
- Determination of oligosaccharide moieties after N-glycopeptidase F digest
- Determination of the structures via sequential enzymatic degradation

Separation and Purification of IgG via SDS-PAGE (SDS Polyacrylamide Gel Electrophoresis) and Electro-Transfer of Immunoglobulins onto an Inert PVDF Membrane[2].

About 50 µg of a glycoprotein are mixed with the same volume of a double concentrated SDS buffer and denatured at 100 °C. The samples are applied onto the prepared SDS polyacrylamide gel, concentrated electrophoretically in the collecting gel, and separated to purity in the separation gel. After this step, the separation gel is put onto a previously activated PVDF membrane (Immobilon

---

2) PVDF is supposed to be inert and does not bind carbohydrates [366].

9.9 Oligosaccharide Analysis of Membrane Coupled Glycoproteins

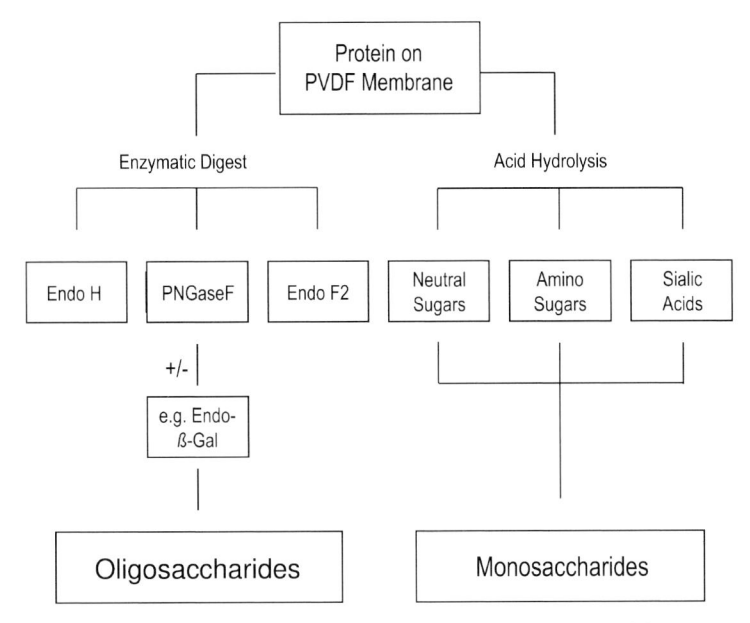

```
                    Protein on
                    PVDF Membrane
         ┌──────────────┴──────────────┐
   Enzymatic Digest                Acid Hydrolysis
   ┌─────┬──────┬──────┐       ┌────────┬────────┬────────┐
 Endo H PNGaseF Endo F2      Neutral   Amino    Sialic
                             Sugars   Sugars    Acids
          +/-|
       e.g. Endo-
         ß-Gal
          |
   Oligosaccharides                Monosaccharides
```

Fig. 9-218. Experimental strategy for analyzing carbohydrates derived from membrane coupled glycoproteins.

PSQ, 0.1 µm pore size, Millipore, USA) and enclosed in a Western Blot cassette. After electro-transfer, the protein on the membrane is stained with a 0.1% Coomassie Brilliant Blue solution. The membrane is washed several times to remove residues from the SDS gel electrophoresis. The completeness of the transfer can be checked by staining the gel in the Coomassie Brilliant Blue solution. A detailed description of this procedure can be found in [369, 370].

Determination of Monosaccharides after Acid Hydrolysis.
For investigating monosaccharides, the carbohydrate moiety of the protein is hydrolyzed with acid, which results in a quantitative cleavage of the monosaccharides. For this purpose, the stained protein is cut out of the membrane, the membrane is activated with methanol, 400 µL of trifluoroacetic acid ($c = 3$ mol/L) or hydrochloric acid ($c = 6$ mol/L) is added, and the sample is incubated for about 4 h at 100°C. The cleavage of N-acetylneuraminic acid and N-glycolylneuraminic acid requires milder hydrolysis conditions (HCl, $c = 0.01$ mol/L, 80°C, 60 min). After hydrolysis, the membranes are removed. In order to remove the acids, the supernatant is dried and dissolved again in an amount of water suitable for HPAEC-PAD. Figure 9-219 shows the analysis of monosaccharides of a membrane-immobilized Mab, which predominantly contains fucose, N-acetylglucosamine, galactose, and mannose. The monosaccharide ratios in IgG monoclonal antibodies are summarized in Table 9-16; the different signals in amperometric detection are taken into account. The values are related to the predicted

number of three mannose residues of the N-connected oligosaccharide core. Odd values can be attributed to oligosaccharides of different composition being present at one or several glycosylation sites of the protein. The results indicate a bi-antennary structure of the complex type with the sequence $FucGlcNAc_2$-Man_3Gal and a ratio of less than 2:3 between galactose and mannose. Thus, the bi-antennary oligosaccharide is not completely galactosylated. Because galactosamine has not been detected in this analysis, IgG's do not seem to contain O-glycosidic bonded carbohydrates.

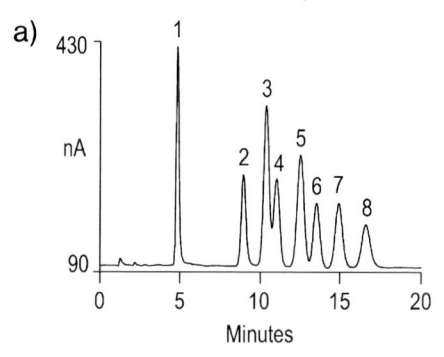

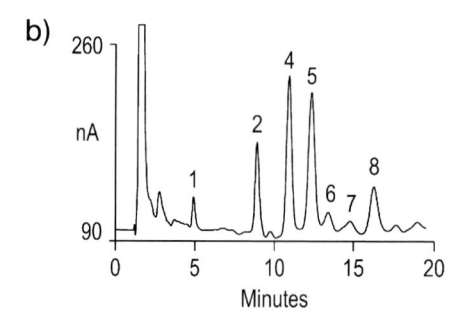

Fig. 9-219. Separation of monosaccharides of a standard (a) and a monoclonal antibody (b). — Separator column: CarboPac PA1; eluant: 16 mmol/L NaOH; flow rate: 1 mL/min; detection: pulsed amperometry on a gold working electrode; peaks: fucose (1), rhamnose as internal standard (2), galactosamine (3), mannosamine as internal standard (4), glucosamine (5), galactose (6), glucose (7), and mannose (8), (taken from [369]).

Table 9-16. Monosaccharide ratios of IgG antibodies (taken from [369]).

IgG	GlcN	Monosaccharide ratio			
		Man*	Gal	Fuc	GalN
Polyclonal human IgG	4.30	3.00	1.92	1.08	0.00
Monoclonal MAK MY9-6	3.22	3.00	0.85	0.89	0.00
Monoclonal MAK M115	3.35	3.00	1.30	1.05	0.00

*) Values related to Man = 3.00.
(Samples were hydrolyzed in 2 mol/L TFA at 100 °C for 5 h.)

Determination of the Oligosaccharide Moieties after N-Glycopeptidase F Digest
Oligosaccharide mapping starts with a chemical or enzymatic cleavage of the oligosaccharides from the glycoprotein. In order to cleave N-connected oligosaccharides, the stained glycoprotein is cut out of the dried PVDF membrane, pre-

treated and dipped in 200 μL of a sodium phosphate buffer (pH 7.6). The enzymatic digest is carried out in the presence of reduced 0.1% Triton-X-100 after adding 1 to 10 units of N-glycopeptidase F at 37 °C over 48 h. The cleavage with enzymes of different specificity is carried out in a similar way. Carbohydrate analysis is performed eventually with the HPAEC-PAD method. In many cases, a very simple "glyco fingerprint" of a protein is thus obtained. Identification of carbohydrates is performed via comparison of the chromatographic behaviour with that of known oligosaccharide standards. Such a procedure is outlined below, taking a polyclonal antibody (Poly) of the immunoglobulin class G as an example.

After their cleavage with N-glycopeptidase F, the glycans are separated with HPAEC-PAD. The profiles of the investigated IgG glycans show a remarkable similarity in the form of three prominent peaks eluting in the retention range of neutral oligosaccharides (Fig. 9-220, chromatogram A). Oligosaccharide 1 has a retention time corresponding to a fucosylated, bi-antennary carbohydrate that does not contain any galactose residues (see Table 9-17, structure 1). Oligosaccharide 3 elutes together with a bi-antennary oligosaccharide containing fucose as well as two galactose residues (see Table 9-17, structure 3). Oligosaccharide 2 elutes between 1 and 3 and thus could possess a monogalactosylated, bi-antennary structure, which again is fucosylated at the core (see Table 9-17, structure 2). Chromatogram A in Fig. 9-220 shows another peak which, according to its retention time, can be attributed with high probability to a monosialylated structure (see Table 9-17, structure 4). These structures have already been identified in other IgG molecules [368, 371].

An indication of the presence of a complex bi-antennary structure is obtained by treating proteins with endoglycanases such as EndoF2 or EndoH. While EndoF2 predominantly cleaves bi-antennary structures between the two GlcNAc residues of the core, EndoH prefers structures of the mannose- or hybrid type [372]. In the case of IgG's, only the digest with EndoF2 shows cleavage products, so that it can be assumed that the oligosaccharides have bi-antennary structures of the complex type (Fig. 9-220, chromatograms B and C).

Structural Elucidation via Sequential Enzymatic Degradation
A sequential exoglycosidase treatment of immunoglobulins allows a more detailed identification of oligosaccharide structures [373]. For this purpose, immobilized proteins are treated with carbohydrate-specific enzymes such as neuraminidase, which specifically cleaves neuraminic acid residues at the terminal end of the oligosaccharide. Other enzymes – β-galactosidase, hexosaminidase, fucosidase, etc. – cleave terminal galactose, N-acetylglucosamine, and fucose from the protein, respectively. Thereafter, the resulting oligosaccharide shows a correspondingly different chromatographic behaviour. Again taking polyclonal IgG as an example, Fig. 9-221 shows the disappearance of peak 4 and the rise of peak 3 as a result of a neuraminidase digest.

Table 9-17. Suggested oligosaccharide structures of immunoglobulin G.

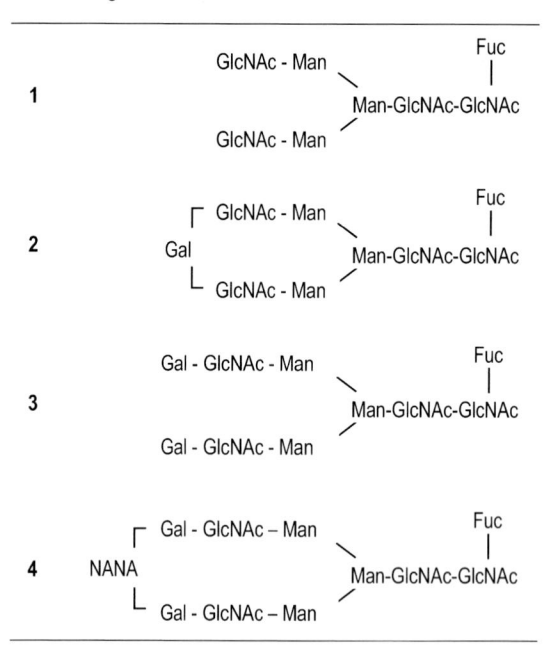

1	GlcNAc - Man \ / GlcNAc - Man	Fuc | Man-GlcNAc-GlcNAc
2	⌈ GlcNAc - Man \ Gal / ⌊ GlcNAc - Man	Fuc | Man-GlcNAc-GlcNAc
3	Gal - GlcNAc - Man \ / Gal - GlcNAc - Man	Fuc | Man-GlcNAc-GlcNAc
4	⌈ Gal - GlcNAc – Man \ NANA / ⌊ Gal - GlcNAc – Man	Fuc | Man-GlcNAc-GlcNAc

Next to it, a peak appears that co-elutes with *N*-acetylneuraminic acid. Thus, the monosialylated structure is found again in structure 3. Enzymatic cleavage with β-galactosidase should lead to a complete degalactosylation with the postulated structures **2** and **3**, so that peak **1** becomes the prominent peak in the HPAEC-PAD chromatogram. Figure 9-222 shows that the peaks **2** and **3** contain terminal galactose residues that results in a strong signal at about four minutes. Further investigations with hexosaminidases confirm the assumption that the investigated oligosaccharides are indeed the compounds listed in Table 9-17.

This sequential oligosaccharide mapping after enzymatic digest with subsequent anion exchange chromatography using pulsed amperometric detection is transferable to numerous other glycoproteins carrying oligosaccharides with known sequences. In the case of more complex oligosaccharide mixtures, other methods for structural elucidation that were described in Section 3.3.8 are used in addition.

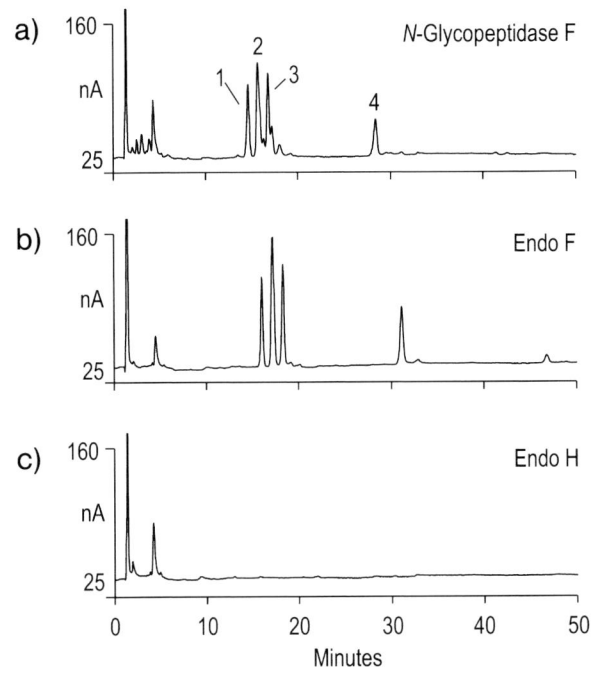

Fig. 9-220. Separation of oligosaccharides of a polyclonal antibody after enzymatic digest with N-glycopeptidase F (a), EndoF2 (b), and EndoH (c). – Separator column: CarboPac PA-100; eluant: (a) 0.1 mol/L NaOH, (b) 0.1 mol/L NaOH + 0.25 mol/L NaOAc; gradient: 0% B to 100% B in 110 min; flow rate: 1 mL/min; detection: pulsed amperometry on a gold working electrode; peaks 1 to 4 correspond to the structures 1 to 4 in Table 9-17; (taken from [369]).

9.10 Other Applications

In addition to the main application areas described in detail, ion chromatography is of significant importance in other fields, too. From the multitude of applications summarized in Table 9-18, only those with especially complex matrices will be dealt with here.

This includes the determination of anions, cations, and amines in samples from the *Petrochemical Industry*. At first sight, this might be surprising, but in the exploration and production of crude oil and natural gas aqueous samples also turn up. Oil and gas reservoirs are often associated with so-called formation waters that differ in composition depending on the geological formation. All formation waters have in common an extremely high chloride content between 10 and 70 g/L. In comparison to sea water, formation waters exhibit higher contents of bromide, iodide, and nitrate as well as lower contents of sulfate and

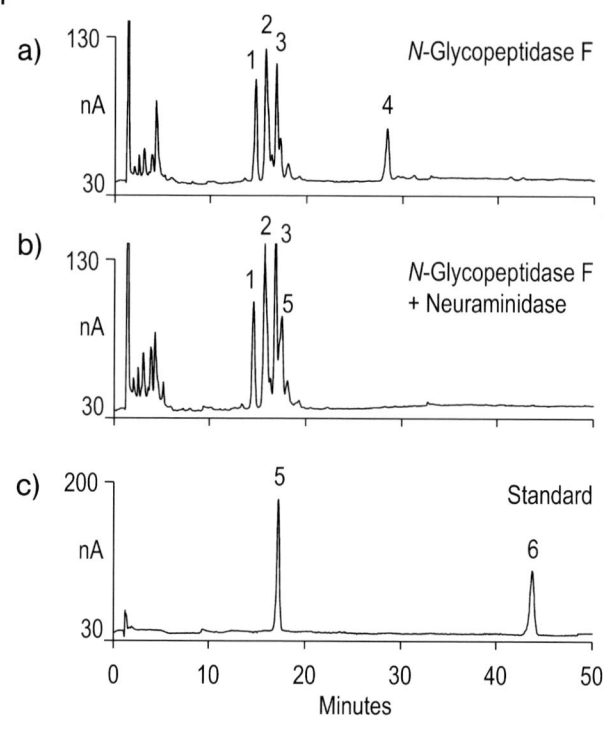

Fig. 9-221. Separation of oligosaccharides of a polyclonal antibody after enzymatic digest without (a) and with neuraminidase (b) as compared to a standard (c). — Chromatographic conditions: see Fig. 9-220; peaks: 1 to 4 correspond to the structures 1 to 4 in Table 9-17, N-acetylneuraminic acid (5) and N-glycolylneuraminic acid (6); (taken from [369]).

magnesium. Formation waters are characterized by the contents of bromide, sulfate, and iodide in addition to chloride; in some cases, nitrate, orthophosphate, fluoride, thiocyanate, and organic acids are also of interest. Application areas in the field of anion analysis include the analysis of aqueous samples that result from exploitation drillings, such as filtrates of drill sludge. Another important application is the analysis of water that was formed together with oil and gas, the composition of which allows conclusions about the original formation water. In addition, tracer studies are carried out, with which the influence of injection waters and drill sludge on the composition of the formation waters can be investigated. The high chloride excess in all those samples is no longer an insuperable problem for modern ion chromatography. Years ago, Kadnar et al. [374] published a method with which bromide, sulfate, and iodide could be determined in the presence of a high chloride excess. They chose an acrylate-based latexed anion exchanger as a stationary phase, from which mineral acids together with polarizable anions such as iodide and thiocyanate can be eluted with a sodium carbonate eluant (see Fig. 3-90 in Section 3.7.2). As an example, Fig. 9-223 shows

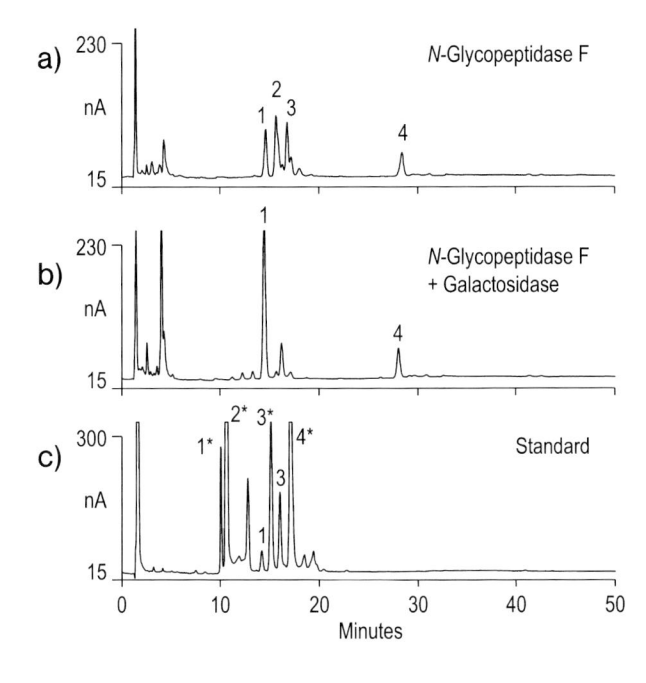

Fig. 9-222. Separation of oligosaccharides of a polyclonal antibody after enzymatic digest without (a) and with β-galactosidase (b). Chromatographic conditions: see Fig. 9-220; peaks **1** to **4** correspond to the structures 1 to 4 in Table 9-17, peaks: **1*** to **4*** correspond to the structures 1, 2, 4, and 8 in Table 3-28 in Section 3.3.8.2 (taken from [369]).

Table 9-18. Survey of other application areas and corresponding analytical examples.

Application area	Analytical examples
Petrochemistry Scrubber solutions Crude oil Process liquors Environmental toxins	Determination of sulfur compounds, cyanide, mineral acids, and transition metals
Pulp & Paper Industry Dressing Bleaching processes Paper manufacturing Wastewater	Determination of sulfur- and chlorine species, alkali metals, and ammonium
Mining and Metal Processing Potash mining Coal processing Phosphatation	Determination of inorganic anions, alkali- and alkaline-earth metals, transition metals, and metal complexes
Agriculture Fertilizers Soil extracts	Determination of inorganic anions, alkali- and alkaline-earth metals, transition metals, and ammonium

the anion chromatogram of a water sample from a crude oil production site. The sample was diluted 1:25 with de-ionized water and injected through a reversed-phase extraction cartridge in order to avoid contaminating the separator column with traces of hydrocarbons.

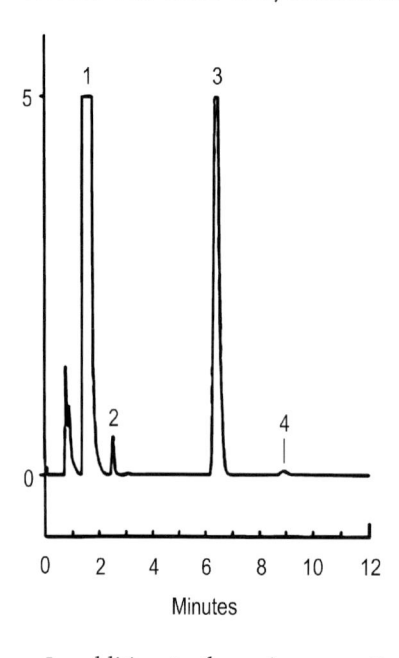

Fig. 9-223. Anion analysis of a water sample from a crude oil production site. – Separator column: IonPac AS9-SC; eluant: 3 mmol/L Na_2CO_3; flow rate: 2 mL/min; detection: suppressed conductivity; injection volume: 25 µL; sample: diluted 1:25 with de-ionized water; solute concentrations: 17650 mg/L chloride (1), 28.1 mg/L bromide (2), 590 mg/L sulfate (3), and 15.1 mg/L iodide (4); (taken from [374]).

In addition to the anions mentioned above, formation waters contain alkali- and alkaline-earth metals, of which sodium is in high excess. Until recently, cation analysis was carried out via flame AAS; higher contents of calcium and magnesium are usually determined via complexometric titration. With the introduction of weak acid cation exchangers for simultaneous analysis of mono- and di-valent cations, ion chromatography became a welcomed alternative to these conventional methods. However, in a more recent paper on the ion chromatographic analysis of alkali- and alkaline-earth metals in formation waters, Kadnar [375] mentioned problems with the determination of ammonium. Because the sodium content in formation waters can easily reach 50 g/L, samples have to be strongly diluted to avoid column overloading. After dilution, the ammonium content is usually less than 1 mg/L, so that the ammonium determination in the presence of a high sodium excess can only be performed with special separator columns such as the IonPac CS16. Alternatively, gradient elution can be applied when using a suppressor system to obtain higher resolution between the individual cations. Figure 9-224 illustrates this, taking a formation water as an example; the cationic components were separated on IonPac CS12A in the microbore format with a methanesulfonic acid gradient. Total analysis time is not much affected by this procedure. Again, the ammonium content in this sample is too low to be detected under these conditions.

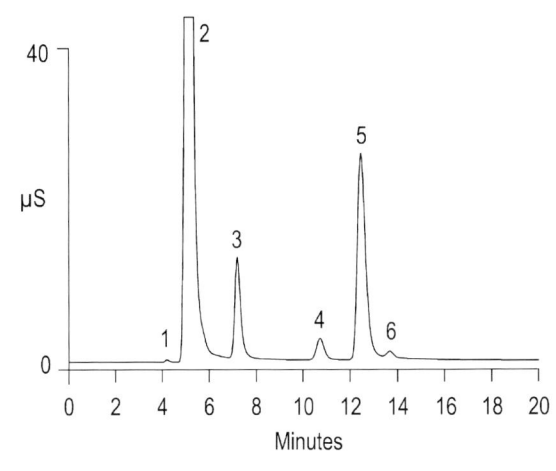

Fig. 9-224. Analysis of alkali- and alkaline-earth metals in a formation water. — Separator column: IonPac CS12A (2-mm); eluant: 18-36 mmol/L methanesulfonic acid; flow rate: 0.25 mL/min; detection: suppressed conductivity; injection volume: 2 µL; sample: diluted 1:100 with de-ionized water; solute concentrations: lithium (1), 49 g/L sodium (2), 4.4 g/L potassium (3), 710 mg/L magnesium (4), 6.2 g/L calcium (5), and 870 mg/L strontium (6).

Natural gas often contains high concentrations of H_2S and CO_2, which are called acid gases in petrochemistry. Because they are very corrosive, acid gases have to be removed from the natural gas before it is pumped into the pipelines. This is usually performed via a gas wash with an amine-containing solution. In most refineries, diethanolamine (DEA), monoethanolamine (MEA), or methyldiethanolamine (MDEA) is used for the absorption of acid gases. According to Fig. 9-225, those ethanolamines are regenerated in a continuous process by the stripping of H_2S and CO_2 in the course of the heat treatment. In a refinery, ethanolamines are cost-intensive chemicals, so that losses due to leakages or degradation reactions increase operating costs. In case of leakages, the amine used for the gas wash gets into the wastewater. Because different amines are often used in various production areas, plant managers can draw conclusions regarding leakages via the type of amine. Nowadays, excellent separations of the various ethanolamines are obtained on IonPac CS15 with a methanesulfonic acid gradient utilizing suppressed conductivity detection (see Fig. 9-226). Due to contamination and ethanolamine degradation, various anions are permanently enriched in the ethanolamine solution, thereby forming heat-stable salts. If the concentration of these salts exceeds 500 mg/L, a number of problems are observed with the gas wash, such as a decrease in absorption capacity of the ethanolamine for acid gases and an increase of corrosion. Thus, the content of iron sulfide increases which, in turn, leads to a foam formation. To diagnose these problems, the amine solution is analyzed at different sites. Kadnar and Rieder [376] used an IonPac AS10 high-capacity latexed anion exchanger with a NaOH eluant for this type of analysis. An example of a 1:10 diluted MDEA solution after heat treatment is illustrated in Fig. 9-227.

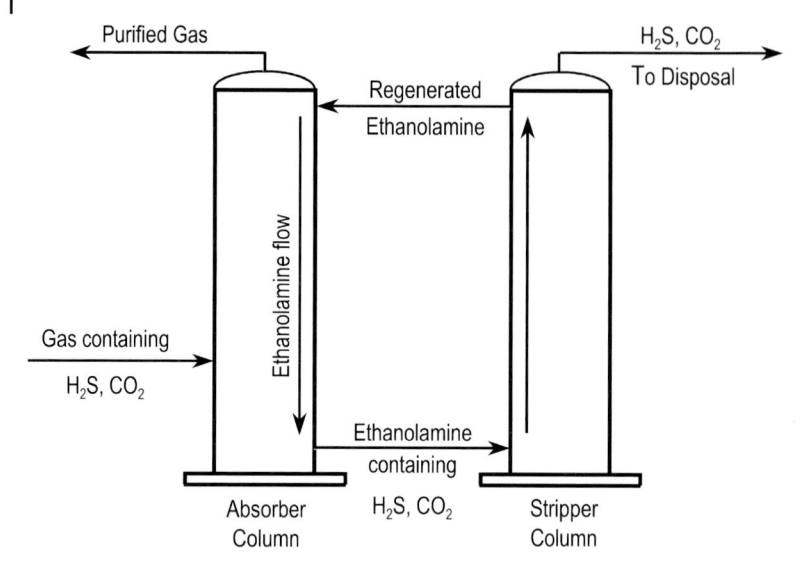

Fig. 9-225. Removal of H$_2$S and CO$_2$ from natural gas.

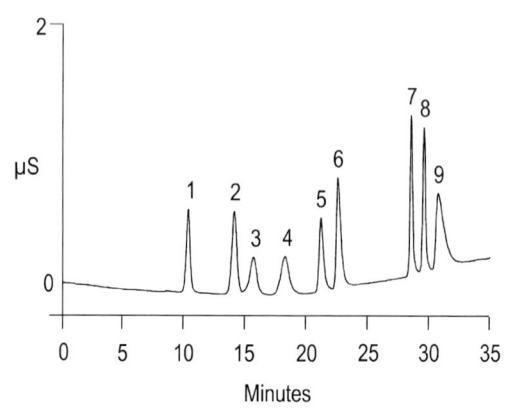

Fig. 9-226. Separation of ethanolamines in presence of alkali- and alkaline-earth metals. – Separator column: IonPac CS15 (2-mm); column temperature: 40 °C; eluant: 2 mmol/L methanesulfonic acid isocratic for 14 min, then to 27 mmol/L in 30 min; flow rate: 0.3 mL/min; detection: suppressed conductivity; injection volume: 2.5 µL; solute concentrations: 0.5 mg/L lithium (1), 2 mg/L sodium (2), 10 mg/L diethanolamine (3), 100 mg/L triethanolamine (4), 5 mg/L monoethanolamine (5), 2.5 mg/L ammonium (6), 1 mg/L magnesium (7), 1.5 mg/L calcium (8), and 5 mg/L potassium (9).

There are four sources for the anions contained in this solution. Chloride, nitrate, and fluoride usually originate from the production water, sulfate and sulfite are oxidation products of endogenous sulfur, thiosulfate is formed as a product of the reaction of H$_2$S with cyanide, and organic acids are products of the thermal degradation of MDEA. The extremely high retention time for thiosulfate of about 50 minutes is a disadvantage of the isocratic procedure shown in Fig. 9-227. However, this problem can be solved by applying a gradient technique on IonPac AS11 (see Fig. 9-228), with which thiosulfate elutes in less than 20 minutes.

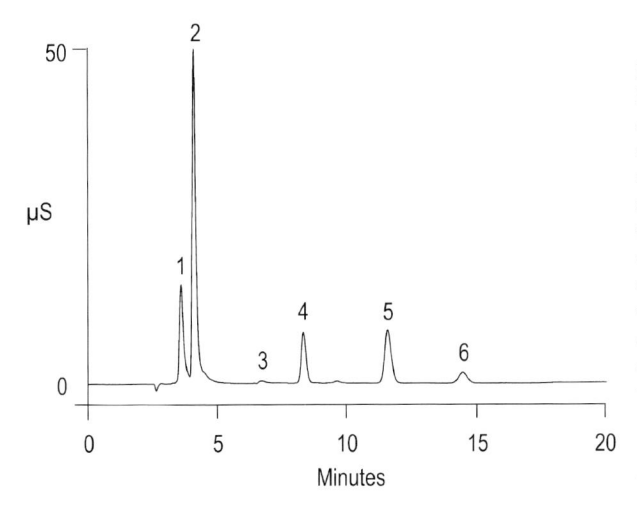

Fig. 9-227. Separation of heat-stable salts in a N-methyldiethanolamine solution. – Separator column: IonPac AS10; eluant: 0.1 mol/L NaOH; flow rate: 1 mL/min; detection: suppressed conductivity; injection volume: 25 µL; solute concentrations: 48 mg/L acetate (1), 54 mg/L formate (2), carbonate (3), 5.9 mg/L chloride (4), 12 mg/L sulfate (5), and 3.5 mg/L oxalate (6).

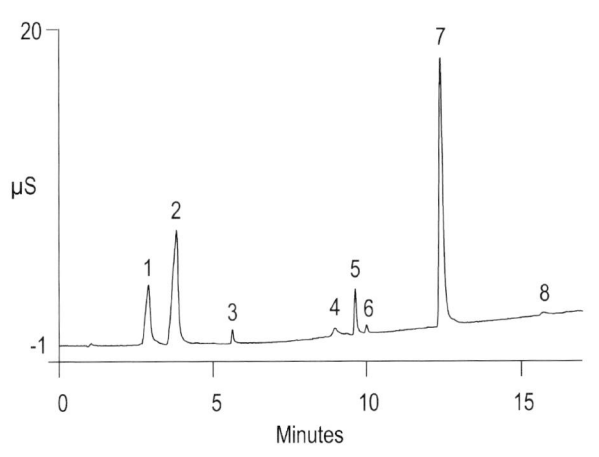

Fig. 9-228. Separation of heat-stable salts in a N-methyldiethanolamine solution applying a gradient elution technique. – Separator column: IonPac AS11; eluant: NaOH gradient; flow rate: 2 mL/min; detection: suppressed conductivity; injection volume: 25 µL; solute concentrations: 50 mg/L acetate (1), 55 mg/L formate (2), 6 mg/L chloride (3), carbonate (4), 12 mg/L sulfate (5), 4 mg/L oxalate (6), 110 mg/L thiosulfate (7), and 0.6 mg/L thiocyanate (8).

N-Methyldiethanolamine is sometimes used in combination with piperazine as an additive. Both amines are usually determined via titration in non-aqueous solution, but can also be eluted from an IonPac CS12A cation exchanger within

20 minutes using a 22 mmol/L sulfuric acid eluant and conductivity detection [377]. However, this does not work for water samples containing alkali- and/or alkaline-earth metals, as they elute within the first few minutes under isocratic conditions and, thus, are poorly resolved.

Toxic anions such as sulfide and cyanide are of high environmental relevance for the petrochemical industry. Process liquors from catalytic cracking plants, for example, sometimes contain very high concentrations of sulfide and cyanide. For ion chromatographic determination, the sample to be analyzed has to be digested with boiling sulfuric acid to release cyanide from metal complexes. The resulting gases, H_2S and HCN, are trapped in an alkaline absorber solution, which can be directly injected onto CarboPac PA1 (see Fig. 3-86 in Section 3.7.2). Detection is carried out via DC amperometry.

Process liquors from the *Pulp & Paper Industry* represent an extremely difficult matrix for ion chromatography. In the manufacturing of paper, wood is processed to small scraps which are then boiled in a Holländer vessel (trough with a rotating blade) under pressure in an alkaline solution. During this process, lignins get into the solution. The cellulose and hemi-cellulose moieties of the wood fibres are left behind; the remainder is subsequently refined with dyes, fillers, and glue. The pulp, which is bleached when required, is passed through a sieve to remove course constituents and then drained, pressed through rollers, and finally dried on a large number of partially heated cylinders. Pulping techniques include the so-called Kraft process [378], in which delignification of wood fibres is carried out by boiling in an aqueous solution of sodium hydroxide and sodium sulfide, which is termed White Liquor. After digestion, the White Liquor turns into the Black Liquor due to dissolved and decomposed wood constituents. After separating the fibres, the Black Liquor is transferred into a storage tank. Sodium sulfate is added to compensate for losses of inorganic elements during the digest. After thickening to a solid content of 60-70% and burning up the organic components in an oven, a melt of sodium carbonate and sodium sulfide is left over; it contains sodium sulfate, sodium sulfite, and sodium thiosulfate impurities. When dissolving this melt in water, the so-called Green Liquor is obtained, which is treated with a calcium hydroxide suspension to transform sodium carbonate back into sodium hydroxide. The White Liquor thus recovered can be fed back into the circuit.

Today, all these process liquors are analyzed by means of ion chromatography [379, 380]. Chloride, oxalate, and sulfur-containing anions are the most important, analytically. Chloride is primarily analyzed because of its corrosive effect, whereas the content of sulfur-containing anions allows conclusions on the efficiency of the process liquors during the pulping process and their subsequent recovery. The presence of oxalate can lead to the formation of salts of low solubility. Using ion chromatography, all these anions can be analyzed in the same run. The TAPPI (Technical Association of the **P**ulp and **P**aper **I**ndustry) method [381] still represents the industrial standard, although its total analysis time of about 35 minutes does not meet today's requirements. The long analysis time is caused by thiosulfate, which is strongly retained on an anion exchanger due to its polarizability. To lower analysis time, Utzman et al. [379] developed a column switching technique, in which thiosulfate only passes a 50-mm long guard column, while all the other anions have to pass the guard column and the analytical separator. Utzman et al. used an OmniPac PAX-100 with the corresponding guard column as a stationary phase; the order is reversed after 3.1 minutes using a column switching valve. As can be seen from the respective chromatogram of a dilute White Liquor sample in Fig. 9-229, thiosulfate already elutes between chloride and sulfite under these conditions, lowering the total analysis time to 10 minutes. However, p-cyanophenol was added to the carbonate/hydroxide eluant to diminish thiosulfate tailing, an eluant component which people nowadays try to avoid. Another disadvantage of this method is the co-elution of thiosulfate with nitrate and orthophosphate when switching the valve after 3.1 minutes. Therefore, gradient elution with a hydroxide eluant on a hydroxide-selective stationary phase such as IonPac AS16 is now favoured for solving this separation problem. It is very important that a 100% solvent-compatible stationary phase is used, so that humic acids and other strongly retained components can be rinsed off the separator column if necessary.

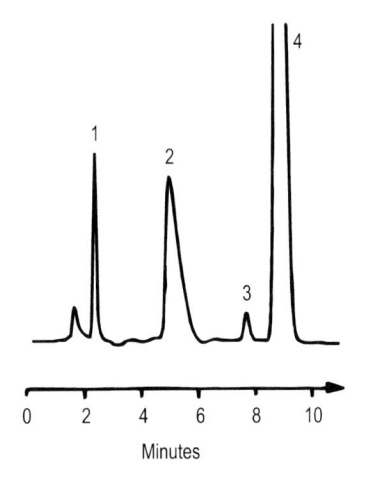

Fig. 9-229. Analysis of a White Liquor utilizing a column switching technique. − Separator column: OmniPac PAX-100 with guard column; eluant: 1.3 mmol/L Na_2CO_3 + 6 mmol/L NaOH + 1.58 mmol/L p-cyanophenol − methanol (95:5 v/v); switching time: 3.1 min after injection; flow rate: 1 mL/min; detection: suppressed conductivity; injection volume: 50 µL; sample: White Liquor, diluted 1:250 with de-ionized water; peaks: (1) chloride, (2) thiosulfate, (3) sulfite, and (4) sulfate; (taken from [379]).

In recent years, the bleaching process has increasingly been carried out with chlorine dioxide (ClO_2). Nevertheless, little is known about the respective side products in the wastewater such as chlorite, chlorate, and chloride. Also, there are very few confirmed findings about the stability of these species in the various effluents of the pulping process. Because the various chlorine species are involved in corrosion processes and are environmentally important, the monitoring of bleaching effluents is of great importance. However, the determination of chlorite in bleaching effluents is interfered by the high concentrations of acetic acid, monochloroacetic acid, and especially formic acid. This problem can be solved with an IonPac AS12A [382], a latexed anion exchanger specifically developed for separating oxyhalides, from which all anions of interest can be eluted within 30 minutes with a mixture of boric acid and sodium tetraborate [383]. A combination of a conductivity detector and a UV detector is recommended. Because the response factors of organic acids relative to chlorite are markedly lower in UV detection as compared with conductivity detection, even small amounts of chlorite in the presence of an excess of formate can be detected with a UV detector. ClO_2 is not dissociated in solution and, consequently, can only be analyzed with this method after reduction to chlorite. Hydrogen peroxide in alkaline solution is a suitable reducing agent:

$$2ClO_2 + H_2O_2 + 2OH^- \rightarrow 2H_2O + O_2 + 2ClO_2^- \tag{325}$$

However, standards and samples cannot be prepared in eluant or diluted with eluant, as chlorite is complexed by tetraborate. Because chlorine dioxide is also complexed by perborate, it is recommended that standards and samples be prepared with water containing 0.1 mmol/L NaOH and 12 mmol/L H_2O_2. As an example, Fig. 9-230 shows the anion chromatogram of a bleaching effluent, which was diluted 1:25 with de-ionized water after sparging with helium.

Because large amounts of wastewater are produced during the bleaching process followed by an alkaline extraction, alternative bleaching techniques are wanted, with which the formation of chlorinated organic compounds can be eliminated. A very promising candidate is ozone, whose bleaching properties are well known. While lignins are immediately degraded by ozone, changes are observed with cellulose that negatively affect the wood extraction process. This problem is still not solved and, thus, represents one of the big challenges for the pulp and paper industry.

Fundamental investigations of the ozonation of cellulose were performed by Nifterik et al. [384] who separated and identified reaction products utilizing the HPAEC-PAD method. Previous experiments revealed that gluconic acid and D-glucuronic acid are ozonation products of D-glucose. Because D-glucose is the main ozonation product of cellobiose, the above mentioned organic acids are formed. In addition, formaldehyde is cleaved, which can be eluted together with carbohydrates from the same stationary phase and detected very sensitively via pulsed amperometry. This is due to the fact that in alkaline solution, formaldehyde exists as *gem*-diol bearing two hydroxide groups. The degradation of cellulose by ozone is temperature-dependent and leads mainly to the formation of cellutriose. At a temperature of 0 °C, degradation is least intensive; this can be attributed to the low decomposition rate of ozone and, consequently, to a reduced attack by hydroxide radicals. Among the monosaccharides the yield of D-xylose is larger than that of D-glucose, which presently cannot be explained.

Moreover, the HPAEC-PAD method is used for analyzing carbohydrates in wood, in pulp, and in the various process liquors [385]. One of the major goals in the pulping process is to minimize polysaccharide degradation. For this pur-

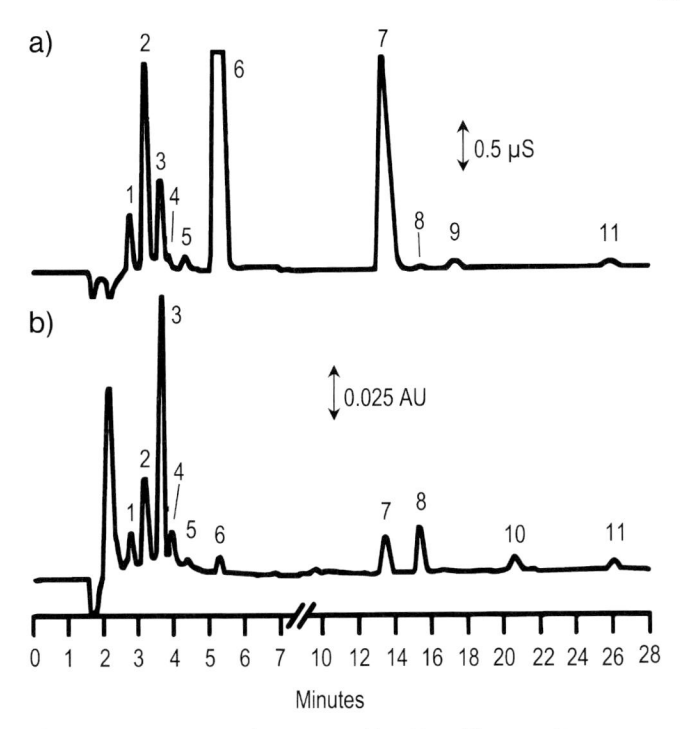

Fig. 9-230. Separation of anions in a bleaching effluent. — Separator column: IonPac AS12A; eluant: 25 mmol/L H_3BO_3 + 25 mmol/L $Na_2B_4O_7$ (pH 8.7); flow rate: 1 mL/min; detection: (a) suppressed conductivity, (b) UV (205 nm); injection volume: 50 µL; peaks: (1) acetate, (2) formate, (3) 21 mg/L chlorite, (4) monochloroacetate, (5) carbonate, (6) 253 mg/L chloride, (7) 205 mg/L chlorate, (8) nitrate, (9) 25 mg/L sulfate, (10) trichloroacetate, and (11) oxalate; (taken from [383]).

pose, a reliable and sensitive method for determining wood sugars in complex matrices is required. Using such a method, influential parameters in the pulping and bleaching processes can be investigated. While acid wood hydrolysates and pulping solutions are relatively free of contaminations, process liquors contain high concentrations of lignins, carboxylic acids, sulfonic acids, and humic acids, which can be irreversibly adsorbed on an anion exchanger and thus damage the stationary phase. For this reason, no other publication on carbohydrate analysis in process liquors from the paper industry is found apart from the work of Sullivan et al. To take care of the separator column Sullivan et al. injected the diluted process liquors through extraction cartridges containing a strong acid anion exchanger and a polyvinylpyrrolidone resin. A large fraction of the above-mentioned compounds is retained on these materials, which increases column lifetime significantly. As an example, Fig. 9-231 shows the determination of monosaccharides in a cellulose hydrolysate. To obtain optimal resolution between the three monosaccharides the hydroxide eluant concentration has to be lowered to 1 mmol/L. Reproducible retention times can only be obtained when using an Eluant Generator™ or when rinsing the separator column post-chromatographically with a high hydroxide concentration. Post-column addition of a NaOH concentrate to enhance detector sensitivity is no longer necessary due to an optimized pulse sequence.

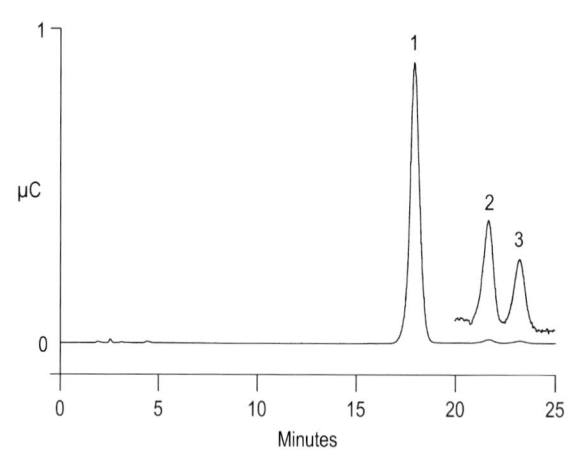

Fig. 9-231. Determination of monosaccharides in a cellulose hydrolysate. – Separator column: CarboPac PA1; eluant: 1 mmol/L KOH (EG40); flow rate: 1 mL/min; detection: pulsed amperometry on a gold working electrode; peaks: (1) glucose, (2) xylose, and (3) mannose.

Reactive dyes, in which chloride and sulfate have to be determined, represent a difficult matrix for ion chromatography in the *Chemical Industry*. At first sight, this task does not seem to be that difficult. However, if aqueous solutions of such dyes are injected without any sample preparation, the peak areas or peak heights of chloride and sulfate decrease continuously with the number of injections. This phenomenon is not completely understood, because reactive dyes are anionic compounds, too, and will eventually be eluted from the separator column. However, it might have something to do with their high affinity towards the stationary phase due to their aromatic character and the fact that they are blocking the ion-exchange groups. In any case, reproducible results are only obtained when the dyes are removed from the samples to be analyzed. Again, OnGuard-P solid-phase extraction cartridges are used for this purpose. They contain a polyvinylpyrrolidone resin on which interfering compounds are retained. Thus, the sample to be analyzed is injected through such a cartridge. It is important to flush these cartridges with de-ionized water before use to avoid chloride and sulfate contaminations. Furthermore, it has to be noted that the sample should be pushed slowly through the cartridge to ensure complete extraction. If these limiting conditions are observed, the chromatogram shown in Fig. 9-232 is obtained for the dye Royal Blue. An aqueous 0.01 % solution was investigated, in which chloride, sulfate, and traces of nitrate and orthophosphate could be detected.

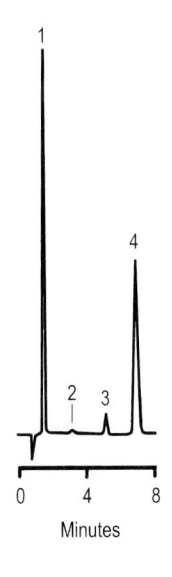

Fig. 9-232. Determination of chloride and sulfate in a reactive dye. — Separator column: IonPac AS4A; eluent: 1.7 mmol/L NaHCO$_3$ + 1.8 mmol/L Na$_2$CO$_3$; flow rate: 2 mL/min; detection: suppressed conductivity; injection: 50 µL of a 0.01 % solution of Royal Blue; peaks: (1) chloride, (2) nitrate, (3) orthophosphate, and (4) sulfate.

One of the most difficult applications in the chemical industry is the analysis of ionic impurities in all kinds of chemicals, whose absolute purity is of fundamental importance in some industries — above all in the semiconductor industry. The degree of difficulty to detect and quantitate ionic impurities in chemicals increases exponentially with the required degree of purity. The separation of fluoride and sulfate in raw phosphoric acid [386] containing these anions in per cent amounts is somewhat simple. Transition metal analysis in this matrix is also possible without any problem. Although in orthophosphoric acid, iron exists predominantly in the oxidation state +2, traces of iron(III) can be transferred to the reduced form by adding 1 mmol/L ascorbic acid to the mobile phase. In this way, a determination of total iron can also be achieved by ion chromatography. In contrast, the analysis of ionic contaminations in ultra-pure chemicals represents one of the big challenges for modern ion chromatography. Attempts at determining ionic impurities in orthophosphoric acid, hydrofluoric acid, and nitric acid are described in Section 9.3.

An interesting application in the field of *Mineralogy* is the determination of ammonium in cement [387]. Ammonium ions are a natural admixture in raw products of the cement industry. In the finished product they should not exceed a certain amount, because cement is used, among other things, to line drinking water pipes. For the ion chromatographic analysis of ammonium ions, 0.1 g to 0.5 g cement is extracted with 100 mL hydrochloric acid ($c = 0.1$ mol/L) for about 10 minutes by shaking and stirring. After a short sedimentation, the extract can be injected directly through a membrane filter (0.22 µm). Sample preparation is not a problem in this case, but the extremely high calcium content of such samples is. To obtain a satisfactory separation of ammonium ions from the alkali- and alkaline-earth metals, a high-performance latexed cation exchanger is used as a stationary phase. Depending on the number of samples to be analyzed, it must be flushed every working day with concentrated hydrochloric acid to remove the more strongly retained alkaline-earth metals from the column, which increasingly occupy ion-exchange groups and thus, lower the ion-exchange capacity. The chromatogram of such a cement extract with an ammonium content of 193 mg/kg shown in Fig. 9-233 demonstrates that even small amounts of ammonium can be detected in cement.

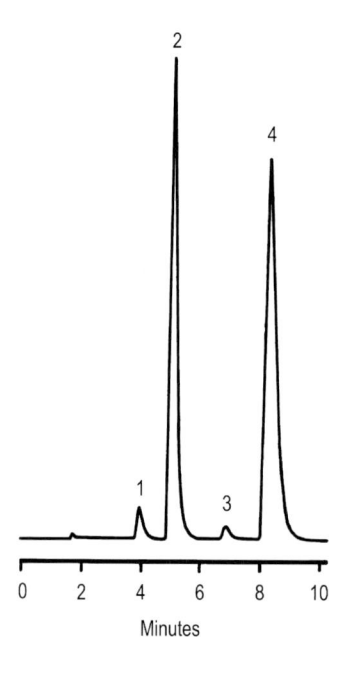

Fig. 9-233. Analysis of ammonium in cement. – Separator column: IonPac CS3; eluant: 30 mmol/L HCl; flow rate: 1 mL/min; detection: suppressed conductivity; injection: 50 µL of a HCl extract; peaks: (1) lithium, (2) sodium, (3) ammonium, and (4) potassium; (taken from [387]).

For the ion chromatographic analysis of chloride in cement, which cancels out the passivation of steel surfaces in concrete and, thus, is only admitted up to a maximum content of 0.1%, Maurer et al. [388] developed a procedure in which the cement sample is extracted with nitric acid. Because a direct chloride determination is impossible due to the high nitrate concentration in the extract, a three-fold excess of silver nitrate is added to the nitric acid extract. The precipitated silver chloride is filtered off and subsequently dissolved in 100 mL of a 0.25% ammonium hydroxide solution. Depending on the chloride concentration, this solution can be further diluted with de-ionized water or injected directly. If methanesulfonic acid is used instead of nitric acid, the extract can be injected directly without any sample preparation when applying a gradient elution technique. A high-capacity, hydroxide-selective IonPac AS15 anion exchanger has proved to be a suitable stationary phase. The chromatogram in Fig. 9-234 reveals that a separation between the extraction agent and chloride is possible without any problem. Moreover, gradient elution with a simple linear hydroxide gradient offers the ability to detect mono-, di-, and tri-carboxylic acids together with mineral acids within 30 minutes. In the present case, an Eluant Generator™ was used for contaminant-free hydroxide preparation, which helps to avoid a baseline drift during the gradient run.

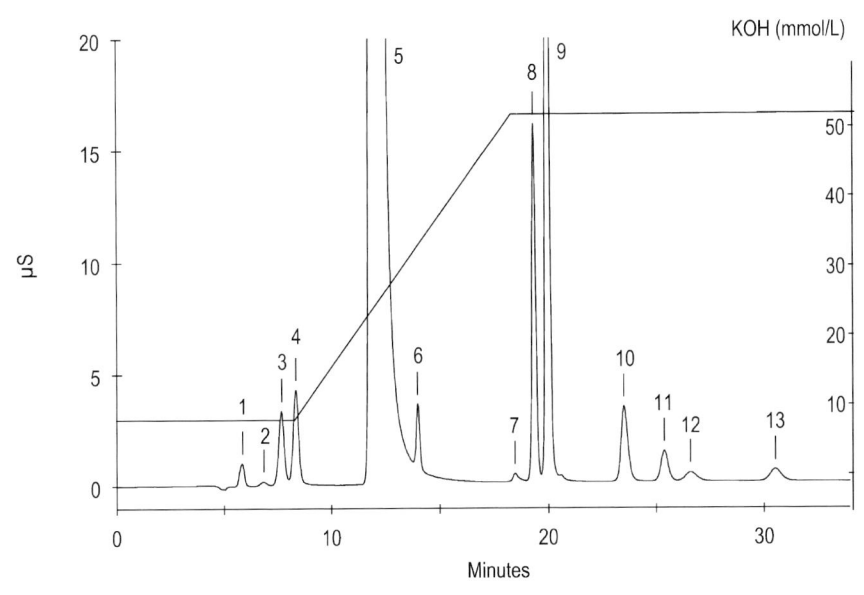

Fig. 9-234. Gradient elution of inorganic and organic anions in cement. – Separator column: IonPac AS15; eluant: KOH gradient as drawn; flow rate: 1.2 mL/min; detection: suppressed conductivity; injection volume: 25 μL; peaks: (1) unknown, (2) gluconate, (3) acetate, (4) formate, (5) methanesulfonate, (6) chloride, (7) carbonate, (8) tartrate, (9) sulfate, (10) nitrate, (11) orthophosphate, (12) unknown, and (13) citrate.

In the gypsum industry [389], ion chromatography complements classical chemical analysis. One of the main application areas is the investigation of easily soluble accompanying compounds in the gypsum raw materials. Of special interest are alkali- and alkaline-earth metals, ammonium ions, and simple inorganic anions. Another very important application is the analysis of technical-grade gypsums, especially flue-gas gypsums. These are formed in an aqueous reaction medium and, by their very nature, contain small amounts of all water-soluble accompanying compounds, which are present in the flue-gas or in the absorption medium. Essentially, these are the cations sodium, magnesium, and calcium, and the anions fluoride, chloride, and sulfate. Ammonium and nitrate ions may be present in the case of flue-gas denitrification. Upon application of ion chromatography, these accompanying compounds can be analyzed within a short time, thus providing a fast quality control method for the power plant.

Another challenge for ion chromatography is the determination of ionic impurities such as bromide, iodide, sulfate, magnesium, and calcium in brine. If the brine is used in electrolytic processes, it has to be extremely pure because, for example, alkaline-earth metals reduce the lifetime of the membranes via interactions with iodide on the membrane surface. The separation of non-polarizable anions such as bromide and sulfate in brine is carried out on a conventional anion exchanger. In the presence of extremely high chloride concentrations, the commonly employed conductivity detection is not suitable for determining bromide. In those cases, detection is performed via measuring the light absorption at 200 nm. For the sake of simplicity, both detection methods should be used simultaneously, as demonstrated in Fig. 9-235 with a brine sample containing about 140 g/L chloride. The separation of iodide is carried out on an anion exchanger with hydrophilic ion-exchange groups such as IonPac AS11 to minimize the adsorption of the polarizable iodide on the stationary phase. With a dilute nitric acid eluant, iodide can be eluted in less than five minutes, thus obtaining maximal sensitivity. Several application modes exist for amperometric detection of iodide. High sensitivity is obtained by using an iodized platinum working electrode. Conditioning of the electrode with saturated potassium iodide solution is necessary to reach the sensitivity required for investigating brines. Alternatively, a silver working electrode can be used. However, the high chloride excess requires the application of a pulse sequence, because baseline stability is severely affected at constant oxidation potential. An example chromatogram that illustrates amperometric iodide determination in ultra-pure brine is shown in Fig. 9-236. Because the formation of silver iodide as part of the oxidation process is reversible, a slight swing at the end of the iodide peak is observed, which can be attributed to the dissolution of silver iodide left at the electrode together with a reduction of silver. In pulsed amperometry this swing

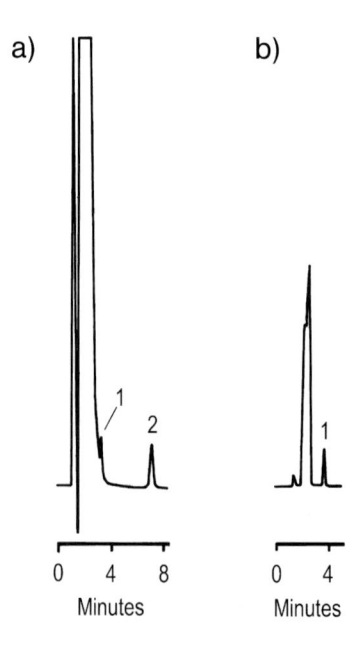

Fig. 9-235. Analysis of bromide and sulfate in brine. — Separator column: IonPac AS4A; eluant: 1.7 mmol/L NaHCO$_3$ + 1.8 mmol/L Na$_2$CO$_3$; flow rate: 1 mL/min; detection: (a) suppressed conductivity, (b) UV (200 nm); injection: 50 µL sample (diluted 1:100); sample: (1) bromide and (2) sulfate.

is less pronounced than in DC amperometry. Despite the high chloride excess iodide detection is not inhibited, even at trace level. However, to obtain high sensitivity, thorough online degassing of the mobile phase and temperature constancy of the measuring cell are of critical importance.

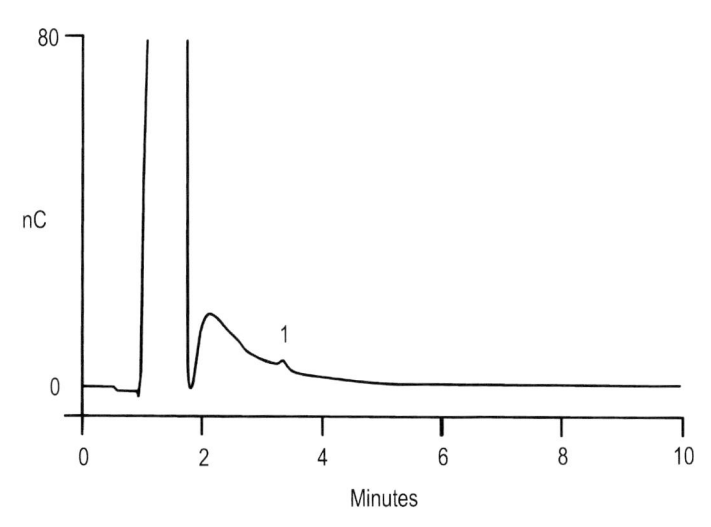

Fig. 9-236. Amperometric determination of iodide in brine (30%). – Separator column: IonPac AS11; eluant: 50 mmol/L HNO_3; flow rate: 1.5 mL/min; detection: DC amperometry at a silver working electrode; injection: 50 µL sample (diluted 1:10) with 16 µg/L iodide (1).

Trace analysis of alkaline-earth metals such as magnesium and calcium in ultra-pure brine is not possible without pre-concentration, because manufacturers of the membranes used for electrolytical processes specify the total concentration of alkaline-earth metals to be 50 µg/L at the most. The current ion chromatographic method is based on the work of Hildebrand [390]. According to his method, the alkaline-earth metals present in brine are selectively enriched on a MetPac column and subsequently separated on a weak acid cation exchanger. The required valve switching for this type of matrix elimination was discussed in detail in Section 9.3 (Fig. 9-77) and exemplified by the analysis of anions in organic solvents. In contrast, the carrier liquid is not pure de-ionized water but dilute hydrochloric acid (c = 1 mmol/L). The hydrochloric acid concentration is dimensioned in such a way that after the successful pre-concentration of the alkaline-earth metals, the high sodium excess is eluted off the column within 20 minutes without affecting the recovery of magnesium and calcium. To obtain optimal recovery rates for magnesium and calcium sample pH has to be adjusted to 11.8, because at this pH value the selectivity of the MetPac column for divalent cations is significantly higher than for monovalent cations. Separation of the pre-concentrated alkaline-earth metals on the analytical column

(IonPac CS12A) is carried out with 20 mmol/L methanesulfonic acid. As can be seen from the respective chromatogram of a 1:10 diluted ultra-pure brine sample (30%) in Fig. 9-237, significant amounts of sodium are eluted from the analytical column despite the intensive rinsing process. However, this does not interfere with the evaluation of the two alkaline-earth metal peaks. The minimum detection limits for alkaline-earth metals in this matrix are on the order of 5 µg/L.

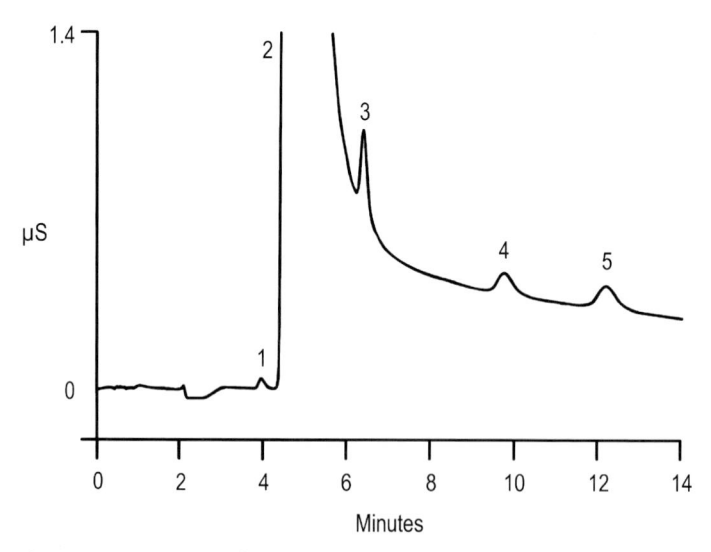

Fig. 9-237. Separation of magnesium and calcium in ultra-pure brine (30%). – Separator column: IonPac CS12A; eluant: 20 mmol/L methanesulfonic acid; flow rate: 1 mL/min; pre-concentrator column: MetPac CC-1; carrier and rinsing liquid: 1 mmol/L HCl; carrier flow rate: 2 mL/min; rinsing time: 20 min; detection: suppressed conductivity; injection volume: 100 µL; solute concentrations: lithium (1), sodium (2), potassium (3), 25 µg/L magnesium (4), and 25 µg/L calcium (5).

Highly concentrated salt solutions are also found as inclusions in minerals such as Halite (NaCl), Sylvite (KCl), and Carnallite ($KMgCl_3 \cdot 6H_2O$). The inclusions have diameters between 1 and >100 µm. According to Knipping at al. [391] the main components of these saturated solutions are the elements Na, K, Mg, and Ca as well as chloride and sulfate. But minor components such as lithium and bromide also have to be determined in order to characterize the composition of these solutions and to interpret the genesis of the liquid inclusions. Due to their small diameter, the extraction of the liquid (about 0.5 µL) from the inclusions has to be monitored with an optical stereo microscope. A needle or a micro-driller made of tungsten carbide is used for opening the cavity. The liquid can then be charged with a micro-burette made of borosilicate glass. To avoid partial evaporation of the liquid a small volume of oil is charged before and after the liquid. The charged liquid volume, which can be calculated based on the geometry of the liquid column inside the capillary, is diluted 1:104 with ultrapure water prior to analysis. Nevertheless, only small volumes are available for the ion chromatographic analysis, so the used ion chromatograph has to be equipped with short connective tubing to limit extra-column band broadening. The separations themselves are relatively simple and carried out on conventional anion- or cation exchangers. Again, a conductivity detector and a UV detector (200 nm) are used in series due to the large concentration difference between chloride and bromide. In combination with other geological and geochemical studies the investigations of Knippling et al. reveal whether and where the salt dome was subject to a metamorphosis in the geological past. Such findings are important in rating the long term stability of a salt dome for the final storage of radioactive waste.

The above-mentioned pre-concentration of transition metals on MetPac CC-1 was originally used by Kingston et al. [392] for the AAS analysis of transition metals in sea water. It is based on the selective concentration of transition metals on a macroporous iminodiacetic acid resin, where divalent cations are retained by chelation. Their affinity towards the stationary phase decreases in the order

$$Hg > Cu > Ni > Pb > Zn > Co > Cd > Fe > Mn > Ba > Ca > Sr > Mg \gg Na$$

After pre-concentration, the transition metals are eluted with acid and can then be analyzed either by atomic spectrometry (AAS, ICP-OES, ICP-MS) or chromatography. Using such a pre-concentration technique as a starting point, Riviello [393] developed an ion chromatographic method for trace analysis of transition metals such as iron, copper, nickel, and manganese in saline samples. The schematic setup of such a system is depicted in Fig. 9-238. In the first step, 5 mL of the 1:10 diluted solution, neutralized with nitric acid (suprapure) and adjusted to pH 5.2-5.3 with 2 mol/L ammonium acetate solution is pumped through a MetPac CC-1 concentrator column, which is connected to a 3-port valve instead of an injection loop. Anions and monovalent cations are not retained at this column and are discharged. The MetPac CC-1 column is then rinsed for 1.5 minutes with ammonium acetate solution (pH 5.5) to remove alkaline-earth metals. However, the metal ions thus concentrated cannot be directly passed onto the analytical separator column, as the eluant required for this (0.4 mol/L nitric acid) is not suitable to separate the metal ions on the analytical separator column. Therefore, they are first passed through a high-capacity cation exchanger TMC-1 connected to a 4-port valve, where they are retained as a narrow band. After the TMC-1 has been converted from the hydronium form to the ammonium form with a 0.1 mol/L ammonium nitrate solution (pH 3.5), the metal ions are passed onto the analytical separator column with the usual PDCA eluant (PDCA: pyridine-2,6-dicarboxylic acid) and are separated. In the meantime, the concentrator column can be prepared for the next sample by cleaning with 1.5 mol/L nitric acid for a short time and then conditioning with ammonium acetate solution.

822 | 9 Applications

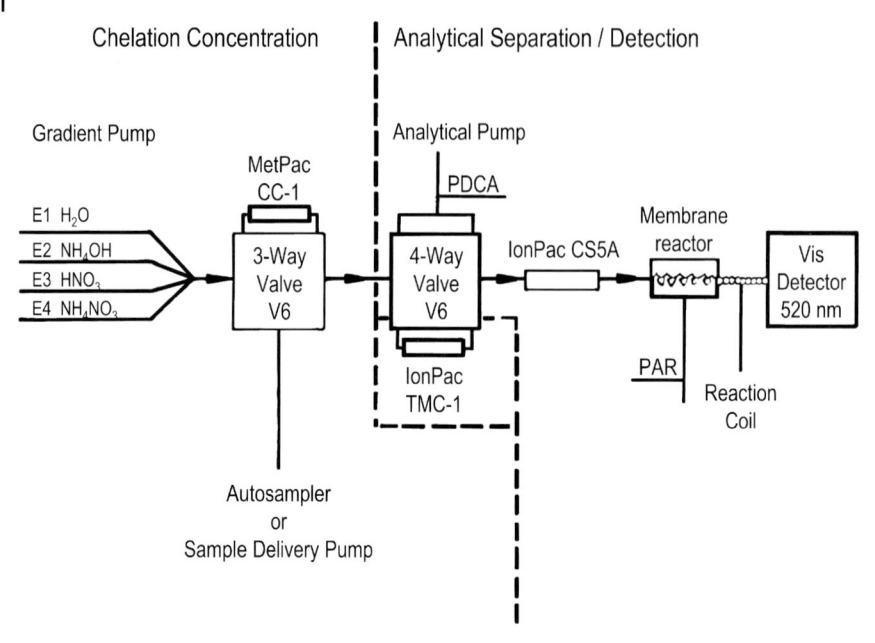

```
                    Chelation Concentration          Analytical Separation / Detection

       Gradient Pump                            Analytical Pump
                              MetPac
                              CC-1                   PDCA
       E1  H₂O                                                                    Membrane
       E2  NH₄OH                                                      IonPac CS5A  reactor       Vis
                              3-Way             4-Way                                            Detector
       E3  HNO₃                Valve             Valve                                           520 nm
       E4  NH₄NO₃               V6                V6
                                                                                          PAR
                                                IonPac                                         Reaction
                                                TMC-1                                            Coil

                            Autosampler
                                or
                         Sample Delivery Pump
```

Fig. 9-238. Schematic setup of an ion chromatographic system for the analysis of transition metals in saline samples.

Although this technique appears to be very complicated at first sight, reproducible results are obtained by automating rinsing processes with a gradient programmer. Figure 9-239 shows the application of this procedure to the analysis of 50 % sodium hydroxide [394], which was chromatographically impossible up to this point. In this figure, the chromatogram of a 1:10 diluted concentrate is compared to that of a blank sample. The level of the blank values for the analyte metal ions depends primarily on how carefully the chromatographic system is prepared. At the beginning, it should be flushed for some time with 0.2 mol/L oxalic acid to remove any metallic impurities from the system. This holds true for eluant containers, all capillary tubing and the two concentrator columns MetPac CC-1 and TMC-1. If all these requirements are met, only traces of zinc – as seen in Fig. 9-239 – are detected in the blank sample. A comparison of sample and standard reveals that the iron content of more than 100 µg/L is markedly higher than that of copper, nickel, and manganese, which are in the lowest µg/L range. The linearity of the method in the investigated concentration range of 0.5 mg/L to 2.5 mg/L for iron, and 0.025 mg/L to 0.2 mg/L for copper, nickel, and manganese is excellent. The recovery rate was determined by Adams [394] to be between 98 % and 106 %; the relative standard deviation with values between 2 % and 8 % is surprisingly low.

As mentioned above, the procedure introduced here for analyzing transition metals is not only applicable to the investigation of strong bases such as sodium hydroxide, potassium hydroxide, ammonium hydroxide, and quaternary ammonium bases, but to all those matrices in which traces of transition metals are

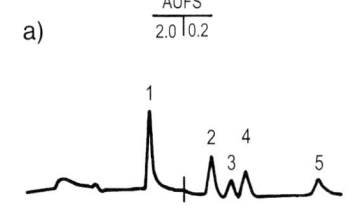

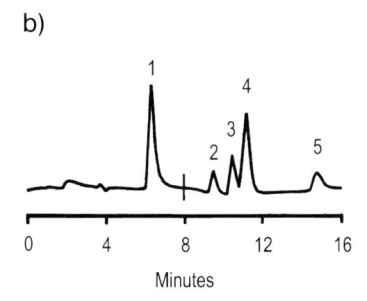

Fig. 9-239. Analysis of transition metals in 50% sodium hydroxide. — Separator column: IonPac CS5; eluant: 6 mmol/L pyridine-2,6-dicarboxylic acid + 10 mmol/L NaOH + 40 mmol/L NaOAc + 50 mmol/L HOAc; flow rate: 1 mL/min; detection: photometry at 520 nm after reaction with PAR; pre-concentrated volume: 5 mL; (a) standard with 0.1 mg/L Fe^{3+} (1) and 5 µg/L each of Cu^{2+} (2), Ni^{2+} (3), Zn^{2+} (4), and Mn^{2+} (5), (b) sample: 50% NaOH (diluted 1:10); (taken from [394]).

to be determined in the presence of a high alkali metal excess. These matrices include alkali salts, saline waters, sea water [395], and also salt-rich tissue samples and urine.

Finally, Heithmar et al. [396] combined this pre-concentration procedure with ICP/MS to overcome interferences in ICP/MS caused by the matrix when environmental samples are investigated. These could be both of spectral nature [397] and of a physical-chemical nature [398]. While the former may be mathematically eliminated by introducing appropriate correction terms in the calibration function, physical-chemical interferences are best removed by the above-described separation of the transition metal analytes from interfering alkali- and alkaline-earth metal salts.

9.11
Sample Preparation and Matrix Problems

Sample preparation embraces all operations which help to bring the samples into a form appropriate for analysis. These processes include a possible crushing of the sample, its homogenization, digestion, dissolution, stabilization, and filtration. All these steps are important and are usually interconnected. However, in most cases difficulties are caused by sample pre-purification [399]. At this point, various techniques will be discussed in more detail, with emphasis on procedures that are typical for ion chromatographic analyses.

After the analyte sample has been dissolved, a number of working steps are often required before the sample can be injected into the chromatograph. In some cases, a simple filtration may suffice. However, there are numerous application problems that require removal of interfering matrix components from the sample. Also, a modification of the chemical nature of the analyte species is sometimes necessary for obtaining a better separation or detection.

In most cases, sample preparation steps take up most of the required analysis time and, thus, contribute substantially to the analysis cost. It should also be mentioned that each manipulation of a sample can falsify the analytical result; therefore, the care taken in sample preparation directly affects the quality of the analytical result.

Sample preparation is usually performed prior to the chromatographic analysis, but it is often coupled directly to it in the form of an intermediate first step. Sample preparation aims to avoid overloading effects by appropriately diluting of the sample, removing interfering matrix constituents, and making ions which are present in very low concentrations accessible to analysis via pre-concentration.

As with all liquid chromatographic techniques, the solutions to be injected for ion chromatography must be free of particulate matter to avoid plugging of the capillary tubings and, above all, the frits in the separator column head. Disposable filters are generally used for the filtration of sample solutions. Owing to their Luer connector, they are also suited for manual injection, for which they are attached to the injection syringe. While membrane filters with a pore diameter of 0.45 µm are sufficient in most cases, aseptic filters with a pore diameter of 0.22 µm should be used for samples with biological activity in order to avoid, as much as possible, a change in sample composition by bacteriological oxidation or reduction. To preclude sample contamination, membrane filters should be rinsed with de-ionized water prior to their use. Depending on the filter brand, nitrate ions may be released from the filters [400]. While ions such as chloride, bromide, sulfate, and alkali metals are chemically stable, nitrite, nitrate, orthophosphate, and ammonium *cannot* be preserved. Excluding oxygen during sampling and storing samples in a refrigerator under light exclusion are the only measures to stabilize samples. In contrast what many assume, chlorinated hydrocarbons and mercury salts have *no* stabilizing effect on the above-mentioned non-preservable ions. In some cases, sample stabilization can be carried out by adding chemicals. Known examples are the addition of formaldehyde solution for stabilizing sulfite and the acidification of samples if alkaline-earth metals are to be measured. In complex matrices such as blood sera, ultrafiltration is recommended to remove proteins, a procedure in which the sample is pressed through a membrane under pressure. Such samples have to be de-proteinated, because proteins exhibit such a high affinity towards the stationary phase of an ion exchanger that cleaning the separator column causes severe difficulties.

One of the most frequent chemical sample modifications for ion chromatography is the neutralization of strongly acidic or strongly alkaline samples. This applies particularly to digestion solutions. They should also not be injected directly, as high concentrations of acid or base in the sample can result in severe baseline instabilities, which are attributed to the huge pH difference between sample and eluant. The neutralization of samples may be accomplished in several ways. A simple addition of acid or base is only rarely possible, as the sample would be contaminated with acid anions or base cations, respectively. Even if a species is chosen which does not have to be determined, the contamination risk by impurities contained in these chemicals is nevertheless very high. Alternatively, ion-exchange resins may be used for sample neutralization. The pH value of strongly alkaline samples, for example, may be lowered by adding a high-capacity cation exchange resin in the hydronium form. However, the ion-exchange resin used must be rinsed thoroughly with de-ionized water to prevent sample contamination by ions which can be washed out from the resin. Although this procedure seems to be easy to perform at first sight, two disadvantages should be noted. First, a relatively large sample volume is required, and this is not always available; second, the sample volume may change upon addition of an ion-exchange resin, as the latter is subject to swelling and shrinking processes in the presence of solvents.

Ion-exchange resins are also used as packed columns for pre-purification of samples. A characteristic example is the sample preparation of brine with the formerly used ICE suppressor (see Section 5.4), a cation exchanger in the silver form [401].

Today, the most convenient form of sample preparation is the use of small disposable cartridges equipped with a Luer connector that can be attached to the disposable syringe. Depending on the application, they contain different packing materials (see Table 9-19). Disposable cartridges can be employed both for the pre-purification of samples and for pre-concentration of solute ions. In the former case, it is important that only matrix components are retained, while in the latter case only solute ions should exhibit a high affinity towards the stationary phase.

Hydrophobic packing materials such as divinylbenzene or chemically modified silica (C_8, C_{18}) [402] are recommended for removing aromatics, long-chain fatty acids, hydrocarbons, and surfactants from the sample. In contrast to ODS materials, divinylbenzene has the advantage of being stable over a broad pH range (pH 0-14). Also, divinylbenzene shows a higher selectivity for aromatic and unsaturated compounds.

Polyvinylpyrrolidone (PVP) is suited for the preparation of samples that contain humic acids, lignins, tannins, and azo dyes. In addition, PVP exhibits a high selectivity for phenolic compounds as well as for aromatic carboxylic acids and aldehydes. PVP is compatible with all common HPLC solvents and stable in the pH range between 1 and 10.

Table 9-19. Commercially available packing materials for solid-phase extraction cartridges.

Type	Packing material	Application
OnGuard-P	Polyvinylpyrrolidone	Removal of: Phenols, humic acids, lignins, azo dyes
OnGuard-RP	Divinylbenzene polymer	Removal of: Surfactants, aliphatic and aromatic hydrocarbons, long-chain carboxylic acids, aromatic dyes
OnGuard-Ag	Strong acid cation exchanger in the silver form	Removal of: Halide ions (excl. fluoride), arsenate, chromate, cyanide, thiosulfate, sulfide
OnGuard-H	Strong acid cation exchanger in the hydrogen form	Removal of: Alkali- and alkaline-earth metals, transition metals, neutralization of basic solutions
OnGuard-Ba	Strong acid cation exchanger in the barium form	Removal of: Sulfate
OnGuard-A	Strong basic anion exchanger in the bicarbonate form	Removal of: Anions, peptides, proteins, neutralization of acidic solutions

Strong acid cation exchangers are also available as disposable cartridges. Typically, these are materials based on polystyrene/divinylbenzene with a high degree of cross-linking (about 10%) that are stable over the entire pH range. Strong acid cation exchangers are offered in two forms. In the hydrogen form they serve to neutralize strongly alkaline samples. Typical applications are soda/potash digestions or concentrated lyes, which may be analyzed for inorganic anions after treatment with such a cartridge [403]. Cation exchangers in the silver form are predominantly employed for removing high amounts of chloride from sea water, brine, and other saline samples [404]. The only disadvantage is that other halide ions such as bromide and iodide are also precipitated. Moreover, silver ions are released which could foul the separator column. Therefore, it is recommended that a cation exchanger in the silver form always be used in combination with a cation exchanger in the hydrogen form.

In general, when using sample preparation cartridges, care must be taken to ensure that they are rinsed thoroughly with de-ionized water prior to their use in order to avoid sample contamination by ionic constituents that possibly remained in the cartridges from the production process. In addition, hydrophobic packing materials require rinsing with organic solvents such as methanol to activate the material's surface and to increase the binding ability for medium-polar to non-polar organics from aqueous solutions. Furthermore, it should be noted that the sample to be analyzed is passed through the cartridge as slowly as possible. Due to their low packing densities, the cartridges are actually designed for flow rates up to 50 mL/min, but it is widely recognized that the required pretreatment is much more efficient at significantly lower flow rates

(<10 mL/min). For the practical execution of the sample pretreatment it is advisable to fill the injection syringe with the solution to be analyzed. This solution is then pushed continuously through the cartridge that has been activated and rinsed in advance. However, the first 2-3 cartridge volumes should be discarded, so that the sample does not become diluted with liquid remaining in the cartridge. The subsequent effluent can be collected or injected directly into the chromatograph.

In elemental analysis of leaves, fruits, vegetable extracts, and plant material of botanical or nutrition-physiological relevance, the organic matrix with its complexing properties represents a strong interference. Classical methods for decomposing the organic matrix include dry ashing and wet digests. While in dry ashing often very low recovery rates are observed, the oxidative wet digest with concentrated acids is characterized by a relatively high consumption of chemicals, which consequently implies a significant contamination risk for the sample. As an alternative to these microwave-based digestion techniques, Buldini et al. [405] had a good experience with oxidative UV photolysis. The oxidation of organically bound carbon as well as the release of organically bound metals and non-metals under the influence of UV radiation was first described by Armstrong et al. [406]. Further investigations by Kolb et al. [407] revealed that the UV oxidation of organic materials in the presence of an oxidation agent is as efficient as chemical oxidation. Many of the above-mentioned botanical materials are completely degraded in less than two hours, while inorganic constituents with the exception of nitrate, iodide, and manganese are not affected by UV radiation. A 500-W mercury high pressure lamp proved to be suitable for irradiation; the sample temperature is kept at 85 °C with a combined air/water cooling. Figure 9-240 shows a chromatogram resulting from photolysis of a certified pine needle sample. For this, 250 mg of the homogenized and pulverized sample are mixed with 1-2 mL of hydrogen peroxide solution (30%) and irradiated for 120 minutes at maximum. After cooling down, de-ionized water is used to fill up to a volume of 10 mL, of which 50 µL are directly injected. The recovery rates for the two certified anions bromide and orthophosphate were 102.2% and 101.6%, respectively.

An oxidative combustion with subsequent absorption of the combustion products in an aqueous medium has proved to be suitable for the simultaneous determination of sulfur and halides [408]. Without a doubt, ion chromatography is the most suitable analysis method for the quantification of these analytes, especially when the sample to be analyzed contains other halogens such as fluorine, bromine, and iodine in addition to chlorine and sulfur. It is important that the analytes are present in a uniform ionic form. However, during the combustion of halogen- and sulfur-containing materials, a number of other ionic and non-ionic compounds with varying concentrations are formed in addition to HX and SO_3. In practice, combustion devices with a special coating on the inner wall are used. This coating helps to catalyze the kinetically inhibited oxidation of the sulfur compounds to SO_3.

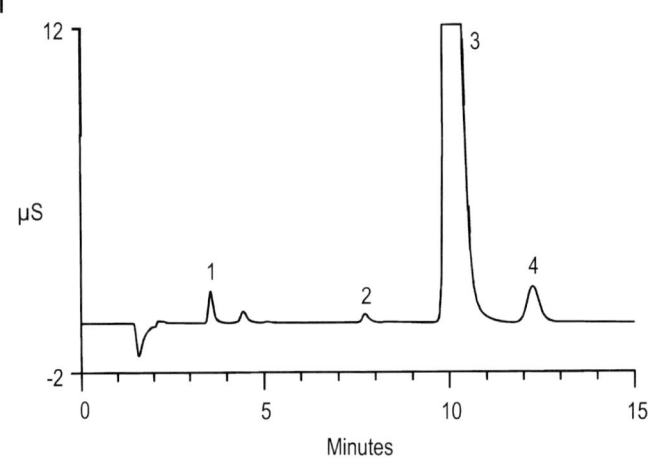

Fig. 9-240. Anion analysis in a certified pine needle sample after UV photolysis. — Separator column: IonPac AS12A; eluant: 2.7 mmol/L Na_2CO_3 + 0.3 mmol/L $NaHCO_3$; flow rate: 1.5 mL/min; detection: suppressed conductivity; injection volume: 50 µL; sample: 250 mg NIST SRM 1575; solute concentrations: 0.05 mg/L (not certified) chloride (1), 0.22 mg/L bromide (2), 93 mg/L orthophosphate (3), and 0.45 mg/L (not certified) sulfate (4); (taken from [405]).

In comparison with the Wickbold method, the combustion in a bomb offers significant advantages. Larger amounts of sample can be weighed in (up to 1 g), which accounts for the inhomogeneity problem with many samples, and smaller absorption volumes (10-20 mL as compared to 200 mL with Wickbold) ensure minimum detection limits below 10 mg/L when using ion chromatography as method of determination. Moreover, liquid, liquid-puffy, and solid samples can be analyzed. To carry out the combustion [409], 100-200 mg of sample (depending on the halogen content) placed between two benzoic acid die-pressed parts (to support combustion) are weighed in an acebutyrate capsule and transferred into a quartz crucible (see Fig. 9-241). The absorber solution contains 10 mL NaOH solution (c = 0.25 mol/L), to which 200 µL H_2O_2 (30%) is added. This composition ensures a quantitative transfer of oxyhalides formed during the combustion to halide ions detectable by ion chromatography [408]. After closing the combustion device, an oxygen pressure of 34 bar is applied. Ignition of the sample is carried out electrically. After cooling down and slueing around the vessel, the combustion gases are expanded in a gas wash bottle containing the same absorber solution. Both absorber solutions are combined and filled up to a defined volume. A typical application is the oxidative combustion of coal for determining chlorine and sulfur. As can be seen from the chromatogram in Fig. 9-242, the ion chromatographic analysis of the combustion products on IonPac AS9 does not pose any problem.

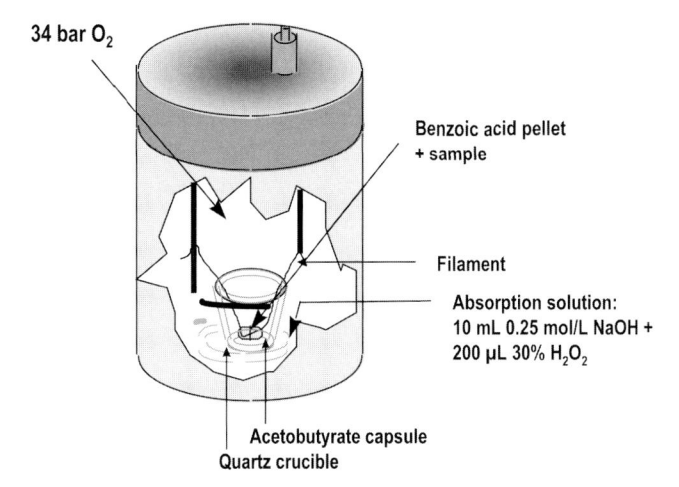

Fig. 9-241. Schematic representation of a vessel for an oxidative bomb combustion (IKA Heitersheim, Germany); (taken from [409]).

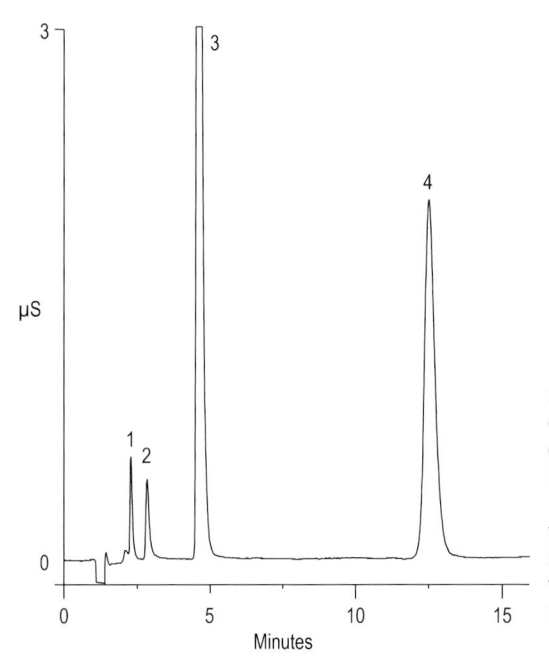

Fig. 9-242. Anion analysis in a coal sample after oxidative bomb combustion. – Separator column: IonPac AS9; eluant: 1.8 mmol/L Na_2CO_3 + 1.7 mmol/L $NaHCO_3$; flow rate: 1.5 mL/min; detection: suppressed conductivity; injection volume: 50 µL; sample: 0.1-0.2 g; peaks: (1) 0.16% Cl, (2) nitrite, (3) nitrate, and (4) 0.67% S.

Another interesting example is the use of oxidative bomb combustion for determining iodine in mixed animal feed [410]. To ensure an optimal iodine supply, 0.1-0.6 mg iodine per kg dry material in form of NaI or $Ca(IO_3)_2$ is added to animal feed. Because iodine exists in animal feed as inorganic species in various oxidation states and as organic species covalently bonded to natural compounds, the goal for a suitable sample preparation must be the measurement of the total iodine content through the transfer of the element into a uniform oxidation state. Traditionally, iodine in animal feed is determined via alkaline ashing with subsequent sulfite reaction; however, long digestion times have to be observed [411]. For the ion chromatographic determination of iodide completely different chromatographic conditions can be employed. However, suppressed conductivity detection is ruled out due to the lack of sensitivity. But when applying UV detection, 100 mg/kg iodide in animal feed can be detected without any problem. As an example, Fig. 9-243 shows the analysis of iodide in a birdseed sample on an IonPac AS14 anion exchanger after oxidative bomb combustion.

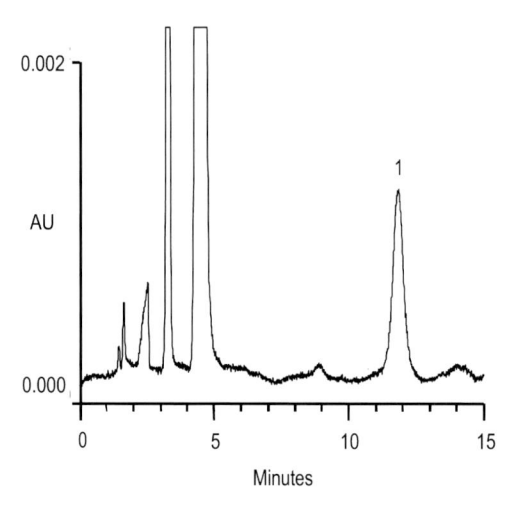

Fig. 9-243. Iodide analysis in a birdseed sample after oxidative combustion. — Separator column: IonPac AS14; eluant: 30 mmol/L NaCl; flow rate: 1.5 mL/min; detection: UV (236 nm); sample: birdseed with 266 mg/kg iodide (1); (taken from [410]).

The determination of phosphorus in organophosphorus compounds has also been simplified with an oxidative combustion in a modified Erlenmeyer flask as a sample preparation. The corresponding absorber solution contains inorganic phosphate. Numerous analytical methods such as gravimetry, titration, spectrophotometry, polarography, AAS, ICP-AES, ^{31}P-NMR, and ion chromatography were suggested for its determination. IC is routinely applied for determining orthophosphate in the presence of other anions. If speciation between orthophosphate and other phosphorus species such as hypophosphite or orthophosphite is a subject, there is practically no cheaper option than ion chromatography with conductivity detection. However, some authors report on inadequate recovery rates in the digestion of organophosphorus compounds. Busman et al. [412] and Senior et al. [413] attribute this to the formation of polyphosphates such as pyrophosphate and cyclic metaphosphate, which cannot be analyzed together with orthophosphate on conventional anion exchangers under isocratic conditions. For this reason, Umali et al. [414] investigated the combustion products of phenylphosphonic acid as a test substance for developing a sample preparation technique, with which phosphorus can be reliably determined in organophosphorus compounds. Umali et al. used a dilute H_2O_2 solution (0.15%) as an absorber solution. Identification of the inorganic phosphorus species was carried out via ^{31}P-NMR, a method successfully used by Gard et al. [415] for the simultaneous determination of orthophosphate, pyrophosphate, and trimetaphosphate in commercial tripolyphosphate. With ^{31}P-NMR, orthophosphate (singlet, δ_p = 3.2 ppm) and pyrophosphate (singlet, δ_p = −6.9 ppm) could be clearly identified in the absorber solution. A third very small peak with a chemical shift of δ_p = −20.74 ppm indicates the presence of trimetaphosphate, which is in good agreement with the statement of Senior et al. [413]. Their statement, however, was not corroborated by positive identification. Because it is now clear that condensed phosphates are formed during oxidative bomb combustion of organophosphorus compounds, they have to be hydrolyzed to orthophosphate prior to the ion chromatographic analysis. Usually, high concentrations of sulfuric acid, hydrochloric acid, nitric acid, or perchloric acid are employed for this, as polyphosphates hydrolyze faster in an acidic environment than in an alkaline one. However, when using one of these acids large amounts of the respective counter ion are introduced to the ion chromatographic system, which can lead to severe interferences in orthophosphate detection depending on the selectivity of the separator column. Optimal hydrolysis results, especially for pyrophosphate, are obtained with a dilute H_2O_2 solution (0.6% w/v), in which pH value is adjusted to pH < 3 with tartaric acid and which is boiled after the digestion for at least 20 minutes.

Finally, alkaline fusion is mentioned for the determination of chlorine and bromine in rock [416]. For this purpose, 0.25 g of the pulverized rock sample is put into a crucible and 2 mL of 1 mol/L NaOH is added. This mixture is evaporated to dryness at 125 °C. Then 1.4 g of sodium peroxide are added and the mixture is heated to about 650-700 °C (red heat) for five minutes until it is melted. After five minutes, the temperature is increased above red heat to destroy excessive sodium peroxide. For better leaching, the warm melt is mixed with 15 mL of de-ionized water. After about 30 minutes, the total content of the crucible is transferred into a polymeric flask, which is filled up to the mark with water. The solution that results from this alkaline fusion contains approximately 0.75 mol/L NaOH and, consequently, cannot be injected directly into the ion chromatograph. For sample neutralization, the above-mentioned OnGuard-H cartridges containing a cation exchanger in the hydrogen form are recommended. Figure 9-244 shows the resulting chromatogram of a Japanese reference material spiked with 200 μg chloride and 20 μg bromide. IonPac AS12A was used as a stationary phase; using this column, the two interesting peaks could be well separated from the matrix. For sensitivity reasons, chloride is detected via suppressed conductivity, while bromide is detected via UV. Blank analyses revealed that the chemicals used contain small amounts of chloride, which have to be considered when reporting results.

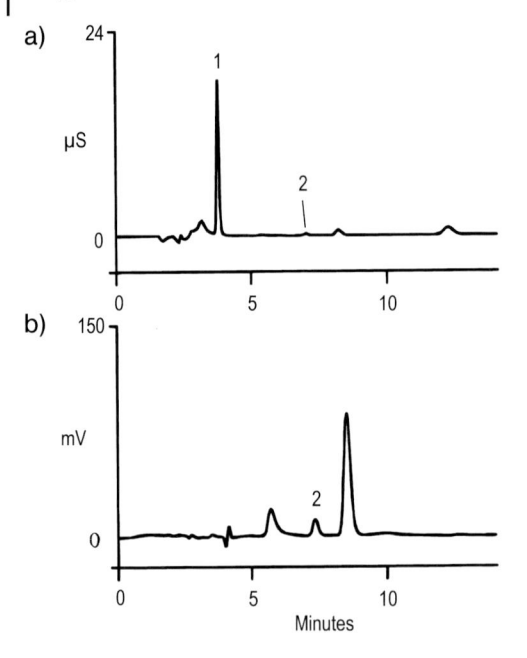

Fig. 9-244. Analysis of chloride and bromide in a basalt sample after alkaline fusion. — Separator column: IonPac AS12A; eluant: 2.7 mmol/l Na_2CO_3 + 0.3 mmol/L $NaHCO_3$; flow rate: 1.5 mL/min; detection: (a) suppressed conductivity, (b) UV (210 nm); sample: JB-2, GSJ reference material, spiked with 200 µg chloride (1) and 20 µg bromide (2); (taken from [416]).

Dialysis techniques are more frequently employed in sample preparation. The diffusion of specific sample components through a membrane is common in all these techniques. If this membrane is neutral and hence permeable for components in a specific molecular weight range, the technique is called *passive* dialysis. In contrast, the term *active* dialysis (also denoted as Donnan dialysis) refers to the diffusion of ions of identical charge through the membrane of an ion exchanger. If this diffusion occurs under the influence of an electrical field, the technique is called *electrodialysis* [417]. In general, all three techniques can be used to prepare samples for ion chromatography. While passive dialysis is a comparatively slow process that requires a larger sample volume and leads to a substantial dilution of the sample, Donnan dialysis is used to selectively remove from the sample or add to the sample a specific species. A detailed description of these techniques is found in the paper by Haddad [399] cited earlier. In this light, the paper published by Pettersen et al. [418] concerning the analysis of anionic impurities in highly concentrated sodium hydroxide via electrodialysis is interesting. In their experimental setup, the authors used an electrolysis cell with two compartments, in which the anode and cathode compartments are separated by a cation exchange membrane. The anode compartment is filled with the sodium hydroxide solution to be analyzed, the cathode compartment with dilute sodium hydroxide solution. During the electrolysis, the following reactions occur at both electrodes:

$$\text{Anode:} \quad 4OH^- \rightleftharpoons O_2 + H_2O + 4e^- \quad (326)$$

$$\text{Cathode:} \quad 4H_2O + 4e^- \rightleftharpoons 2H_2 + 4OH^- \quad (327)$$

This means that the hydroxide ion concentration decreases in the anode compartment while it increases in the cathode compartment. Because diffusion of hydroxide ions through the cation exchange membrane is impossible, the sodium hydroxide concentration in the sample is lowered so much that it may be injected without any further pretreatment.

Finally, sample preparation techniques that are based on a chemical modification of the sample should be mentioned. Modification is sometimes necessary to convert the analyte component into a form suitable for separation or detection. A characteristic example is the oxidation of cyanide to cyanate with sodium hypochlorite [419], which enables indirect cyanide determination by electrical conductivity. In contrast to cyanide with its low dissociation (pK = 9.2), the more strongly dissociated cyanate ion (pK = 3.66) may be detected with both forms of conductivity detection. Because cyanate elutes on a conventional anion exchanger in the retention range of chloride, modification of the chromatographic conditions, which would be necessary for amperometric detection of cyanide, is not required. However, a simultaneous analysis of mineral acids and cyanate is only possible if a separator column is used that allows cyanate to be separated from all other inorganic anions. Nitrite, which elutes immediately before cyanate on a conventional IonPac AS4A anion exchanger, represents a potential interference. If a satisfactory separation of both ions is impossible, nitrite may be degraded by adding amidosulfonic acid at pH 7. (Nitrite is not oxidized to nitrate by hypochlorite!) It should also be noted that the determination of bromide and nitrate is not affected by the presence of chlorate resulting from hypochlorite. Chlorate used to be eluted together with nitrate, but may be resolved to baseline on IonPac AS9 (-SC) and other modern stationary phases.

A similar example represents the determination of borate, which is usually analyzed by ion-exclusion chromatography. However, when borate is converted into tetrafluoroborate by reacting it with hydrofluoric acid [420], an indirect borate determination together with other inorganic anions becomes possible upon application of anion exchanger. Because tetrafluoroborate is a polarizable anion, an anion exchanger with extremely hydrophilic functional groups should be used to elute tetraborate as a symmetrical peak.

A suitable sample preparation technique for determining ionic impurities in concentrated acids and bases is AutoNeutralization [421], an automated neutralization of acidic or basic samples with a special membrane suppressor. This is done in order to analyze traces of anions in strongly basic samples and traces of cations in strongly acidic samples. The automated neutralization is carried out with a special module (SP10, Dionex Corporation) which can be coupled to an ion chromatograph. The instrumental setup of such coupling is schematically depicted in Fig. 9-245. With the integrated metal-free pump, the content of the sample loop is transported with de-ionized water through the collection loop into the SRN neutralization unit, which is connected to a valve, so that the sample to be analyzed can be passed several times through the neutralization unit via time control of the valve. As in AutoSuppression, neutralization is based on the electrolysis of water yielding hydronium- and hydroxide ions, which act as a regenerant for the ion-exchange membranes. The chemical reactions in such a neutralization unit were illustrated in Fig. 3-77 (Section 3.6.4) and Fig. 4-35 (Section 4.3.4). Instead of using a membrane-based self-regenerated suppressor, Montgomery et al. [422] describe the use of a neutralization unit based on two packed columns (ERIN, Alltech, Deerfield, USA) for the same purpose. The two columns are alternated for neutralization and electrochemically regenerated after every run. Because packed columns are used, the dead volume is much larger than in a membrane-based unit. Important applications for this sample preparation technique include anion analysis in a concentrated sodium hydroxide solution as formed during chloralkali electrolysis and purity control of amines in the semiconductor industry. In the area of cation analysis, the determination of alkali- and alkaline-earth metals in high-purity acids such as hydrofluoric acid, hydrochloric acid, sulfuric acid, and orthophosphoric acid is the most important application (again, for the semiconductor industry) [423].

A detailed description of sample preparation techniques is found in a book published by Frei and Zech [424].

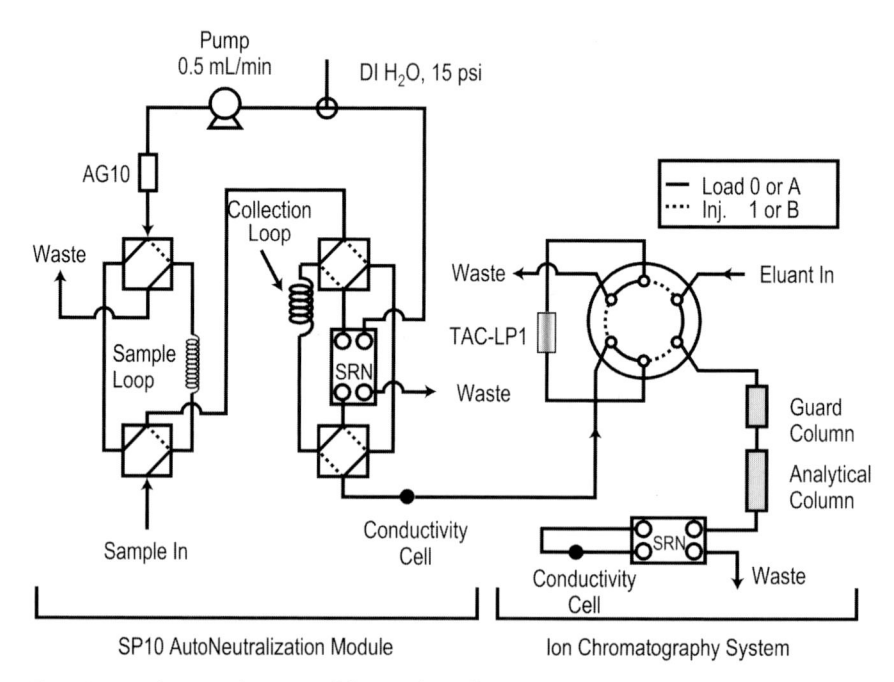

Fig. 9-245. Schematic depiction of the coupling of an AutoNeutralization unit to an ion chromatograph.

Column Maintenance

One of the most effective methods for protecting separator columns is the use of guard columns which normally contain the same stationary phase as the analytical columns. Compounds with high affinities towards the stationary phase being used are retained on the guard column, and therefore poisoning of the analytical separator column is avoided. Fouling of the analytical column is typically indicated by a loss of separation efficiency. Because the capacity of guard columns is limited, they must be rinsed occasionally.

The type of rinsing agents mainly depends on the type of stationary phase. Details are found in the guidelines provided by the column manufacturer. In general, guard and separator columns should be stored sealed when they are removed from the chromatograph. Drying out of the column packing can lead to alterations in the column bed and result in a reduction of separation efficiency. Furthermore, guard and separator columns should be protected from direct light and should be stored free from vibration. Fats, oils, surfactants, humic acids, and lignins as well as cellulose, proteins, and other high molecular compounds are considered – particularly in ion exchangers – to be column poisons. The poisons must be removed by one of the preparation techniques described above. Organic polymer-based ion exchangers should only be rinsed with organic solvents if the materials are 100% solvent compatible. Otherwise, swelling and shrinking phenomena may result, depending on the kind of separation material. In general, only water-soluble solvents such as methanol and acetonitrile should be used. In any case, the guidelines of the manufacturers are to be followed.

Outlook

When comparing the level of development of ion chromatography in the year of publication of the second edition of this book with that at the beginning of this new millennium, one cannot help seeing another quantum step in the development of this method. Pioneering developments such as the contamination-free electrolytic preparation of hydroxide eluants with the related simplification of gradient techniques, the new trend towards hydroxide-selective stationary phases with their superior separation properties, and the rapid development of hyphenated techniques do not seem to leave any wishes unfulfilled. Nevertheless, as in all forms of liquid chromatography, in ion chromatography, eluants have to be prepared throrougly and eluant containers are to be refilled from time to time. If gradient elution is required, two or even three different eluants have to be delivered, usually by employing an expensive gradient pump. In the early 1990s, Dasgupta et al. [425, 426] pioneered electrodialysis for eluant clean-up in IC, thus circumventing the necessity of a gradient pump. But even their techniques still required the preparation of an eluant, although this can be done less precisely as eluant concentration is adjusted electrically.

The author of this book does not presume to be able to look into the future, but with a little bit of fantasy one can imagine an ion chromatographic system that does not require manual eluant preparation. At first sight, this might appear strange, but in suppressor technology, evolution has already culminated in the development of a continuous, electrochemically regenerated membrane-based device that is extremely easy to use and, therefore, practically "invisible". When extrapolating this development onto eluant preparation, one realizes that ion chromatography could indeed be revolutionized. A higher degree of ease-of-use and an even higher degree of automation towards remote control are advantages which easily come to mind. We are not far away from this dream. In 1998, Hamish Small, the pioneer of ion chromatography, introduced an idea which he termed "Ion Reflux" [427, 428] that could change this method very effectively. It is based on classical ion chromatographic elements; i.e., analytes are separated on ion exchangers and detected via conductivity utilizing continuously regenerated suppressor systems. But pure de-ionized water is fed into the system instead of an electrolyte solution. In this case, water is not an eluant but is used as a carrier, a small but important difference. The electrolyte solution that is still required is generated *in situ* with eluant generation and suppression being closely related to each other. The key of this process is an electrically polarized ion-exchange bed operated in the ion reflux mode. This becomes comprehensible when considering that ion exchange resins are good conductors due to the mobile counter ions. The simplest application of an electrically polarized ion-exchange bed is a column with porous electrodes on both sides. When filling this column with cation exchange particles and pumping water towards the anode, an electromigration of potassium ions towards the cathode is observed under the influence of an electric field; hydroxide ions and hydrogen gas are formed at the cathode via electrolysis (Fig. 9-246). Potassium ions and hydroxide ions then recombine to form potassium hydroxide, which is not removed from the system but flows with water in the opposite direction. At the anode, hydronium ions and oxygen are formed, so that a boundary layer between the two ionic forms of the cation exchanger is formed in the cation exchange bed. If potassium hydroxide reaches this boundary layer, the known neutralization reaction occurs:

$$KOH + Resin\text{-}H^+ \rightarrow Resin\text{-}K^+ + H_2O \tag{328}$$

Under the influence of the electric field, the number of hydronium ions generated at the anode is the same as that of hydroxide ions generated at the cathode, so the flow of hydronium ions in one direction is in equilibrium with the flow of KOH into the other direction. Thus, the boundary layer between hydronium and potassium ions is spatially fixed. If the applied current and, consequently, the flow of hydronium ions to the cathode are changed, the flow of hydroxide ions to the anode is changed in the same way. Thus, the electrochemical split of water is the motor for this circulating pool of potassium ions.

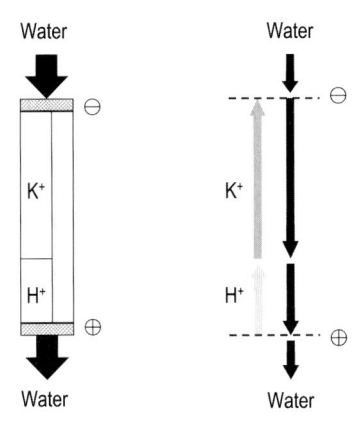

Fig. 9-246. Electromigration of ions and flow directions in an "Ion Reflux" device.

How can such a device, which elegantly combines eluant generation and suppression, be combined with the third element of ion chromatography, the separation on an ion exchanger? For this purpose, only an oppositely charged anion exchange colloid has to be electrostatically agglomerated on the negatively charged surface of the cation exchanger. Small calls this the monolithic version of "Ion Reflux". While the anions of a sample will leave this column at any time, the sample cations remain in the column and add to the pool of potassium ions. Because the ion-exchange capacity of the resin used is a fixed quantity, the hydronium area of the device becomes smaller and smaller. Although the problem can be solved by passing the sample through a cation exchanger prior to injection, it is recommended that the separation of ions be divided from the other two functions. A potential application form is illustrated in Fig. 9-247. In contrast to Fig. 9-246, the cathode at the column inlet is replaced by porous frit. The cathode is now outside the column and connected to the resin material via a cation exchange membrane. This membrane permits the electromigration of cations, but precludes an exchange of liquid between the cathode chamber and the resin material. Now, any separator column can be placed between the outlet of the cathode chamber and the column inlet. The circulation of potassium ions

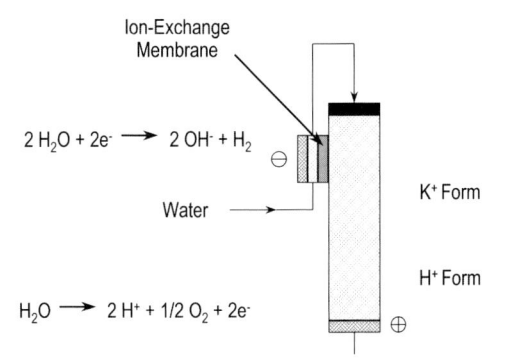

Fig. 9-247. "Ion Reflux" device for the combination with a separator of any kind.

in the manner described above remains unaffected. The small amount of gas generated during electrolysis can be compressed by coupling a restriction capillary to the conductivity cell outlet, so that it does not interfere with the detection. Finally, Fig. 9-248 shows an anion chromatogram that was obtained on a short hydroxide-selective anion exchanger utilizing the "Ion Reflux" device. The applied current was increased stepwise during the run in order to demonstrate that ions with different affinities to the stationary phase can also be separated with this device, even though the optimization of the separation was not the primary objective.

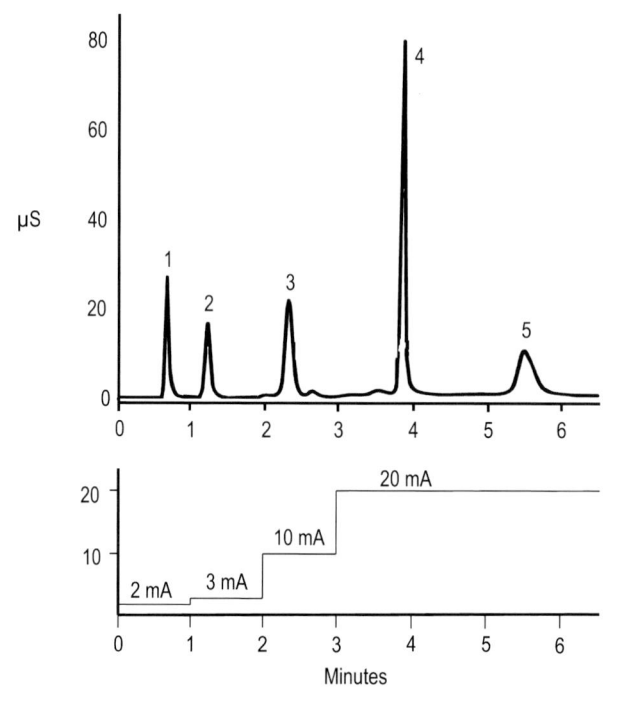

Fig. 9-248. Separation of selected inorganic anions with an electrical gradient according to the "Ion Reflux" principle. — Separator column: IonPac AG11; eluant: KOH; flow rate: 1 mL/min; detection: suppressed conductivity; solute concentrations: 1 mmol/L each of fluoride (1), chloride (2), nitrate (3), sulfate (4), and orthophosphate (5); (taken from [427]).

Although it currently cannot be assessed whether and how those fascinating ideas can be commercialized, there are good reasons to incorporate the "Ion Reflux" principle into the design of future ion chromatography systems. Thus, even 29 years after its introduction, ion chromatography remains an exciting analytical method.

Concluding Remarks

A multitude of separation and detection methods developed in the past years for analyzing ionic compounds have been introduced within the scope of this book. Their advantages and disadvantages have been discussed. However, it should be noted that none of the described methods are *universally* applicable; none can be considered the method of choice in every case. Hence, for any given analytical problem in this field, the following factors should be taken into account when selecting an appropriate analysis method.

- Type and concentration of the species to be analyzed
- Required resolution and precision
- Speed and cost of analysis
- Ease-of-use
- Automation

It was the objective of the third edition of this book to provide a survey of the method of ion chromatography, which has rapidly evolved since the publication of the second edition, and to assist as much as possible in choosing individual techniques.

Literature

Chapter 1

1.1 M. Tswett, *Trav. Soc. Nat. Var.* **14** (1903) 1903.

1.2 M. Tswett, *Ber. Deut. Botan. Ges.* **24** (1906) 385.

1.3 N.A. Izmailov and M.S. Schraiber, *Farmatsiya* **3** (1938).

1.4 E. Stahl, *Pharmazie* **11** (1956) 633.

1.5 E. Stahl, *Chemiker Ztg.* **82** (1958) 323.

1.6 A.J.P. Martin and R.L. Synge, *Biochem. J.* **35** (1941) 1358.

1.7 A.T. James and A.J.P. Martin, *Analyst* **77** (1952) 915.

1.8 C. Horvath, W. Melander and I. Molnar, *J. Chromatogr.* **125** (1976) 129.

1.9 J.H. Knox: "Theory of HPLC, Part II: Solute Interactions with the Mobile Phase and Stationary Phases in Liquid Chromatography". In: C.F. Simpson (ed.). *Practical High Performance Liquid Chromatography*. Heyden and Son, Chichester 1976.

1.10 R.P.W. Scott: "Theory of HPLC, Part II: Solute Interactions with the Mobile Phase and Stationary Phases in Liquid Chromatography". In: C.F. Simpson (ed.). *Practical High Performance Liquid Chromatography*. Heyden and Son, Chichester 1976.

1.11 L.R. Snyder, *Chromatogr. Rev.* **7** (1965) 1.

1.12 G. Guiochon: "Optimization in Liquid Chromatography". In: C. Horvath (ed.). *High Performance Liquid Chromatography*, Vol. 2. Academic Press, New York 1980.

1.13 H.J. Möckel, Lecture: "Instrumentelle Analytik I". Technical University Berlin 1974 to 1984.

1.14 H. Small, T.S. Stevens and W.C. Bauman, *Anal. Chem.* **47** (1975) 1801.

1.15 D.T. Gjerde, J.S. Fritz and G. Schmuckler, *J. Chromatogr.* **186** (1979) 509.

1.16 R.M. Wheaton and W.C. Bauman, *Ind. Eng. Chem.* **45** (1953) 228.

1.17 J. Weiss, S. Reinhard, C.A. Pohl, C. Saini and L. Narayaran, *J. Chromatography* **706** (1995) 81.

1.18 D. Jensen, J. Weiss, M.A. Rey and C.A. Pohl, *J. Chromatography* **640** (1993) 65.

1.19 R.D. Rocklin and C.A. Pohl, *J. Liq. Chromatogr.* **6** (9) (1983) 1577.

1.20 D.A. Martens and W.T. Frankenberger, *J. Liq. Chromatogr.* **15** (1992) 423.

1.21 P. Jandik and G. Bonn, *Capillary Electrophoresis of Small Molecules and Ions*. VCH Publishers Inc., New York 1993.

1.22 P.K. Dasgupta and L. Bao, *Anal. Chem.* **65** (1993) 1003.

1.23 N. Avdalovic, C.A. Pohl, R.D. Rocklin and J.R. Stillian, *Anal. Chem.* **65** (1993) 1470.

1.24 E.L. Johnson and B. Stevens. *Basic Liquid Chromatography*. Varian Associates Inc., Palo Alto, CA 1978, p. 92.

1.25 U. Leuenberger, R. Gauch, K. Rieder and E. Baumgartner, *J. Chromatogr.* **202** (1980) 461.

1.26 J.R. Stillian and C.A. Pohl, *J. Chromatogr.* **499** (1990) 249.

1.27 DIN 38 405, Part 19 (1988) *Die Bestimmung der Anionen Fluorid, Chlorid, Nitrit, Bromid, Nitrat, Orthophosphat und Sulfat in wenig belasteten Wässern mit der Ionenchromatographie*.

Chapter 2

2.1 A.J.P. Martin and R.L.M. Synge, *Biochem. J.* **35** (1941) 1358.

2.2 J.J. van Deemter, F.J. Zuiderweg and A. Klinkenberg, *Chem. Eng. Sci.* **5** (1956) 271.

2.3 E.L. Johnson and B. Stevenson, *Basic Liquid Chromatography*. Varian Associates Inc., Palo Alto, CA 1978.
2.4 A.I.M. Keulemans and A. Kwantes, in: D.K. Desty and C.L.A. Harbourn (eds.). *Vapor Phase Chromatography*. Butterworths, London 1956; p. A10.
2.5 J.C. Giddings, *J. Chromatogr.* **5** (1961) 46.
2.6 J.F.K. Huber and J.A.R.J. Hulsman, *Anal. Chim. Acta* **38** (1967) 305.
2.7 G.J. Kennedy and J.H. Knox, *J. Chromatogr. Sci.* **10** (1972) 549.
2.8 J.N. Done and J.H. Knox, *J. Chromatogr. Sci.* **10** (1972) 606.
2.9 C.S. Horvath and H.-J. Lin, *J. Chromatogr.* **126** (1976) 401.
2.10 C.S. Horvath and H.-J. Lin, *J. Chromatogr.* **149** (1978) 43.
2.11 E. Katz, K.L. Ogan and R.P.W. Scott, *J. Chromatogr.* **270** (1983) 51.
2.12 R.E. Majors, *Anal. Chem.* **44** (1972) 1722.

Chapter 3

3.1 *The Bible*, 2. Ms. **15**, 23-25.
3.2 W. Rieman and H.F. Walton, *Ion Exchange in Analytical Chemistry*. Pergamon Press 1970.
3.3 H. Small, T.S. Stevens and W.C. Bauman, *Anal. Chem.* **47** (1975) 1801.
3.4 H.J. Möckel, Lecture: "Instrumentelle Analytik I". TU Berlin 1974 to 1984.
3.5 H.J. Möckel, *J. Chromatogr.* **317** (1984) 589.
3.6 H.J. Möckel, T. Freyholdt, J. Weiß and I. Molnar: "The HPLC of Divalent Sulphur". In: I. Molnar (ed.) *Practical Aspects of Modern HPLC*. Walter de Gruyter & Co. Berlin – New York 1982.
3.7 C. Eon and G. Guiochon, *J. Colloid. Interface Sci.* **45** (1973) 521.
3.8 C. Eon, *Anal. Chem.* **47** (1975) 1871.
3.9 Y. Takata and G. Muto, *Bunseki Kagaku* **28** (1979) 15.
3.10 N. Yoza, K. Ito, Y. Hirai and S. Ohashi, *J. Chromatogr.* **196** (1980) 471.
3.11 H. Terada, T. Ishihara and Y. Sakabe, *Eisei Kagaku* **26** (1980) 136.
3.12 G.A. Sherwood and D.C. Johnson, *Anal. Chim. Acta* **129** (1981) 101.
3.13 R.G. Gerritse, *J. Chromatogr.* **171** (1979) 527.
3.14 F.A. Buytenhuys, *J. Chromatogr.* **218** (1981) 57.
3.15 N. Vonk, *European Spectroscopy News* **53** (1984) 25.
3.16 D.T. Gjerde, J.S. Fritz and G. Schmuckler, *J. Chromatogr.* **186** (1979) 509.
3.17 D.T. Gjerde, G. Schmuckler and J.S. Fritz, *J. Chromatogr.* **187** (1980) 35.
3.18 J.S. Fritz, D.T. Gjerde and C. Pohlandt, *Ion Chromatography*. Dr. Alfred Hüthig Verlag Heidelberg – Basel – New York 1982; p. 102.
3.19 D.T. Gjerde and J.S. Fritz, *J. Chromatogr.* **176** (1979) 199.
3.20 K.M. Roberts, D.T. Gjerde and J.S. Fritz, *Anal. Chem.* **53** (1981) 1691.
3.21 D.P. Lee, *J. Chromatogr. Sci.* **22** (8) (1984) 327.
3.22 P. Walser, *LaborPraxis*, July/August (1985) 878.
3.23 J. Weiss, S. Reinhard, C. Pohl, C. Saini and L. Narayanan, *J. Chromatogr.* **706** (1995) 81.
3.24 C. Saini, C. Pohl and L. Narayanan, "An Improved Ion Exchange Phase for the Determination of Fluoride and Other Common Anions by Ion Chromatography", Presentation at Pittsburgh Conference 1995, New Orleans, LA, USA.
3.25 D. Jensen and C. Rattmann, *GIT Fachz. Lab. Chromatographie Spezial* **6** (1996) 115.
3.26 P.R. Haddad and A.L. Heckenberg, *J. Chromatogr.* **300** (1984) 357.
3.27 T. Okada and T. Kuwamoto, *Anal. Chem.* **55** (1983) 1001.
3.28 R. Saari-Nordhaus, I. Henderson and J.M. Anderson, Jr., *J. Chromatogr.* **546** (1991) 89.
3.29 F. Vlacil and I. Vins, *J. Chromatogr.* **391** (1987) 119.
3.30 J.R. Benson and D.J. Woo, *J. Chromatogr. Sci.* **22** (1984) 386.
3.31 P.R. Haddad, P.E. Jackson and A.L. Heckenberg, *J. Chromatogr.* **346** (1985) 139.
3.32 T.S. Stevens and M.A. Langhorst, *Anal. Chem.* **54** (1982) 950.
3.33 J.S. Fritz and J.N. Story, *J. Chromatogr.* **90** (1974) 267.

3.34 J. Weiß, *GIT Spezial, Chromatographie* **2** (1992) 67.

3.35 DIN 38405, Teil 13 (1981) *Bestimmung von Cyaniden*.

3.36 C. Pohl, V. Summerfelt and J. Stillian, Presentation at *Pittsburgh Conference* 1987, Atlantic City, N.J., USA.

3.37 J.E. Girard and J.A. Glatz, *Am. Lab.* **13** (1981) 26.

3.38 R.L. Stevenson and K. Harrison, *Am. Lab.* **13** (1981) 76.

3.39 S. Matsushita, Y. Tada, N. Baba and K. Hosako, *J. Chromatogr.* **259** (1983) 459.

3.40 D.R. Jenke and G.K. Pagenkopf, *Anal. Chem.* **55** (1983) 225.

3.41 K. Harrison, W.C. Beckham jr., T. Yates and C.D. Carr, *Am. Lab.* May 1985.

3.42 T. Braumann, personal communication.

3.43 E. Blasius, K.P. Janzen, W. Adrian and G. Klautke, *J. Chromatogr.* **96** (1974) 89.

3.44 G.A. Melson, *Coordination Chemistry of Macrocyclic Compounds*. 1st ed. Plenum Press New York 1979.

3.45 L.R. Sousa, D.H. Hoffmann, L. Kaplan and D.J. Cram, *J. Am. Chem. Soc.* **96** (1974) 7100.

3.46 J.D. Lamb and P.A. Drake, *J. Chromatogr.* **482** (1989) 367.

3.47 K. Kimura, H. Harino, E. Hajatta and T. Shono, *Anal. Chem.* **58** (1986) 2233.

3.48 E. Blasius and K.P. Janzen, *Top. Curr. Chem.* **98** (1981) 163.

3.49 E. Blasius, K.P. Janzen, W. Klein, H. Klotz, V.B. Nguyen, T. Nguyen-Tien, R. Pfeiffer, G. Scholten, H. Simon, H. Stockemer and A. Toussaint, *J. Chromatogr.* **201** (1980) 147.

3.50 E. Blasius and K.P. Janzen, *Israel. J. Chemistry* **26** (1985) 25.

3.51 E. Blasius, K.P. Janzen, H. Luxenburger, V.B. Nguyen, H. Klotz and J. Stockemer, *J. Chromatogr.* **167** (1978) 307.

3.52 E. Blasius, K.P. Janzen, H. Simon and J. Zender, *Fresenius Z. Anal. Chem.* **320** (1985) 435.

3.53 E. Blasius, K.P. Janzen and J. Zender, *Fresenius Z. Anal. Chem.* **325** (1986) 126.

3.54 M. Igawa, K. Saito, J. Tsukamoto and M. Tanaka, *Anal. Chem.* **53** (1981) 1942.

3.55 M. Nakajima, K. Kimura and T. Shono, *Anal. Chem.* **55** (1983) 463.

3.56 J.D. Lamb and R.G. Smith, *J. Chromatogr.* **640** (1993) 33.

3.57 R.M. Izatt, J.S. Bradshaw, S.A. Nielsen, J.D. Lamb and J.J. Christensen, *Chem. Rev.* **85** (1985) 271.

3.58 R.M. Izatt, K. Pawlak, J.S. Bradshaw and R.L. Bruening, *Chem. Rev.* **91** (1991) 1721.

3.59 J.D. Lamb, R.G. Smith, R.C. Anderson and M.K. Mortensen, *J. Chromatogr. A* **671** (1994) 55.

3.60 A. Woodruff, C. Pohl, A. Bordunov and N. Avdalovic, Presentation at International Ion Chromatography Symposium, Chicago, IL, USA 2001.

3.61 J.D. Lamb, R.G. Smith and J. Jagodzinski, *J. Chromatogr.* **640** (1993) 33.

3.62 A. Clearfield, *Inorganic Ion Exchange Materials*. CRC Press Boca Raton, FL, USA 1982.

3.63 C. Laurent, H. Billiet and L. de Galan, *Chromatographia* **17** (1983) 394.

3.64 C. Laurent, H. Billiet and L. de Galan, *J. Chromatogr.* **285** (1984) 161.

3.65 G.M. Schwab and A.N.Z. Ghosh, *Z. Angew. Chem.* **53** (1940) 39.

3.66 L.R. Snyder, *Principles of Adsorption Chromatography*. Marcel Dekker New York 1968; p. 163.

3.67 G.L. Schmitt and D.J. Pietrzyk, *Anal. Chem.* **57** (1985) 2247.

3.68 L.G. Sillen, *Stability Constants of Metal-Ion Complexes*. The Chemical Society: London 1971, Supplement 1, No. 25.

3.69 O.A. Shpigun, I.N. Voloshik and Yu.A. Zolotov, *Anal. Sci.* **8** (1985) 335.

3.70 K. Irgum, *Anal. Chem.* **59** (1987) 358.

3.71 K. Irgum, *Anal. Chem.* **59** (1987) 363.

3.72 D.T. Gjerde and J.S. Fritz, *Anal. Chem.* **53** (1981) 2324.

3.73 J. Hertz and U. Baltensperger, *LC Magazin* **2** (1984) 600.

3.74 R.P.W. Scott, *Liquid Chromatography Detectors*. Elsevier New York 1977; p. 89.

3.75 C. Dengler, M. Kolb and M. Läubli, *GIT Fachz. Lab.* **6** (1996) 609.

3.76 J. Stillian, R. Slingsby and C.A. Pohl: "A Revolutionary New Suppressor For Ion Chromatography". Presentation at the *Pittsburgh Conference* 1985, New Orleans, LA, USA.

3.77 Z.W. Tian, R.Z. Hu, H.S. Lin and J.T. Wu, *J. Chromatogr.* **439** (1988) 159.

3.78 K.-H. Jansen, K.-H. Fischer and B. Wolf, *US Pat.* 4 459 357 (1984).

3.79 T. Ban, T. Murayama, S. Muramoto and Y. Hanaoka, *US Pat.* 4 403 039 (1983).

3.80 D.L. Strong and P.K. Dasgupta, *Anal. Chem.* **61** (1989) 939.

3.81 P.W. Atkins, *Physical Chemistry*. W.H. Freeman, San Francisco, USA 1978.

3.82 C. Pohl, R.W. Slingsby, J.R. Stillian and R. Gajek, *US Pat.* 4 999 098 (1991).

3.83 A. Henshall, S. Rabin, J. Statler and J. Stillian, *Am. Lab.* **24** (1992) 20R.

3.84 S. Rabin, J. Stillian, V. Barreto, K. Friedman and M. Toofan, *J. Chromatogr.* **640** (1993) 97.

3.85 G. Ackermann, W. Jugelt, H.-H. Möbius, H.D. Suschke and G. Werner, *Elektrolytgleichgewichte und Elektrochemie*. VEB Verlag für Grundstoffindustrie Leipzig 1974.

3.86 DIN 38405, Part 19 (1988) *Bestimmung der Anionen Fluorid, Chlorid, Nitrit, Orthophosphat, Bromid, Nitrat und Sulfat in wenig belasteten Wässern mit der Ionenchromatographie*.

3.87 Holleman-Wiberg, *Lehrbuch der anorganischen Chemie*. 71.–80. ed., Walter de Gruyter & Co., Berlin 1971, p. 425.

3.88 R.D. Rocklin and E.L. Johnson, *Anal. Chem.* **55** (1983) 4.

3.89 K. Han and W.F. Koch, *Anal. Chem.* **59** (1987) 1016.

3.90 C. Pohlandt-Watson: "A Revised Ion-Chromatographic Method for the Determination of Free Cyanide". Randburg, Council for Mineral Technology, Report M 283 1986; p. 1–3.

3.91 P.K. Dasgupta, K. DeCesare and J.C. Ullrey, *Anal. Chem.* **52** (1980) 1912.

3.92 M. Lindgren, A. Cedergren and J. Lindberg, *Anal. Chim. Acta* **141** (1982) 279.

3.93 H. Beyer, *Lehrbuch der organischen Chemie*. 17th ed. S. Hirzel Verlag Stuttgart 1973; p. 177.

3.94 J.E. Tong, K. Schertenleib and R.A. Carpio, *Solid State Technology* **27** (1984) 161.

3.95 R.M. Merril, *LC-GC Magazin* **6** (1988) 416.

3.96 S. Reinhard, Applications Report No. 09/93/03, Dionex GmbH Idstein, Germany.

3.97 Holleman-Wiberg, *Lehrbuch der anorganischen Chemie*. 71.–80. Ed., Walter de Gruyter & Co., Berlin 1971, p. 403f.

3.98 J. Stillian and C. Pohl, *J. Chromatogr.* **499** (1990) 249.

3.99 S. Rabin and J. Stillian, *J. Chromatogr. A* **671** (1994) 63.

3.100 J. Dean (ed.), *Lange's Handbook of Chemistry*. McGraw Hill, New York, 11th edition 1973.

3.101 J.S. Fritz, D.L. DuVal and R.E. Barron, *Anal. Chem.* **56** (1984) 1177.

3.102 D.T. Gjerde and J.S. Fritz, *Ion Chromatography*. 2nd ed. Dr. Alfred Hüthig Verlag Heidelberg – Basel – New York 1987; p. 148.

3.103 J.G. Tarter (ed.), *Ion Chromatography*, Vol. 37. Marcel Dekker Inc. New York and Basel 1987; p. 37.

3.104 J. Chang and J.S. Fritz, Iowa State University, Ames, Iowa, USA 1981.

3.105 J.S. Fritz, D.T. Gjerde and R.M. Becker, *Anal. Chem.* **52** (1980) 1519.

3.106 R.M. McCormick and B.L. Karger, *J. Chromatogr.* **199** (1980) 259.

3.107 B.A. Bidlingmeyer, *J. Chromatogr. Sci.* **18** (1980) 525.

3.108 R.W. Melander, J.F. Erard and Cs. Horvath, *J. Chromatogr.* **282** (1983) 229.

3.109 T. Okada and T. Kuwamoto, *Anal. Chem.* **56** (1984) 2073.

3.110 S. Levin and E. Grushka, *Anal. Chem.* **58** (1986) 1602.

3.111 S. Levin and E. Grushka, *Anal. Chem.* **59** (1987) 1157.

3.112 D.R. Jenke and G.K. Pagenkopf, *Anal. Chem.* **56** (1984) 88.

3.113 D.R. Jenke and G.K. Pagenkopf, *Anal. Chem.* **56** (1984) 85.

3.114 H. Small and T.E. Miller jr., *Anal. Chem.* **54** (1982) 462.

3.115 A. Ringbom, *Complexation in Analytical Chemistry*. Interscience New York 1963; p. 198 f.

3.116 J. Behnert, P. Behrend and A. Kipplinger, *LaborPraxis* **10** (8) (1986) 872.

3.117 Y. Sekiguchi, personal communication.

3.118 E. Vaeth, P. Sladek and K. Kenar, *Fresenius Z. Anal. Chem.* **329** (1987) 584.

3.119 Y. Baba, N. Yoza and S. Ohashi, *Chromatographia* **350** (1985) 119.

3.120 L.M. Nair, R. Saari-Nordhaus and J.M. Anderson, jr., *J. Chromatogr.* **671** (1994) 309.

3.121 Method 552, *The Determination of Haloacetic Acids in Drinking Water by Liquid/Liquid Extraction and Gas Chromatography with Electron Capture Detection*. US Environmental Protection Agency, Cincinnati, OH, USA 1989.

3.122 P.D. Perrin and B. Dempsey, *Buffers for pH and Metal Ion Control*. Chapman and Hall London 1974.

3.123 J.D. Lamb and Y.K. Ye, *J. Chromatogr.* **602** (1992) 189.

3.124 H.R. Christen, *Grundlagen der allgemeinen und anorganischen Chemie*. 4th ed., Verlag Sauerländer – Salle Aarau 1973, p. 565.

3.125 A. Gaedcke, Dissertation, Universität Düsseldorf 1986.

3.126 U. Fischer, Dissertation, Universität Düsseldorf 1986.

3.127 *Eur. Pat.* 61106 (7.8.85) Henkel KGaA.

3.128 *D. P.* 3111152.1 Henkel KGaA.

3.129 *U.S. Pat.* 4,572,807 (25.2.86) Henkel KGaA.

3.130 H. Waldhoff and P. Sladek, *Fresenius Z. Anal. Chem.* **320** (1985) 163.

3.131 A.W. Fitchett and A. Woodruff, *LC Magazin* **1**, No. 1 (1983).

3.132 F. Pacholec, D.T. Rossi, L.D. Ray and S. Vazopolos, *LC Magazin* **3**, No. 12 (1985) 1068.

3.133 J. Weiß, *Fresenius Z. Anal. Chem.* **320** (1985) 679.

3.134 G. Tschäbunin, P. Fischer and G. Schwedt, *Fresenius Z. Anal. Chem.* **333** (1989) 111.

3.135 G. Tschäbunin, P. Fischer and G. Schwedt, *Fresenius Z. Anal. Chem.* **333** (1989) 117.

3.136 J. Weiß and G. Hägele, *Fresenius Z. Anal. Chem.* **328** (1987) 46.

3.137 S. Reinhard and J. Weiß, Applications Report No. 07/95/07, Dionex GmbH Idstein, Germany.

3.138 G. Hägele and J. Weiß, "Ionenchromatographie und NMR – Einige vergleichende Studien am Beispiel der 1-Phosphonopropan-1,2,3-tricarbonsäure". In: A. Kettrup, J. Weiß and D. Jensen (eds.), *Spurenanalytische Bestimmung von Ionen*. Ecomed Verlagsgesellschaft Landsberg 1997.

3.139 Literature citation 7 in [142]

3.140 Literature citation 9 in [142]

3.141 Literature citation 10 in [142]

3.142 L.R. Snyder, *Chromatogr. Reviews* **7** (1965) 1.

3.143 T. Sundén, H. Lindgren, A. Cedergren and D.D. Siemer, *Anal. Chem.* **55** (1983) 2.

3.144 J.G. Tarter, *Anal. Chem.* **56** (1984) 1264.

3.145 R.D. Rocklin, C.A. Pohl and J.A. Schibler, *J. Chromatogr.* **411** (1987) 107.

3.146 H. Schwab, W. Rieman and P.A. Vaughan, *Anal. Chem.* **29** (1957) 1357.

3.147 R.D. Rocklin, C.A. Pohl and R.W. Slingsby, *J. Liq. Chromatogr.* **9** (1986) 757.

3.148 Y. Liu, N. Avdalovic, H. Small, R. Matt and H. Dhillon: "On-Line Large Capacity High Purity Acid and Base Generation Devices and Their Applications in Ion Chromatography". Presentation No. 1179, Pittsburg Conference 1998, New Orleans, LA, USA.

3.149 H. Kurth, Dionex (Switzerland) AG, personal communication.

3.150 N.E. Good, G.D. Winget, W. Winter, T.N. Conolly, S. Izawa and R.M.M. Singh, *Biochemistry* **5** (1966) 467.

3.151 J.P. Ivey, *J. Chromatogr.* **287** (1984) 128.

3.152 W.R. Jones, P. Jandik and A.L. Heckenberg, *Anal. Chem.* **60** (1988) 1977.

3.153 P.R. Knapp, *Handbook of Analytical Derivatization Reactions*. John Wiley New York 1979; p. 539.

3.154 H.D. Scobell, K.M. Brobst and E.M. Steele, *Cereal Chem.* **54** (1977) 1905.

3.155 J. Simmer and J. Puls, *J. Chromatogr.* **156** (1978) 197.

3.156 E. Rajakylä, *J. Chromatogr.* **353** (1986) 1.

3.157 R.T. Yong, L.P. Milligan and G.W. Mathison, *J. Chromatogr.* **209** (1981) 316.

3.158 D.L. Hendrix, R.E. Lee, J.G. Baust and H. James, *J. Chromatogr.* **210** (1981) 45.

3.159 R.D. Rocklin and C.A. Pohl, *J. Liq. Chromatogr.* **6** (1983) 1577.

3.160 J.A. Rendleman (1971): "Ionization of Carbohydrates in the Presence of Metal Hydroxides and Oxides". In: *Carbohydrates in Solution*. Advances in Chemistry, Series 117, Am. Chem. Soc., Washington DC.

3.161 P. Edwards and K. Haak, *Am. Lab.* April 1983.

3.162 R.E. Smith and R.A. MacQuarrie, *Anal. Biochem.* **170** (1988) 308.

3.163 J. Kerth, Applications Report No. 01/93/01, Dionex GmbH Idstein, Germany.

3.164 D.C. Johnson and T.Z. Polta, *Chromatogr. Forum* **1** (1986) 37.

3.165 J. Puls, personal communication.

3.166 T.R.I. Cataldi, I.G. Casella, and D. Centonze, *Anal. Chem.* **69** (1997) 4849.

3.167 T.R.I. Cataldi, D. Centonze, and G. Margiotta, *Anal. Chem.* **69** (1997) 4842.

3.168 T.R.I. Cataldi, C. Campa, G. Margiotta and S.A. Bufo, *Anal. Chem.* **70** (1998) 3940.

3.169 S. Honda, S. Suzuki and K.J. Kakehi, *J. Chromatogr.* **291** (1984) 317.

3.170 G.J. Bonn, *J. Chromatogr.* **350** (1985) 381.

3.171 M. Stefansson and D.J. Westerlund, *J. Chromatogr. A* **720** (1996) 127.

3.172 S.J. Angyal, *Adv. Carbohydr. Chem. Biochem.* **47** (1989) 1.

3.173 J. Thayer, P. Jandik, Y. Liu and N. Avdalovic, "Improved-Throughput Monosaccharide Analyses by HPAE-PAD with Electrochemically-Produced, Carbonate-Free Eluant", Presentation P220, Pittsburgh Conference 1999, New Orleans, LA, USA.

3.174 B.N. Mathur, P. Whalen, R. Shahani and K.M. Shahani, *Ind. J. Dairy Sci.* **45** (1992) 190.

3.175 J.A. Kynaston, S.C. Fleming, M.F. Laker and A.D.J. Pearson, *Clin. Chem.* **39** (1993) 453.

3.176 C. Bruggink and P. Bastiaanse, Dionex B.V. Breda (The Netherlands), personal communication.

3.177 H. Schiweck, *Lebensmittelchem. Gerichtl. Chem.* **41** (1987) 49.

3.178 *DE-PS* 3446380, 22.5.1986, Pfeifer & Langen, Köln.

3.179 K. Thielecke, H.-P. Lieker and T. Paskach, *Zuckerind.* **114** (1989) 953.

3.180 C. Jansen, Applications Report No. 02/88/08, Dionex GmbH Idstein, Germany.

3.181 B.B. Wheals and P.C. White, *J. Chromatogr.* **176** (1979) 421.

3.182 E. Rajakylä and M. Palopaski, *J. Chromatogr.* **282** (1983) 595.

3.183 L.A.Th. Verhaar, B.F.M. Kuster and H.A. Clearsens, *J. Chromatogr.* **284** (1984) 1.

3.184 C. Jansen, personal communication.

3.185 K. Koizumi, Y. Kubota, T. Tanimoto and Y. Okada, *J. Chromatogr.* **464** (1989) 365.

3.186 K. Koizumi, M. Fukuda and S. Hizukuri, *J. Chromatogr.* **585** (1991) 233.

3.187 D. Voet and J.G. Voet, *Biochemistry*. John Wiley & Sons, New York 1990.

3.188 T. Feizi, *Nature* **314** (1985) 53.

3.189 T.N. White and R.P. Mecham (eds.) *Biology of Proteoglycans*. Academic Press 1987.

3.190 D.G. George, W.C. Barker and L.T. Hunt, *Nucl. Acids Res.* **14** (1986) 11.

3.191 N. Sharon and H. Lis, *Spektrum der Wissenschaft* **3** (1993) 66.

3.192 G.P. Reddy and C.A. Bush, *Anal. Biochem.* **198** (1991) 278.

3.193 C. Kunz and S. Rudloff, *Acta Paediatr.* **82** (1993) 903.

3.194 T.W. Rademacher, R.B. Parekh and D.A. Dwek, *Ann. Rev. Biochem.* **57** (1988) 785.

3.195 J.E. Sadler. In: *Biology of Carbohydrates*. V. Ginsburg, P.W. Robbins (eds.), John Wiley & Sons, New York 1984.

3.196 J.C. Paulson, *Trends in Biochem. Sci.* **14** (1989) 272.

3.197 P. Hermentin, R. Witzel, R. Doenges, R. Bauer, H. Haupt, T. Patel, R.B. Parekh and D. Brazel, *Anal. Biochem.* **206** (1992) 419.

3.198 S. Honda, S. Suzuki, A. Nose, K. Yamamoto and K. Kakehi, *Carbohydrate Research* **215** (1991) 238.

3.199 P. Jackson, *Anal. Biochem.* **196** (1991) 238.

3.200 A. Haselbeck, E. Schickaneder, H. von der Eltz and W. Hoesel, *Anal. Biochem.* **191** (1990) 25.

3.201 S. Takasaki, T. Mitsuochi and A. Kobata, *Methods Enzymol.* **83** (1981) 263.

3.202 W.T. Wang, N.C. LeDonne jr., B. Ackerman and C.C. Sweeley, *Anal. Biochem.* **141** (1984) 366.

3.203 N. Tomija, N. Kurono, H. Ishihara, S. Teijima, S. Endo, Y. Arata and N. Takahashi, *Anal. Biochem.* **163** (1987) 489.

3.204 Dionex Corporation, Sunnyvale, USA, Technical Note 40: "Glycoprotein Monosaccharide Analysis Using HPAEC-PAD".

3.205 Dionex Corporation, Sunnyvale, USA, Technical Note 41: "Analysis of Sialic Acids Using HPAEC-PAD".

3.206 Dionex Corporation, Sunnyvale, USA, Technical Note 42: "Glycoprotein Oligosaccharide Analysis Using HPAEC-PAD".

3.207 M. Weitzhandler, C. Pohl, J. Rohrer, L. Narayanan, R. Slingsby and N. Avdalovic, *Anal. Biochem.* **241** (1996) 128.

3.208 J. Rohrer, J. Thayer, M. Weitzhandler and N. Avdalovic, *Glycobiology* **8** (1998) 35.

3.209 P. Hermentin and J. Seidat, "Microscale analysis of *N*-acetylneuraminic acid". In: H.S. Conradt (ed.) *Protein Glycosylation: Cellular, Biotechnological, and Analytical Aspects. GBF Monographs* **15** (1991) 185, VCH Verlagsgesellschaft Weinheim Cambridge.

3.210 M.R. Hardy, R.R. Townsend and Y.C. Lee, *Anal. Biochem.* **170** (1988) 54.

3.211 M. Weitzhandler, J. Rohrer, J.R. Thayer and N. Avdalovic. In: *Methods in Molecular Biology*, Glycoanalysis Protocol, E.F. Hounsell (ed.), Humana Press. Inc., in press.

3.212 N. Tomija, N. Kurono, H. Ishihara, S. Teijima, S. Endo, Y. Arata and N. Takahashi, *Anal. Biochem.* **163** (1987) 489.

3.213 V.K. Dua and C.A. Bush, *Anal. Biochem.* **137** (1984) 33.

3.214 W.M. Blanken, M.L.E. Bergh, P.L. Kappen and D.H. Van den Eijnden, *Anal. Biochem.* **145** (1985) 322.

3.215 E.F. Hounsell, J.M. Rideout, N.J. Pickering and C.K. Lim, *J. Liq. Chromatogr.* **7** (1984) 661.

3.216 M.R. Hardy, R.R. Townsend, T.C. Wong and Y.C. Lee, *Biochemistry* **25** (1986) 5716.

3.217 M.R. Hardy and R.R. Townsend, *Proc. Natl. Acad. Sci. USA* **85** (1988) 3289.

3.218 R.R. Townsend, M.R. Hardy, J.D. Olechno and S.R. Carter, *Nature* **335** (1988) 379.

3.219 A. Neuberger and B.M. Wilson, *Carbohydr. Res.* **17** (1971) 89.

3.220 J.S. Rohrer, *Glycobiology* **5** (1995) 359.

3.221 L.J. Basa and *M.W. Spellman, J. Chromatogr.* **499** (1990) 205.

3.222 L.M. Hernandez, L. Ballou and C.E. Ballou, *Carbohydr. Res.* **203** (1990) 1.

3.223 G. Pfeiffer, H. Geyer, R. Geyer, I. Kalsner and P. Wendorf, *Biomed. Chromatogr.* **4** (1990) 193.

3.224 R.R. Townsend, P.H. Atkinson and R.B. Tremble, *Carbohydr. Res.* **215** (1991) 211.

3.225 R.R. Townsend, M.R. Hardy, O. Hindsgaul and Y.C. Lee, *Anal. Biochem.* **174** (1988) 459.

3.226 R.R. Townsend, M.R. Hardy, D.A. Cumming, J.P. Carver and B. Bendiak, *Anal. Biochem.* **182** (1989) 1.

3.227 P. Hermentin, R. Witzel, J.F.G. Vliegenhart, J.P. Kamerling, M. Nimtz and H.S. Conradt, *Anal. Biochem.* **203** (1992) 281.

3.228 M.R. Hardy and R.R. Townsend, *Carbohydr. Res.* **188** (1989) 1.

3.229 E. Watson, A. Bhide, W.C. Kenney and F.-K. Lin, *Anal. Biochem.* **205** (1992) 90.

3.230 M. Weitzhandler and N. Avdalovic. In: A Laboratory Guide to Glycoconjugate Analysis, P. Jackson and J. Gallagher (ed.), Birkhauser Verlag AG (in preparation).

3.231 M. Nimtz and H.S. Conradt. In: Protein Glycosylation – Cellular, Biotechnological and Analytical Aspects, H.S. Conradt (ed.), *GBF Monographs* **15** (1991) 235, VCH Verlagsgesellschaft Weinheim, New York, Cambridge.

3.232 J.R. Barr, K.R. Anumula, M.B. Vettese, P.B. Taylor and S.A. Carr, *Anal. Biochem.* **192** (1991) 181.

3.233 G.D. Roberts, W.P. Johnson, S. Burman, K.R. Anumula and S.A. Carr, *Anal. Chem.* **67** (1995) 3613.

3.234 W.M.A. Niessen, R.A.M. van der Hoeven, J. van der Greef, H.A. Schols, G. Lucas-Lokhorst, A.G.J. Voragen and C. Bruggink, *Rapid Commun. Mass Spectr.* **6** (1992) 474.

3.235 R.A.M. van der Hoeven, W.M.A. Niessen, H.A. Schols, C. Bruggink, A.G.J. Voragen and J. van der Greef, *J. Chromatogr.* **627** (1992) 63.

3.236 W.M.A. Niessen, R.A.M. van der Hoeven, J. van der Greef, H.A. Schols, A.G.J. Voragen and C. Bruggink, *J. Chromatogr.* **647** (1993) 319.

3.237 W.M.A. Niessen, R.A.M. van der Hoeven, J. van der Greef, H.A. Schols, A.G.J. Voragen and C. Bruggink, *Carbohydr. Netherlands* **9** (1993) 7.

3.238 H.A. Schols, M. Mutter, A.G.J. Voragen, W.M.A. Niessen, R.A.M. van der Hoeven, J. van der Greef and C. Bruggink, *Carbohydr. Res.* **261** (1994) 335.

3.239 M.W. Spellman, L.J. Basa, C.K. Leonard J.A. Chakel, and J.V. O'Connor, *J. Biol. Chem.* **264** (1989) 14100.

3.240 J.C. Rouse and J.E. Vath, *Anal. Biochem.* **238** (1996) 82.

3.241 J.R. Thayer, J.S. Rohrer, N. Avdalovic and R.P. Gearing, *Anal. Biochem.* **256** (1998) 207.

3.242 D. Tetaert, B. Soudan, J.-M. Lo-Guidice, C. Richet, P. Degand, G. Boussard, C. Mariller and G. Spik, *J. Chromatogr. B* **658** (1994) 31.

3.243 R.C. Hoffman, H. Andersen, K. Walker, J.D. Krakover, S. Patel, M.R. Stamm and S.G. Osborn, *Biochemistry* **35** (1996) 14849.

3.244 G. Spik, V. Debruyne, J. Montrieul, H. von Halbeek and J.F.G. Vliegenhart, *FEBS Lett.* **183** (1985) 65.

3.245 J.S. Rohrer and N. Avdalovic, *Protein Expression and Purification* **7** (1996) 39.

3.246 K.R. Anumula and P.B. Taylor, *Eur. J. Biochem.* **195** (1991) 269.

3.247 M. Weitzhandler, D. Kadlecek, N. Avdalovic, J.G. Forte, D. Chow and R.R. Townsend, *J. Biol. Chem.* **268** (1993) 5121.

3.248 D.C. Andersen, C.F. Goochee, G. Cooper and M. Weitzhandler, *Glycobiology* **4** (1994) 459.

3.249 U.R. Bhat and E.A. Helgeson, *Am. Biotech. Lab.* **1** (1994) 16.

3.250 A.P. Clarke, P. Jandik, R.D. Rocklin, Y. Liu and N. Avdalovic, *Anal. Chem.* **71** (1999) 2774.

3.251 P. Jandik, A.P. Clarke, N. Avdalovic, D.C. Anderson and J. Cacia, *J. Chromatogr. B* **732** (1999) 193.

3.252 P. Jandik, C. Pohl, V. Barreto and N. Avdalovic, Anion exchange chromatography and integrated amperometric detection of amino acids. In: C. Cooper, N. Packer and K. Williams (eds.), *Methods in Molecular Biology*, Vol. 159; Amino Acid Analysis Protocols, Humana Press, Inc. Totowa, NJ, USA 2001.

3.253 P. Jandik, J. Cheng, D. Jensen, S. Manz and N. Avdalovic, *J. Chromatogr. B* **758** (2001) 189.

3.254 P. Jandik, J. Cheng, D. Jensen, S. Manz and N. Avdalovic, *Anal. Biochem.* **287** (2000) 38.

3.255 J.W. Wimberley, *Anal. Chem.* **53** (1981) 1709.

3.256 M. Weitzhandler, D. Farnan, J. Horvath, J.S. Rohrer, R.W. Slingsby, N. Avdalovic and C. Pohl, *J. Chromatogr. A* **828** (1998) 365.

3.257 R.J. Harris, *J. Chromatogr.* **705** (1995) 129.

3.258 R.J. Harris, A.A. Murnane, S.L. Utter, K.L. Wagner, E.T. Cox, G. Polastri, J.C. Helder and M.B. Sliwkowski, *Bio/Technology* **11** (1993) 1293.

3.259 R.J. Harris, K.L. Wagner and M.W. Spellman, *Eur. J. Biochem.* **194** (1990) 611.

3.260 P. Rao, A. Williams, A. Baldwin-Ferro, E. Hanigan, D. Kroon, M. Makowski, E. Meyer, V. Numsuwan, E. Rubin and A. Tran, *BioPharm* **4** (1991) 38.

3.261 D.A. Lewis, A.W. Guzetta, W.S. Hancock and M. Costello, *Anal. Chem.* **66** (1994) 585.

3.262 A.B. Johnson, J.M. Shirokawa, W.S. Hancock, M.W. Spellman, L.J. Basa and D.W. Aswald, *J. Biol. Chem.* **264** (1989) 14262.

3.263 M.V. Panandi, A.W. Guzetta, W.S. Hancock and D.W. Aswald, *J. Biol. Chem.* **269** (1994) 243.

3.264 A. Tuong, M. Maftough, C. Ponthus, O. Whitechurch, C. Roitsch and C. Picard, *Biochemistry* **31** (1992) 8291.

3.265 J. Cacia, C.P. Quan, M. Vasser, M.B. Sliwkowski and J. Frenz, *J. Chromatogr.* **634** (1993) 229.

3.266 P.K. Tsai, M. Bruner, J.I. Irwin, C.C.Y. Ip, C.N. Oliver, R.W. Nelson, D.B. Volkin and C.R. Middaugh, *Pharm. Res.* **10** (1993) 1580.

3.267 D.W. Aswald. *Deamidation and Isoaspartate Formation in Peptides and Proteins.* CRC Series in Analytical Biotechnology, CRC Press, Boca Raton, FL, USA 1995.

3.268 A.D. Donato, M.A. Ciardello, M.D. Nigris, R. Piccoli, L. Mozzarella and G. D'Alessio, *J. Biol. Chem.* **268** (1993) 4745.

3.269 E. Margoliash, in B. Chance and R. Estabrook (eds.), *Hemes and Hemoproteins.* Academic Press, Inc. New York 1966, p. 373.

3.270 J. Rohrer and N. Avdalovic, personal communication.

3.271 J. Rohrer, *J. Chromatogr. A* **667** (1994) 75.

3.272 M. Rhodes, P. Azari and R. Feeney, *J. Biol. Chem.* **230** (1958) 390.

3.273 J. Rohrer, personal communication.

3.274 R.L. Wiseman, J.E. Fothergill and L.A. Fothergill, *J. Biochem.* **127** (1972) 775.

3.275 G. Zon. In: *HPLC in Biotechnology*, W.S. Hancock (ed.), John Wiley & Sons 1990, p. 301.

3.276 A. Pingoud, A. Fliess and V. Pingoud. HPLC of Oligonucleotides. In: *HPLC of Macromolecules*, R.W.A. Oliver (ed.), Oxford University Press 1989, p. 183.

3.277 A.J. Bourque and A.S. Cohen, *J. Chromatogr. B* **662** (1994) 343.

3.278 C.G. Huber, P.J. Oefner and G.K. Bonn, *Anal. Biochem.* **212** (1993) 351.

3.279 J. Thayer, R.M. McCormick and N. Avdalovic, *Meth. Enzym.* (in press).

Chapter 4

4.1 H. Small, T.S. Stevens and W.C. Bauman, *Anal. Chem.* **47** (1975) 1801.

4.2 P. Hajos and J. Inczédy, *J. Chromatogr.* **201** (1980) 253.

4.3 C.A. Pohl and E.L. Johnson, *J. Chromatogr. Sci.* **18** (1980) 442.

4.4 P. Kolla, J. Köhler and G. Schomburg, *Chromatographia* **23** (1987) 465.

4.5 M.W. Läubli and B. Kampus, *J. Chromatogr. A* **706** (1995) 103.

4.6 M.W. Läubli and B. Kampus, *J. Chromatogr. A* **706** (1995) 99.

4.7 D. Jensen, J. Weiss, M.A. Rey and C.A. Pohl, *J. Chromatogr.* **640** (1993) 65.

4.8 D. Jensen, *Labo* **4** (1992) 64.

4.9 Applications Report No. 04/91/06, Dionex GmbH Idstein, Germany.

4.10 Applications Report No. 01/92/02, Dionex GmbH Idstein, Germany.

4.11 D. Jensen and S. Böhling, *LaborPraxis* **11** (1996) 481.

4.12 M.A. Rey, C.A. Pohl, J.J. Jagoszinski, E.Q. Kaiser and J.M. Riviello, *J. Chromatogr. A* **804** (1998) 201.

4.13 P.E. Jackson, *Ion Chromatography in Environmental Analysis.* In: R.A. Meyers (ed.) Encyclopedia of Analytical Chemistry, John Wiley & Sons Chichester 2000, p. 2779-2801.

4.14 Dionex Corporation, Sunnyvale, USA, Application Update No. 138.

4.15 J. Weil, Diploma thesis, Fachhochschule Fresenius, Wiesbaden 1990.

4.16 G. Schomburg, *LC-GC Magazin* **6** (1988) 36.

4.17 L. Nair, R. Saari-Nordhaus and J.M. Anderson, jr., *J. Chromatogr.* **640** (1993) 41.

4.18 L. Nair, R. Saari-Nordhaus and J.M. Anderson, jr., *J. Chromatogr.* **671** (1994) 43.

4.19 K. Kimura, H. Harino, E. Hajata and T. Shono, *Anal. Chem.* **58** (1986) 2233.

4.20 K. Bächmann and K.-H. Blaskowitz, *Fresenius Z. Anal. Chem.* **333** (1989) 15.

4.21 A. Henshall, S. Rabin, J. Statler and J. Stillian, *Am. Lab.* **24** (1992) 20R.

4.22 S. Rabin, J. Stillian, V. Barreto, K. Friedman and M. Toofan, *J. Chromatogr.* **640** (1993) 97.

4.23 J. Weiss, *Ionenchromatographie*. 2nd ed. VCH Verlagsgesellschaft Weinheim 1991, p. 192f.

4.24 O. Samuelson, *Z. Anal. Chem.* **116** (1939) 328.

4.25 J.S. Fritz and J.N. Story, *Anal. Chem.* **46** (1974) 825.

4.26 S. Elchuk and R.M. Cassidy, *Anal. Chem.* **51** (1979) 1434.

4.27 J.M. Riviello and W.E. Rich, Dionex Corp., internal report 1982.

4.28 S.S. Heberling, J.M. Riviello, M.S. Taylor, S. Papanu and M. Ebenhahn: "New and Versatile IC Columns for Metal Separations". Presentation No. 26, Rocky Mountain Conference 1984, Denver, CO, USA.

4.29 C.A. Pohl and R.M. Riviello, Presentation No. 108, 24. Rocky Mountain Conference 1982, Denver, CO, USA.

4.30 G.J. Sevenich and J.S. Fritz, *Anal. Chem.* **55** (1983) 12.

4.31 J. Weiß, *LaborPraxis*, **4** (1987) 321.

4.32 J. Weiß, *LaborPraxis* **5** (1987) 468.

4.33 A.E. Martell and R.M. Smith, *Critical Stability Constants*, Vol. 3. Plenum New York 1977.

4.34 D.-R. Yan and G. Schwedt, *LaborPraxis* **1** (1987) 48.

4.35 S.S. Heberling and J.M. Riviello: "Advances in High Performance IC of Transition and Post-Transition Metals". Presentation at 27th Rocky Mountain Conference 1985, Denver, CO, USA.

4.36 S.S. Heberling: "Recent Advances in Metals Determination by Ion Chromatography". Presentation at *Pittsburgh Conference* 1986, Atlantic City, N.J., USA.

4.37 S. Somerset, personal communication.

4.38 M.P. Harrods, A. Siriraks and J. Riviello, *J. Chromatogr.* **602** (1992) 119.

4.39 R.R. Greenberg and H.M. Kingston, *Anal. Chem.* **55** (1983) 1160.

4.40 Y. Igarashi, C.K. Kim, Y. Takatu, K. Shiraishi, M. Yamamoto and N. Ikeda, *Anal. Sci.* **6** (1990) 157.

4.41 S.B. Savvin, *Talanta* **8** (1961) 673.

4.42 S.B. Savvin, *Talanta* **11** (1964) 7.

4.43 H.G. Petrov and C.D. Strehlow, *Anal. Chem.* **39** (1967) 265.

4.44 B. Budesinsky, *Talanta* **16** (1960) 1277.

4.45 R.M. Cassidy and S. Elchuk, *Anal. Chem.* **54** (1982) 1558.

4.46 C.H. Knight and R.M. Cassidy, *Anal. Chem.* **56** (1984) 474.

4.47 S.S. Heberling, J.M. Riviello, M. Shifen and A.W. Ip, *Res. Dev.* September (1987) 74.

4.48 A.P. le Roex and R.T. Watkins, *Chem. Geol.* **88** (1990) 151.

Chapter 5

5.1 R.M. Wheaton and W.C. Bauman, *Ind. Eng. Chem.* **45** (1953) 228.

5.2 C. Sarzanini and S. Cavalli, *Cromatografia Ionica*. UTET Libreria Turin 1998.

5.3 G.A. Harlow and D.H. Morman, *Anal. Chem.* **36** (1964) 2438.

5.4 R. Wood, L. Cummings and T. Jupille, *J. Chromatogr. Sci.* **18** (1980) 551.

5.5 D.P. Lee and A.D. Lord, *LC-GC Magazine* **5** (1987) 261.

5.6 V.T. Turkelson and M. Richards, *Anal. Chem.* **50** (1978) 1420.

5.7 T. Jupille, D.W. Togami and D.E. Burge, *Ind. Res. Dev.* **25** (1983) 151.

5.8 W.E. Rich, E.L. Johnson, L. Lois, P. Kabra, B. Stafford and L. Marton, *Clin. Chem.* **26** (1980) 1492.

5.9 K. Tanaka and J.S. Fritz, *J. Chromatogr.* **361** (1986) 151.

5.10 K. Watanabe and T. Ishizaki, Poster No. 66, International Ion Chromatography Symposium 1995, Dallas, TX, USA.

5.11 J.-G. Chen and M. Wu, *Micro* **1** (1997) 31.

5.12 H.-J. Kim and Y.-K. Kim, *J. Food Sci.* **51** (1986) 1360.

5.13 P. Cunniff (ed.), AOAC Official Methods of Analysis of AOAC International, 16th edition, AOAC International Arlington, VA, 1995, Vol. 2, Chapter 47, p. 33-34.

5.14 H.P. Wagner and M.J. McGarrity, *J. Chromatogr.* **546** (1991) 119.

5.15 D.R. Lide (ed.) *CRC Handbook of Chemistry and Physics*. 72nd edition, CRC Press BocaRaton – Ann Arbor – Boston 1991.

5.16 W.E. Rich, F. Smith, L. McNeill and T. Sidebottom, "Ion Exclusion Coupled to Ion Chromatography: Instrumentation and Application". In: *Ion Chromatographic Analysis of Environmental Pollutants,* Vol. 2. Ann Arbor Science, Ann Arbor Michigan 1979; p. 17.

5.17 W.E. Rich, E.L. Johnson, L. Lois, P. Kabra and L. Marton, "Organic Acids by Ion Chromatography". In: L. Marton and P. Kabra (eds.), *Liquid Chromatography in Clinical Analysis*. The Humana Press, Inc. 1981.

5.18 E. Kaiser, J. Rohrer and W. Ausserer, "Determination of Trace Anions in Concentrated Weak Acids by Ion Chromatography", Presentation No. 962, Pittsburg Conference, Orlando 1999.

5.19 A. Siriraks, C.A. Pohl and M. Toofan, *J. Chromatogr.* **602** (1992) 89.

5.20 Dionex Corporation, Sunnyvale, USA, Technical Note No. 44.

5.21 T. Jupille, M. Gray, B. Black and M. Gould, *Am. Lab.* **13** (1981) 80.

5.22 W.R. LaCourse, D.C. Johnson, M.A. Rey and R.W.Slingsby, *Anal. Chem.* **63** (1991) 134.

5.23 R. Rocklin, R.W. Slingsby and A. Woodruff, "Optimization of Conditions for the Determination of Alcohols and Aldehydes Using Separation on High Capacity Ion-Exchange Resins and Pulsed Amperometric Detection." Presentation No. 264, Pittsburgh Conference, New Orleans, 1995.

5.24 D.H. Spackman, W.H. Stein and S. Moore, *Anal. Chem.* **59** (1958) 1190.

5.25 M.C. Roach and M.D. Harmony, *Anal. Chem.* **59** (1987) 411.

5.26 W.D. Hill, F.H. Walters, T.D. Wilson and J.D. Stuart, *Anal. Chem.* **51** (1979) 1338.

5.27 S. Einarsson, B. Josefsson and S. Lagerkvist, *J. Chromatogr.* **282** (1983) 609.

5.28 S. Einarsson, S. Folestad, B. Josefsson and S. Lagerkvist, *Anal. Chem.* **58** (1986) 1638.

5.29 J.L. Glajch and J.J. Kirkland, *J. Chromatogr. Sci.* **25** (1987) 4.

5.30 B.-L. Johansson and K. Isaksson, *J. Chromatogr.* **356** (1986) 383.

5.31 M. Abrahamsson and K. Gröningsson, *J. Chromatogr.* **154** (1978) 313.

5.32 R.L. Heinrikson and S.C. Meredith, *Anal. Biochem.* **136** (1983) 65.

5.33 H. Scholze, *J. Chromatogr.* **350** (1985) 453.

5.34 J.M. Wilkinson, *J. Chromatogr. Sci.* **16** (1978) 547.

5.35 N. Kaneda, M. Sato and K. Yagi, *Anal. Biochem.* **127** (1982) 49.

5.36 G. Ogden, *LC-GC Magazine* **5** (1987) 28.

5.37 K. Dus, S. Lindroth, R. Pabst and R. Smith, *Anal. Biochem.* **14** (1966) 41.

5.38 P.E. Hare, *Space Life Sci.* **3** (1972) 354.

5.39 J.R. Benson, U.S. Patent Nr. 3,686,118 (1972).

5.40 K. Piez and L. Morris, *Anal. Biochem.* **1** (1960) 187.

5.41 M. Roth, *Anal. Chem.* **43** (1971) 880.

5.42 M. Roth and A. Hampai, *J. Chromatogr.* **83** (1973) 353.

5.43 J.R. Benson and P.E. Hare, *Proc. Nat. Acad. Sci. USA* **72** (1975) 619.

5.44 K.S. Lee and D.G. Drescher, *Int. J. Biochem.* **9** (1978) 457.

5.45 P. Bohlen and M. Mellet, *Anal. Biochem.* **94** (1974) 313.

5.46 D.R. Jenke and D.S. Brown, *Anal. Chem.* **59** (1987) 1509.

5.47 T.L. Perry, G.H. Dixon and S. Hansen, *Nature (London)* **206** (1965) 895.

5.48 G.V. Paddock, G.B. Wilson and A.-C. Wang, *Biochem. Biophys. Res. Commun.* **87** (1979) 946.

5.49 H. Godel, Th. Graser, P. Földi and P. Fürst, *J. Chromatogr.* **297** (1984) 49.

5.50 T. Gerritsen, M.L. Rehberg and H.A. Weisman, *Anal. Biochem.* **11** (1965) 460.

5.51 H. Matsubara and R.M. Sasaki, *Biochem. Biophys. Res. Commun.* **35** (1969) 175.

5.52 A.P. Williams, *J. Chromatogr.* **373** (1986) 175.

Chapter 6

6.1 D.P. Wittmer, N.O. Nuessle and W.G. Haney, jr., *Anal. Chem.* **47** (1975) 1422.

6.2 S.P. Sood, L.E. Sartori, D.P. Wittmer and W.G. Haney, *Anal. Chem.* **48** (1976) 796.

6.3 "Paired Ion Chromatography, an Alternative to Ion Exchange". Waters Associates, Milford, Mass. 1975.

6.4 J.H. Knox and G.R. Laird, *J. Chromatogr.* **122** (1976) 17.

6.5 J.H. Knox and J. Jurand, *J. Chromatogr.* **125** (1976) 89.

6.6 C. Horvath, W. Melander, I. Molnar and P. Molnar, *Anal. Chem.* **49** (1977) 2295.

6.7 C. Horvath, W. Melander and I. Molnar, *J. Chromatogr.* **125** (1976) 129.

6.8 J.C. Kraak, K.M. Jonker and J.F.K. Huber, *J. Chromatogr.* **142** (1977) 671.

6.9 N.E. Hoffmann and J.C. Liao, *Anal. Chem.* **49** (1977) 2231.

6.10 P.T. Kissinger, *Anal. Chem.* **49** (1977) 883.

6.11 W.R. Melander and C. Horvath, *J. Chromatogr.* **201** (1980) 211.

6.12 J.H. Knox and J. Jurand, *J. Chromatogr.* **103** (1975) 311.

6.13 B.A. Bidlingmeyer, S.N. Deming, W.P. Price, jr., B. Sachok and M. Petrusek, *J. Chromatogr.* **186** (1979) 419.

6.14 B.A. Bidlingmeyer, *J. Chrom. Sci.* **18** (1980) 525.

6.15 C. Pohl, "Mobile Phase Ion Chromatography (MPIC). Theory and Separation." Dionex Dept. of Research & Development, IC-Exchange No. 2 (1982).

6.16 N. Skelly, *Anal. Chem.* **54** (1982) 712.

6.17 R.N. Reeve, *J. Chromatogr.* **177** (1979) 393.

6.18 R.M. Cassidy and S. Elchuk, *Anal. Chem.* **54** (1982) 1558.

6.19 R.M. Cassidy and S. Elchuk, *J. Chrom. Sci.* **21** (1983) 454.

6.20 R.M. Cassidy and S. Elchuk, *J. Chromatogr.* **262** (1983) 311.

6.21 B.B. Wheals, *J. Chromatogr.* **262** (1983) 61.

6.22 W. Jost, R. Spatz, R. Dietz and F. Eisenbeiss, *LaborPraxis* **11** (1984) 1184.

6.23 W. Jost, R. Spatz, R. Dietz and F. Eisenbeiss, *LaborPraxis* **10** (1984) 1016.

6.24 Z. Iskandarani and D.J. Pietrzyk, *Anal. Chem.* **54** (1982) 2427.

6.25 Z. Iskandarani and D.J. Pietrzyk, *Anal. Chem.* **54** (1982) 2601.

6.26 I. Molnar, H. Knauer and D. Wilk, *J. Chromatogr.* **201** (1980) 225.

6.27 M. Dreux, M. Lafosse and M. Pequinot, *Chromatographia* **15** (1982) 653.

6.28 B.A. Bidlingmeyer, C.T. Santasassia and F.V. Warren, jr., *Anal. Chem.* **59** (1987) 1843.

6.29 J.P. de Kleijn, *Analyst (London)* **107** (1982) 223.

6.30 U. Leuenberger, R. Gauch, K. Rieder and E. Baumgärtner, *J. Chromatogr.* **202** (1980) 461.

6.31 H.J. Cortes, *J. Chromatogr.* **234** (1982) 517.

6.32 J. Weiß and M. Göbl, *Fresenius Z. Anal. Chem.* **320** (1985) 439.

6.33 M. Weidenauer, P. Hoffmann and K.H. Lieser, *Fresenius Z. Anal. Chem.* **331** (1988) 372.

6.34 S.B. Rabin and D.M. Stanbury, *Anal. Chem.* **57** (1985) 1131.

6.35 R. Steudel and G. Holdt, *J. Chromatogr.* **361** (1986) 379.

6.36 J. Weiß, *LaborPraxis* **11** (1987) 321.

6.37 J. Weiß, H.J. Möckel, A. Müller, E. Diemann and H.-J. Walberg, *J. Chromatogr.* **439** (1988) 93.

6.38 A. Müller, E. Diemann, R Jostes and H. Bögge, *Angew. Chem.* **93** (1981) 957.

6.39 D.L. McAlese, *Anal. Chem.* **59** (1987) 541.

6.40 H. Beyer, *Lehrbuch der organischen Chemie*. 17th Edition S. Hirzel Verlag, Stuttgart 1973; p. 691.

6.41 A. Marcomini and W. Giger, *Anal. Chem.* **59** (1987) 1709.

6.42 J. Weiß, *J. Chromatogr.* **353** (1986) 303.

6.43 B. Steinbrech, D. Neugebauer and G. Zulauf, *Fresenius Z. Anal. Chem.* **324** (1986) 154.

6.44 G. Janssen, Diploma Thesis FH Niederrhein 1987/88.

6.45 G. Aced, E. Anklam and H.J. Möckel, *J. Liq. Chromatogr.* **10** (1987) 3321.

6.46 G. Aced, Dissertation Technical University Berlin 1989, D 83.

6.47 H.J. Möckel. In: J.C. Giddings, E. Grushka and P.R. Brown (eds.), *Advances in Chromatography*, Vol. 26. Marcel Dekker, New York 1987.

6.48 R.W. Slingsby, *J. Chromatogr.* **371** (1986) 373.

6.49 J. Weiß and G. Hägele, *Fresenius Z. Anal. Chem.* **328** (1987) 46.

6.50 C. Umile and J.F.K. Huber, *J. Chromatogr. A* **723** (1996) 11.

6.51 C. Umile and J.F.K. Huber, *J. Chromatogr.* **640** (1993) 27.

6.52 C. Umile and J.F.K. Huber, *Talanta* **41** (1994) 1101.

6.53 J.R. Stillian and C.A. Pohl, *J. Chromatogr.* **499** (1990) 249.

6.54 R.W. Slingsby and M. Rey, *J. Liq. Chromatogr.* **13** (1990) 107.

Chapter 7

7.1 G. Ackermann, W. Jugelt, H.-H. Möbius, H.D. Suschke and G. Werner, *Elektrolytgleichgewichte und Elektrochemie*. VEB Verlag für Grundstoffindustrie, Leipzig 1974.

7.2 M. Göbl, *GIT Fachz. Lab.* **27** (1983) 261-65 and 373-75.

7.3 J.S. Fritz and D.T. Gjerde, *Ion Chromatography*. Dr. Alfred Hüthig-Verlag, Heidelberg – Basel – New York 1987.

7.4 D.T. Gjerde and J.S. Fritz, *Anal. Chem.* **53** (1981) 2324.

7.5 D.T. Gjerde, J.S. Fritz and G. Schmuckler, *J. Chromatogr.* **186** (1979) 509.

7.6 J.E. Girard and J.E. Glatz, *Am. Lab.* **13** (1981) 26.

7.7 T. Okada and T. Kuwamoto, *Anal. Chem.* **55** (1983) 1001.

7.8 J.P. Ivey, *J. Chromatogr.* **287** (1984) 128.

7.9 K. Irgum, *Anal. Chem.* **59** (1987) 358.

7.10 H. Small, *Ion Chromatography*. Plenum Press New York 1989, p. 180 ff.

7.11 R.D. Rocklin and E.L. Johnson, *Anal. Chem.* **55** (1983) 4.

7.12 K. Han and W.F. Koch, *Anal. Chem.* **59** (1987) 1016.

7.13 L.K. Tan and J.E. Dutrizac, *Anal. Chem.* **58** (1986) 1383.

7.14 T. Kolb, G. Bogenschütz and J. Schäfer, *LaborPraxis* **5** (1998) 66.

7.15 S. Hughes and D.C. Johnson, *Anal. Chim. Acta* **149** (1981) 1.

7.16 R.D. Rocklin and E.L. Johnson, *Anal. Chem.* **55** (1982) 392.

7.17 J. Weiss, *GIT Supplement Chromatographie* **3** (1986) 65.

7.18 D.C. Johnson and W.R. LaCourse, *Anal. Chem.* **62** (1990) 589A.

7.19 M.B. Jensen and D.C. Johnson, *Anal. Chem.* **69** (1997) 1776.

7.20 R.D. Rocklin, A.P. Clarke and M. Weitzhandler, *Anal. Chem.* **70** (1998) 1496.

7.21 M. Weitzhandler, C. Pohl, L. Narayanan, R. Slingsby and N. Avdalovic, *Anal. Biochem.* **241** (1996) 128.

7.22 D.C. Johnson and W.R. LaCourse, in: *Carbohydrate Analysis: High Performance Liquid Chromatography and Capillary Electrophoresis*. Z. El Rassi (ed.), Elsevier Science Amsterdam 1994, Chapter 10.

7.23 A.P. Tomilov, S.G. Mairanovskij, M.Y. Fioshin and V.A. Smironov, *Electrochemistry of Organic Compounds*. Halsted New York 1972, p. 314.

7.24 S. Hughes, L. Meschi and D.C. Johnson, *Anal. Chim. Acta* **132** (1981) 1.

7.25 W.R. LaCourse and G.S. Owens, *Anal. Chim. Acta* **307** (1995) 301.

7.26 C.O. Dasenbrock and W.R. LaCourse, *Anal. Chem.* **70** (1998) 2415.

7.27 V. Hanko and J. Rohrer, "Determination of Sulfur-Containing Antibiotics by Improved HPLC-Pulsed Electrochemical Detection", Presentation No. 636, Pittsburg Conference, Orlando 1999.

7.28 J.A. Polta and D.C. Johnson, *J. Liq. Chromatogr.* **6** (1983) 1727.

7.29 D.A. Martens and W.T. Frankenberger, *J. Liq. Chromatogr.* **15** (1992) 423.

7.30 L.E. Welch, W.R. LaCourse, D.A. Mead and D.C. Johnson, *Anal. Chem.* **61** (1989) 555.

7.31 A.P. Clarke, P. Jandik, R.D. Rocklin, Y. Liu and N. Avdalovic, *Anal. Chem.* **71** (1999) 2774.

7.32 P. Edwards and K. Haak, "New Pulsed Amperometric Detector for Ion Chromatography", *Am. Lab.* April 1983.

7.33 R.P. Buck, S. Singhadeja and L.B. Rogers, *Anal. Chem.* **26** (1954) 1240.

7.34 R.N. Reeve, *J. Chromatogr.* **177** (1979) 393.

7.35 U. Leuenberger, R. Gauch, K. Rieder and E. Baumgärtner, *J. Chromatogr.* **202** (1980) 461.

7.36 R.J. Williams, *Anal. Chem.* **55** (1983) 851.

7.37 S.H. Kola, K.A. Buckle and M. Wooton, *J. Chromatogr.* **260** (1983) 189.

7.38 P.E. Jackson, P.R. Haddad and S. Dilli, *J. Chromatogr.* **295** (1984) 471.

7.39 R.D. Rocklin, *Anal. Chem.* **56** (1984) 1959.

7.40 S. Shibata, in: Chelates in Analytical Chemistry. H.A. Flaschka and A.J. Barnard (eds.), Marcel Dekker New York 1972, Vol. 4, Chapter 1.

7.41 D.-R. Yan and G. Schwedt, *LaborPraxis* January/February (1987) 48.

7.42 D.-R. Yan and G. Schwedt, *Fresenius Z. Anal. Chem.* **327** (1987) 503.

7.43 C.A. Lucy and H.N. Dinh, *Anal. Chem.* **66** (1994) 793.

7.44 R.M. Smith and A.E. Martell, *Critical Stability Constants, Amines.* 2nd ed., Plenum Press New York 1975.

7.45 A.E. Martell and R.M. Smith, *Critical Stability Constants, First Supplement.* 2nd ed., Plenum Press New York 1975.

7.46 J. Stillian, "Trace Analysis via Post Column Chemistry in Ion Chromatography: Silica andppb Calcium and Magnesium in Brines", Presentation at Pittsburgh Conference 1984, Atlantic City, N.J., USA.

7.47 W. Wang, Y. Chen, and M. Wu, *Analyst (London)* **109** (1984) 281.

7.48 C.H. Knight, R.M. Cassidy, B.M. Recoskie and L.W. Green, *Anal. Chem.* **56** (1984) 474.

7.49 D.P. Hautman and M.J. Bolyard, *J. Chromatogr.* **602** (1991) 7.

7.50 L. Charles and D. Pépin, *Anal. Chem.* **70** (1998) 353.

7.51 R.J. Joyce and H.S. Dhillon, *J. Chromatogr. A* **671** (1994) 165.

7.52 L. Charles, D. Pépin and B. Casetta, *Anal. Chem.* **68** (1996) 2554.

7.53 D.T. Heitkemper, L.A. Kaine, D.S. Jackson and K.A. Wolnik, *J. Chromatogr. A* **67** (1994) 101.

7.54 H.S. Weinberg and H. Yamada, *Anal. Chem.* **70** (1998) 1.

7.55 H.P. Wagner, B.V. Pepich, D.P. Hautmann and D.J. Munch, "Performance Evaluation of EPA Method 302.0 and Validation of EPA Method 303.0." Presentation at the International Ion Chromatography Symposium 1999, San Jose, CA, USA.

7.56 D.P. Hautmann, D.J. Munch and J.D. Pfaff, NERL, USEPA, Method 300.1 (1997).

7.57 C.R. Warner, D.H. Daniels, F.L. Joe and G.W. Diachenko, *Food Additives and Contaminants* **13** (1996) 633.

7.58 USEPA Appendix B to Part 136-*Definition and Procedure for the Determination of the Method Detection Limit*-Revision 1.11, Fed. Reg. 40:136B:141.

7.59 Y. Bichsel and U. von Gunten, *Anal. Chem.* **71** (1999) 34.

7.60 J. Weiß, *Fresenius Z. Anal. Chem.* **328** (1987) 46.

7.61 Y. Baba, N. Yoza and S. Ohashi, *J. Chromatogr.* **295** (1984) 153.

7.62 Y. Baba, N. Yoza and S. Ohashi, *J. Chromatogr.* **348** (1985) 27.

7.63 Y. Baba, N. Yoza and S. Ohashi, *J. Chromatogr.* **350** (1985) 119.

7.64 E. Vaeth, P. Sladek and K. Kenar, *Fresenius Z. Anal. Chem.* **239** (1987) 584.

7.65 H. Small and T.E. Miller jr., *Anal. Chem.* **54** (1982) 462.

7.66 R.A. Cochrane and D.E. Hillman, *J. Chromatogr.* **241** (1982) 392.

7.67 I.M. Kolthoff and E.B. Sandell, *Textbook of Quantitative Inorganic Analysis.* Macmillan, New York 1949; p. 662-68.

7.68 M. Dreux, M. Lafosse, and M. Pequinot, *Chromatographia* **15** (1982) 653.

7.69 P.R. Haddad and A.L. Heckenberg, *J. Chromatogr.* **252** (1982) 177.

7.70 P.R. Haddad and A.L. Heckenberg, *Chem. Aust.* **50** (1983) 275.

7.71 M. Roth and H. Hampai, *J. Chromatogr.* **83** (1973) 353.

7.72 P. de Montigny, J.F. Stobaugh, R.S. Givens, R.G. Carlson, K. Srinivasachar, L.A. Sternson and T. Higuchi, *Anal. Chem.* **59** (1987) 1096.

7.73 Application Report No. 109, Dionex Corporation, Sunnyvale, CA, USA.

7.74 S.H. Lee and L.R. Field, *Anal. Chem.* **56** (1984) 2647.

7.75 A.W. Wolkoff and R.H. Larose, *J. Chromatogr.* **99** (1974) 731.

7.76 S. Katz, W.W. Pitt jr., J.E. Mrochek and S.J. Dinsmore, *J. Chromatogr.* **101** (1974) 193.

7.77 J.H. Sherman and N.D. Danielson, *Anal. Chem.* **59** (1987) 490.

7.78 J.H. Sherman and N.D. Danielson, *Anal. Chem.* **59** (1987) 1483.

7.79 K. Bächmann and K.-H. Blaskowitz, *Fresenius Z. Anal. Chem.* **333** (1989) 15.

7.80 S. Mho and E.S. Yeung, *Anal. Chem.* **57** (1985) 2253.

7.81 T. Takeuchi and E.S. Yeung, *J. Chromatogr.* **370** (1986) 83.

7.82 N. Nimura and T. Kinoshita, *Anal. Letters* **13** (1980) 191.

7.83 T. Nakaya, T. Tomomoto and M. Imoto, *Bull. Chem. Soc. Japan* **40** (1967) 691.

7.84 F.A. Buytenhuys, *J. Chromatogr.* **218** (1981) 57.

7.85 P.R. Haddad und A.L. Heckenberg, *J. Chromatogr.* **300** (1984) 357.

7.86 E.A. Stadlbauer, C. Trieu, H. Hingmann, H. Rohatzsch, J. Weiß and R. Maushart, *Fresenius Z. Anal. Chem.* **330** (1988) 1.

7.87 J.M. Pettersen, *Anal. Chim. Acta* **160** (1984) 263.

7.88 E.A. Woolson and N. Aharonson, *J. Assoc. Off. Anal. Chem.* **63** (1980) 523.

7.89 A. Montaser and D.W. Golightly, *Inductively Coupled Plasmas in Analytical Atomic Spectroscopy*. VCH Publishers New York 1987.

7.90 G. Horlick, S.H. Tan, M.A. Vaughan and Y. Shao, *Inductively Coupled Plasma – Mass Spectrometry, Radiation Chemistry: Principles and Applications*. VCH Publishers New York 1987.

7.91 D. Jensen and W. Blödorn, *GIT Fachz. Lab.* **7** (1995) 654.

7.92 M.J. Powell, D.W. Boomer and D.R. Wiederin, *Anal. Chem.* **67** (1995) 2474.

7.93 R. Roehl, M. Alforque and J. Riviello, "Comparison of the Determination of Hexavalent Chromium by Ion Chromatography coupled with ICP-MS or with Colorimetry", Presentation at the Plasma Winter Conference on Plasma Spectrochemistry 1992, San Diego, CA, USA.

7.94 D.W. Boomer, M.J. Powell and J. Hipfner, *Talanta* **37** (1990) 127.

7.95 M.J. Powell, E.S.K. Quan, D.W. Boomer and D.R. Wiederin, *Anal. Chem.* **64** (1992) 2253.

7.96 R. Roehl and M.M. Alforque, *Atomic Spectrosc.* **11** (1990) 211.

7.97 I.T. Urasa and F. Ferede, *Anal. Chem.* **59** (1987) 1563.

7.98 G.J. De Menna, "The speciation and structure elucidation of transition metal complexes by HPLC-DCP", Presentation at Pittsburgh Conference 1986, Atlantic City, N.J., USA.

7.99 D.T. Heitkemper, L.A. Kaine, D.S. Jackson and K.A. Wolnik, *J. Chromatogr. A* **671** (1994) 101.

7.100 J.M. Riviello, A. Siriraks, R.M. Manabe, R. Roehl and M. Alforque, *LC•GC* **9** (1991) 704.

7.101 D. Schlegel, J. Mattusch and K. Dittrich, *J. Chromatogr. A* **683** (1994) 261.

7.102 J. Mattusch and R. Wennrich, *Anal. Chem.* **70** (1998) 3649.

7.103 W.R. Biggs, J.T. Gano and R.J. Brown, *Anal. Chem.* **56** (1984) 2653.

7.104 W. Buchberger and W. Ahrer, *J. Chromatogr. A* **850** (1999) 99.

7.105 R.C. Simpson, C.C. Fenselau, M.R. Hardy, R.R. Townsend, Y.C. Lee and R.J. Cotter, *Anal. Chem.* **62** (1990) 248.

7.106 J.J. Conboy, J.D. Henion, M.W. Martin and J.A. Zweigenbaum, *Anal. Chem.* **62** (1990) 800.

7.107 R.C. Willoughby and R.F. Browner, *Anal. Chem.* **56** (1984) 2626.

7.108 W.M.A. Niessen, *Liquid Chromatography–Mass Spectrometry*. 2nd ed., Marcel Dekker New York – Basel 1999.

7.109 C.S. Creaser and J.W. Stygall, *Analyst* **118** (1993) 1467.

7.110 A. Cappiello, *Mass Spectrom. Rev.* **15** (1996) 283.

7.111 P.C. Winkler, D.D. Perkins, W.K. Williams and R.F. Browner, *Anal. Chem.* **60** (1988) 15.

7.112 J.D. Kirk and R.F. Browner, *Biomed. Environ. Mass. Spectrom.* **18** (1989) 355.

7.113 J.C. Richardson and R.F. Browner, Proceedings of the 41st ASMS Conference on Mass Spectrometry and Allied Topics, 1993, San Francisco, CA, USA, p. 663.

7.114 J. Hsu, *Anal. Chem.* **64** (1992) 434.

7.115 I.S. Kim, F.I. Sasinos, R.D. Stephens and M.A. Brown, *Environ. Sci. Technol.* **24** (1990) 1832.

7.116 H. Bagheri, J. Slobodnik, R.M.M. Recasens, R.T. Ghijsen and U.A.Th. Brinkman, *Chromatographia* **37** (1993) 159.

7.117 A.L. Yergey, C.G. Edmonds, I.A.S. Lewis and M.L. Vestal, *Liquid Chromatography/Mass Spectrometry, Techniques and Applications.* Plenum Press New York 1990, p. 31 ff.

7.118 P.J. Arpino, *Mass Spectrom. Rev.* **11** (1992) 3.

7.119 J.A. McCloskey and P.F. Crain, *Int. J. Mass Spectrom. Ion Processes* **118/119** (1992) 593.

7.120 J.B. Fenn, M. Mann, C.K. Meng, S.F. Wong and C.M. Whitehouse, *Science* **246** (1989) 64.

7.121 R.D. Smith, J.A. Loo, C.G. Edmonds, C.J. Baringa and H.R. Udseth, *Anal. Chem.* **62** (1990) 882.

7.122 M. Mann, *Org. Mass Spectrom.* **25** (1990) 575.

7.123 W.M.A. Niessen, R.A.M. van der Hoeven, J. van der Greef, H.A. Schols, G. Lucas-Lokhorst, A.G.J. Voragen and C. Bruggink, *Rapid. Commun. Mass Spectrom.* **6** (1992) 474.

7.124 W.M.A. Niessen, R.A.M. van der Hoeven, J. van der Greef, H.A. Schols, A.G.J. Voragen and C. Bruggink, *J. Chromatogr.* **627** (1992) 63.

7.125 W.M.A. Niessen, R.A.M. van der Hoeven, J. van der Greef, H.A. Schols and A.G.J. Voragen, *Rapid Commun. Mass Spectrom.* **6** (1992) 197.

7.127 W.M.A. Niessen, R.A.M. van der Hoeven, J. van der Greef, H.A. Schols, A.G.J. Voragen and C. Bruggink, *J. Chromatogr.* **647** (1993) 319.

7.128 D. Barceló, G. Durand, R.J. Vreeken, G.J. de Jong and U.A.Th. Brinkman, *Anal. Chem.* **62** (1990) 1696.

7.129 M. Yoshida, T. Watabiki, T. Tokiyasu and N. Ishida, *J. Chromatogr.* **628** (1993) 235.

7.130 D. Barceló, G. Durand and R.J. Vreeken, *J. Chromatogr.* **647** (1993) 271.

7.131 R.E.A. Escott und D.S.W. Chandler, *J. Chromatogr. Sci.* **27** (1989) 134.

7.132 H.Fr. Schröder, *J. Chromatogr.* **647** (1993) 219.

7.133 J.M. Ballard and L.D. Betowski, *Org. Mass Spectrom.* **21** (1986) 575.

7.134 D.A. Flory, M.A. McLean, M.L. Vestal and L.D. Betowski, *Rapid Commun. Mass Spectrom.* **1** (1987) 48.

7.135 J. Abián, M.I. Churchwell and W.A. Korfmacher, *J. Chromatogr.* **629** (1993) 267.

7.136 K.L. Tyczkowska, R.D. Voyksner and A.L. Aronson, *J. Chromatogr.* **490** (1989) 101.

7.137 R.D. Voyksner, K.L. Tyczkowska and A.L. Aronson, *J. Chromatogr.* **567** (1991) 389.

7.138 J.B. Fenn, M. Mann, C.K. Meng, S.F. Wong and C.M. Whitehouse, *Mass Spectrom. Rev.* **9** (1990) 37.

7.139 R.B. Cole (ed.), *Electrospray Ionization Mass Spectrometry: Fundamentals, Instrumentation, and Applications.* John Wiley & Sons New York 1997.

7.140 E.D. Lee and J.D. Henion, *Rapid Commun. Mass Spectrom.* **6** (1992) 727.

7.141 J.F. Banks, jr., S. Shen, C.M. Whitehouse and J.B. Fenn, *Anal. Chem.* **66** (1994) 406.

7.142 J. Efer, U. Ceglarek, T. Anspach, W. Engewald and W. Pihan, *Wasser Abwasser* **138** (1997) 577.

7.143 L. Charles, D. Pépin and B. Casetta, *Anal. Chem.* **68** (1996) 2554.

7.144 L. Charles and D. Pépin, *Anal. Chem.* **70** (1998) 353.

7.145 S. Rabin and J. Stillian, *J. Chromatogr. A* **671** (1994) 63.

7.146 S.B. Mohsin, *Anal. Chem.* **71** (1999) 3603.

Chapter 8

8.1 J. Aßhauer and H. Ullner, "Quantitative Analysis". In: H. Engelhardt and I. Halász (eds.). *Handbook of High Performance Liquid Chromatography*. Springer-Verlag, Heidelberg 1985.

8.2 DIN 38402 Part 11: Allgemeine Angaben − Probenahme von Abwasser (June 1985).

8.3 DIN 38402 Part 12: Allgemeine Angaben − Probenahme aus stehenden Gewässern (June 1985).

8.4 DIN 38402 Part 13: Allgemeine Angaben − Probenahme aus Grundwasserleitern (December 1985).

8.5 DIN 38402 Part 14: Allgemeine Angaben − Probenahme von Rohwasser und Trinkwasser (March 1986).

8.6 DIN 38402 Part 15: Allgemeine Angaben − Probenahme aus Fließgewässern (July 1986).

8.7 DIN 38402 Part 16: Allgemeine Angaben − Probenahme aus dem Meer (August 1987).

8.8 DIN 38402 Part 17: Allgemeine Angaben − Probenahme von fallenden, nassen Niederschlägen in flüssigem Aggregatzustand (May 1988).

8.9 DIN 38402 Part 19: Allgemeine Angaben − Probenahme von Schwimm- und Badebeckenwasser (April 1988).

8.10 DIN 38402 Part 20: Allgemeine Angaben − Probenahme von Tidegewässern (August 1987).

8.11 DIN 38402 Part 30/31: Allgemeine Angaben − Vorbehandlung, Teilung und Homogenisierung heterogener Wasserproben für die Bestimmung von Schwermetallen (Deutsche Einheitsverfahren zur Wasser-, Abwasser- und Schlammanalytik (DEV) − 20. Lieferung 1988).

8.12 DIN 38402 Part 30/32: Allgemeine Angaben − Vorbehandlung , Teilung und Homogenisierung heterogener Wasserproben für die Bestimmung der absorbierbaren organisch gebundenen Halogene (AOX) (Deutsche Einheitsverfahren zur Wasser-, Abwasser- und Schlammanalytik (DEV) − 21. Lieferung 1989, VCH).

8.13 E. Kucera, *J. Chromatogr.* **19** (1965) 237.

8.14 E. Grushka, *J. Phys. Chem.* **76** (1972) 2586.

8.15 J.Å. Jönsson, *Chromatographia* **18** (1984) 427.

8.16 J.P. Foley and J.G. Dorsey, *Anal. Chem.* **55** (1983) 730.

8.17 W.W. Yau, *Anal. Chem.* **49** (1977) 395.

8.18 A. Becker, *LaborPraxis* **9** (1999) 16.

8.19 R. Kaiser and G. Gottschalk, *Elementare Tests zur Beurteilung von Meßdaten*. Bibliographisches Institut Mannheim 1972.

8.20 W. Funk, V. Dammann, C. Vonderheid and G. Oehlmann, *Statistische Methoden in der Wasseranalytik*. VCH Verlagsgesellschaft Weinheim 1985.

8.21 M. Doury-Berthod, P. Giampaoli, H. Pitsch, C. Sella and C. Poitrenaud, *Anal. Chem.* **57** (1985) 2257.

8.22 J. Mandel, *The Statistical Analysis of Experimental Data*. John Wiley & Sons, New York 1964.

8.23 C. Vonderheid, V. Damman, W. Dürr, W. Funk and H. Krutz, *Vom Wasser* **57** (1981) 59.

8.24 K. Doerffel, *Statistik in der analytischen Chemie*. VCH Verlagsgesellschaft Weinheim 1987.

8.25 F. Ahmend, G. Bernhart, D. Rinne, M. Rogge and B. Rüdesheim, *LaborPraxis* **5** (1988) 524.

8.26 W. Funk, V. Dammann and G. Donnevert, *Qualitätssicherung in der analytischen Chemie*. VCH Verlagsgesellschaft Weinheim 1992.

8.27 DIN 38402 Part 51: Allgemeine Angaben − Kalibrierung von Analysenverfahren, Auswertung von Analysenergebnissen und lineare Kalibrierfunktionen für die Bestimmung von Verfahrenskenngrößen (May 1986).

8.28 J. Hartung, *Statistik*. Oldenburg Verlag München 1982.

8.29 J.C. Miller and J.N. Miller, *Statistics for Analytical Chemistry*. Second Edition, Ellis Horwood Ltd. Chichester 1988.

8.30 D.G. Mitchell, W.N. Mills and J.S. Garden, *Anal. Chem.* **49** (1977) 1655.

8.31 H. Mager, *Moderne Regressionsanalyse*. Salle & Sauerländer Frankfurt 1982.

8.32 W. Huber, *Fresenius Z. Anal. Chem.* **319** (1984) 379.

8.33 DIN 38405 D19 (1988) *Bestimmung der Anionen Fluorid, Chlorid, Nitrit, Orthophosphat, Bromid, Nitrat und Sulfat in wenig belasteten Wässern mit der Ionenchromatographie.*

8.34 E. Jackwerth, *Chemie für Labor und Betrieb* **33** (1982) 4.

8.35 J.B. Roos, *Analyst* **87** (1962) 832.

8.36 H. Kaiser, *Fresenius Z. Anal. Chem.* **209** (1965) 1.

8.37 H. Kaiser, *Fresenius Z. Anal. Chem.* **216** (1966) 80.

8.38 M.A. Sharaf, D.L. Illmann, and B.R. Kowalski, *Chemometrics.* John Wiley & Sons, New York 1986.

8.39 W. Shewhart, *The Economic Control of Quality of Manufactured Product.* D. van Nostrand Company, Inc. New York 1931.

8.40 R.H. Woodward and P.L. Goldsmith, *Cumulative Sum Techniques. Mathematical and Statistical Techniques for Industry.* Monograph No. 3, Oliver and Boyd for ICI, Edinburgh 1972.

8.41 P.-Th. Wilrich, *QZ 24* **10** (1979) 260.

8.42 E.S. Page, *Biometrika* **41** (1954) 100.

8.43 G.B. Wetherill, *Sampling Inspection and Quality Control.* 2nd ed., Chapman and Hall, London, New York 1982.

8.44 G.A. Barnard, *J.R. Statist. Soc. B* **21** (1959) 239.

8.45 J.D. Juran, F.M. Gryna and R.S. Bingham (eds.), *Quality Control Handbook.* McGraw Hill Book Company New York 1975.

Chapter 9

9.1 T. Darimont, G. Schulze and M. Sonneborn, *Fresenius Z. Anal. Chem.* **314** (1983) 383.

9.2 G. Schwedt, "Ionen-Chromatographie (IC) – die High Performance-LC für Anionen und Kationen", *LaborPraxis* January/February (1984).

9.3 J.A. Mosko, *Anal. Chem.* **56** (1984) 629.

9.4 M.A. Tabatabai and W.A. Dick, *J. Environ. Qual.* **12** (1983) 209.

9.5 O.A. Shpigun, *Trends Anal. Chem.* **4** (1985) 29.

9.6 F. Schöller and F. Ollram, *Österr. Wasserwirtsch.* **35** (1983) 73.

9.7 G. Resch and E. Grünschläger, *Vom Wasser* **62** (1984) 207.

9.8 R. Schwabe, T. Darimont, T. Möhlmann, E. Pabel and M. Sonneborn, *Int. J. Environ. Anal. Chem.* **14** (1983) 169.

9.9 T. Okada and T. Kuwamoto, *Anal. Lett.* **17** (1984) 1743.

9.10 J.P. Wilshire, *LC-Magazin* **1** (1983) 290.

9.11 S.W. Krasner, W.H. Glaze, H.S. Weinberg, P.A. Daniel and I.N. Najm, *J. Am. Water Works Assoc.* **85** (1993) 73.

9.12 F. Sacher, A. Matschi and H.-J. Brauch, *Acta Hydrochim. Hydrobiol.* **23** (1995) 26.

9.13 W.R. Haag, J. Hoigné and H. Bader, *Vom Wasser* **59** (1982) 237.

9.14 W.R. Haag and J. Hoigné, *Environ. Sci. Techn.* **17** (1983) 261.

9.15 U. von Gunten and J. Hoigné, *Environ. Sci. Technol.* **28** (1994) 1234.

9.16 U. von Gunten and Y. Oliveras, *Environ. Sci. Technol.* **32** (1998) 63.

9.17 Y. Kurokawa, Y. Maekawa, M. Takahashi and Y. Hayashi, *Environ. Health Perspect.* **87** (1990) 309.

9.18 World Health Organization: Guidelines for drinking water quality. WHO, Geneve 1993.

9.19 Commission of the European Communities: Proposal for a council directive concerning the quality of water intended for human consumption. Brussels 1994.

9.20 U.S. EPA. "The Determination of Inorganic Anions in Water by Ion Chromatography." Method 300.0. Washington, DC: U.S. EPA, 1993.

9.21 C.Y. Kuo, S. Krasner and H. Weinberg, Proceedings of the American Water Works Association Water Quality Technology Conference, San Diego, CA, Denver, CO: AWWA, 1990.

9.22 D.P. Hautman and M. Bolyard, *J. Chromatogr.* **602** (1991) 7.

9.23 D.P. Hautman and M. Bolyard, *J. Chromatogr.* **602** (1992) 65.

9.24 H. Weinberg, *J. Chromatogr. A* **671** (1994) 141.

9.25 R.J. Joyce and H.S. Dhillon, *J. Chromatogr. A* **671** (1994) 165.

9.26 C. Bruggink, W.J.M. Van Rossum and J.G.M.M. Smeenk, *H₂O* **28** (1995) 343.

9.27 D.T. Heitkemper, L.A. Kaine, D.S. Jackson and K.A. Wolnik, *J. Chromatogr. A* **67** (1994) 101.

9.28 M. Nowak and A. Seubert, *Anal. Chim. Acta* **359** (1998) 193.

9.29 H.S. Weinberg and H. Yamada, *Anal. Chem.* **70** (1998) 1.

9.30 E. Salhi and U. von Gunten, *Wat. Res.* **33** (1999) 3239.

9.31 H.P. Wagner, B.V. Pepich, D.P. Hautman and D.J. Munch, *J. Chromatogr. A* **850** (1999) 119.

9.32 D.P. Hautmann, D.J. Munch and J.D. Pfaff, NERL, USEPA, Method 300.1 (1997).

9.33 H.P. Wagner, B.V. Pepich, D.P. Hautman and D.J. Munch, *J. Chromatogr. A* **882** (2000) 300.

9.34 M.G. Ondrus and G. Gordon, *Inorg. Chem.* **11** (1972) 985.

9.35 A. Iatrou and W.R. Knocke, *AWWA* **5** (1992) 63.

9.36 C.A. Lucy, *LC-GC* **14** (1996) 406.

9.37 M. Legrand, M. De Angelis and R.J. Delmas, *Anal. Chim. Acta* **156** (1984) 181.

9.38 U. Nickus and M. Kuhn, *J. Chromatogr. A* **671** (1994) 225.

9.39 A. Döscher, M. Schwikowski and H.W. Gäggeler, *J. Chromatogr. A* **706** (1995) 249.

9.40 J. Ivask and J. Pentschuk, *J. Chromatogr. A* **770** (1997) 125.

9.41 K. Oikawa, K. Murano, Y. Enomoto, K. Wada and T. Inimata, *J. Chromatogr. A* **671** (1994) 211.

9.42 IC Application Report C-19, Metrohm AG, Herisau, Switzerland.

9.43 P.R. Haddad and P.E. Jackson, *Ion Chromatography: Principles and Applications*. In: J. Chromatogr. Library, Vol. 46, Elsevier, Amsterdam 1990, p. 109.

9.44 P.E. Jackson, M. Laikhtman and J.S. Rohrer, *J. Chromatogr. A* **850** (1999) 131.

9.45 P.E. Jackson, S. Gokhale, T. Streib, J.S. Rohrer and C.A. Pohl, *J. Chromatogr. A* **888** (2000) 151.

9.46 USEPA Appendix B of Part 136-*Definition and Procedure for the Determination of the Method Detection Limit*-Revision 1.11, Fed. Reg. 40:136B:141.

9.47 Ch. Siegert, "Das Chlordioxidverfahren bei der Bekämpfung von Geruch und Geschmack im Trinkwasser sowie die Bestimmung von Cl_2, ClO_2 und $NaClO_2$ nebeneinander". In: *Fortschritte der Wasserchemie*, Vol. 1. Akademie-Verlag 1964; p. 25.

9.48 E. Bandi, "Untersuchungen über die Anwendungsmöglichkeit von Chlordioxid zur Entkeimung von Badewasser". In: *Mitteilungen aus dem Gebiet der Lebensmitteluntersuchung und Hygiene*. Publication by Eidg. Gesundheitsamt Bern **58** (1967) 170.

9.49 Y. Bichsel and U. von Gunten, *Anal. Chem.* **71** (1999) 34.

9.50 K. Weber and M. Pichler, *Ber. Deutsche Chem. Gesellschaft* **73** (1940) 494.

9.51 B. Nowak and U. von Gunten, *J. Chromatogr. A* **849** (1999) 209.

9.52 N.K. Kristiansen, K.T. Aune, M. Froeshaug, G. Becher and E. Lundanes, *Water Res.* **30** (1996) 2155.

9.53 G.R. Fuchs and K. Bächmann, *Fresenius Z. Anal. Chem.* **327** (1987) 205.

9.54 L.M. Nair, R. Saari-Nordhaus and J.M. Anderson, jr., *J. Chromatogr. A* **671** (1994) 309.

9.55 C. Sarzanini, M.C. Bruzzoniti and E. Mentasti, *J. Chromatogr. A* **850** (1999) 197.

9.56 Y. Sekiguchi, Dionex Nippon K.K., personal communication.

9.57 S. Hashimoto and A. Otsuki, *J. High Resolut. Chromatogr.* **21** (1998) 55.

9.58 L. Charles and D. Pépin, *J. Chromatogr. A* **804** (1998) 105.

9.59 IC Application Report S-33, Metrohm AG, Herisau, Switzerland.

9.60 Y. Miura and H. Hamada, *J. Chromatogr. A* **850** (1999) 153.

9.61 C. Pohl, M. Rey, D. Jensen and J. Kerth, *J. Chromatogr. A* **850** (1999) 239.

9.62 M.A. Rey, C.A. Pohl, J.J. Jagodzinski, E.Q. Kaiser and J.M. Riviello, *J. Chromatogr. A* **804** (1998) 201.

9.63 P. Mostbauer and G. Balderas, *GIT Spezial "Chromatographie"* **1** (1991) 4.

9.64 K. Fischer, *Korrespondenz Abwasser* **45** (1998) 676.

9.65 M.A. Rey, J.M. Riviello and C.A. Pohl, *J. Chromatogr. A* **789** (1997) 149.

9.66 Y. Huang, S. Mou and J.M. Riviello, *J. Chromatogr. A* **868** (2000) 209.

9.67 W. Bashir and B. Paull, *J. Chromatogr. A* **910** (2001) 301.

9.68 D.E. Kimbrough and I.H. Suffet, *Analyst* **121** (1996) 309.

9.69 M. Betti and S. Cavalli, *J. Chromatogr.* **538** (1991) 365.

9.70 I.N. Voloschik, M.L. Litvina and B.A. Rudenko, *J. Chromatogr. A* **706** (1995) 351.

9.71 M.J. Shaw, S.J. Hill, P. Jones and P. Nesterenko, *J. Chromatogr. A* **876** (2000) 127.

9.72 IC Application Report C-18, Metrohm AG, Herisau, Switzerland.

9.73 H.L. Tucker and R.W. Flack, *J. Chromatogr. A* **804** (1998) 131.

9.74 W.D. Vermillion and M.D. Crenshaw, *J. Chromatogr. A* **770** (1997) 253.

9.75 A.F. Kingery and H.E. Allen, *Anal. Chem.* **66** (1994) 155.

9.76 H. Matusiewicz and D.F.S. Natusch, *Int. J. Environ. Anal. Chem.* **8** (1980) 227.

9.77 U. Baltensperger and J. Hertz, *J. Chromatogr.* **324** (1985) 153.

9.78 M.J. Willison and A.G. Clarke, *Anal. Chem.* **56** (1984) 1037.

9.79 J. Forrest, D.J. Spandau, R.L. Tanner and L. Newman, *Atmos. Environ.* **16** (1982) 1473.

9.80 J.M. Lorrain, C.R. Fortune and B. Dellinger, *Anal. Chem.* **53** (1981) 1302.

9.81 T.W. Dolzine, G.G. Esposito and D.S. Rinehart, *Anal. Chem.* **54** (1982) 470.

9.82 J.H. Margeson, J.E. Knoll, M.R. Midgett, G.B. Oldaker and W.E. Reynolds, *Anal. Chem.* **59** (1985) 1586.

9.83 H. Velásquez, H. Ramzrez, J. Dízas, M. Gonzáles de Nava, B. Sosa de Borrego and J. Morales, *J. Chromatogr. A* **739** (1996) 295.

9.84 M. Nonomura and T. Hobo, *J. Chromatogr. A* **804** (1998) 151.

9.85 D.V. Vinjamoori and C.S. Ling, *Anal. Chem.* **53** (1981) 1689.

9.86 M. Nonomura, T. Hobo, E. Kobayashi, T. Murayama and M. Satoda, *J. Chromatogr. A* **739** (1996) 301.

9.87 US EPA Method 7A: Determination of Nitrogen Oxide Emissions from Stationary Sources − Ion Chromatographic Method, Environmental Protection Agency, Washington DC, 1994, p. 626.

9.88 US EPA Method 7D: Determination of Nitrogen Oxide Emissions from Stationary Sources − Alkaline Permanganate/Ion Chromatographic Method, Environmental Protection Agency, Washington DC, 1994, p. 637.

9.89 R. Cee and J.C. Ku, *Analyst* **119** (1994) 57.

9.90 H. Sontheimer and M. Schnitzler, *Vom Wasser* **59** (1982) 169.

9.91 M. Schnitzler, G. Lévay, W. Kühn and H. Sontheimer, *Vom Wasser* **61** (1983) 263.

9.92 M. Schnitzler, *GIT Supplement Chromatographie* **4** (1985) 32.

9.93 E. Kaiser, J. Statler and S. Heberling, "Improved Detection Limits for the Ion Chromatographic Determination of Anions in High Purity Water with Direct Injection." Presentation Pittsburgh Conference 1996, Chicago, IL, USA.

9.94 Y. Liu, N. Avdalovic, J. Rohrer, M. Laikhtman, E. Kaiser and H, Dhillon, "Determination of Trace Level Anions in High Purity Water by Ion Chromatography with On-Line Eluant Generation." Presentation at the International Ion Chromatography Symposium, Osaka, Japan 1998.

9.95 T. Wacker and M. Rziha, *PowerPlant Chemistry* **2** (2000) 27.

9.96 G. Resch and E. Grünschläger, *VGB Kraftwerkstechnik* **62** (1982) 127.

9.97 Th. Ehmann, L. Fabry, L. Kotz and S. Pahlke, "Ion Chromatography and Capillary Electrophoresis in the Semiconductor Industry." Presentation at 2. Fachtagung Ionenchromatographie, Idstein 1999.

9.98 D. Bostic, G. Burns and S. Harvey, *J. Chromatogr.* **602** (1992) 163.

9.99 A. Siriraks and J. Stillian, *J. Chromatogr.* **640** (1993) 371.

9.100 M.D.H. Amey and D.A. Bridle, *J. Chromatogr.* **640** (1993) 323.

9.101 G. Boyle, B. Handy and A. van Geen, *Anal. Chem.* **59** (1982) 1499.

9.102 W.R. Seitz and D.M. Hercules, *Chemiluminescence and Bioluminescence.* Plenum Press New York 1973, p. 427 ff.

9.103 P. Jones, T. Williams and L. Ebdon, *Anal. Chim. Acta* **237** (1990) 291.

9.104 S. Charbonneau, R. Gilbert and L. Lépine, *Anal. Chem.* **67** (1995) 1204.

9.105 M.A. Rey, J. Riviello, C. Pohl and J. Jagodzinski, "Dealing with High-to-Low ConcentrationRatios of Sodium and Ammonium in Ion Chromatography." Presentation at Pittsburgh Conference 1997, Atlanta, GA, USA.

9.106 M.A. Rey, C. Pohl and J. Riviello, "Column Switching for the Determination of Trace Cations in Waters Containing Amines." Presentation at the Power/Semiconductor Process Analysis Workshop 1995, Dallas, USA.

9.107 E. Kaiser, H. Dhillon and S. Heberling, "Application of AutoNeutralization to the Determination of Sub-Parts-Per-Billion Concentrations of Anions in Power Plant Water Treated with Amines." Presentation at Pittsburgh Conference 1996, Chicago, USA.

9.108 IC Application Report S-5, Metrohm AG, Herisau, Switzerland.

9.109 J. Weiß and G. Hägele, *Fresenius Z. Anal. Chem.* **328** (1987) 46.

9.110 E. Vaeth, P. Sladek and K. Kenar, *Fresenius Z. Anal. Chem.* **329** (1987) 584.

9.111 H. Gutberlet, *VGB Kraftwerkstechnik* **1** (1994) 119.

9.112 J. Weiß and M. Göbl, *Fresenius Z. Anal. Chem.* **320** (1985) 439.

9.113 M Geißler and R. van Eldik, *Anal. Chem.* **64** (1992) 3004.

9.114 D.A. Howell, "The Use of AutoNeutralization for the Analysis of High Purity Amine Products." Presentation at the Power/Semiconductor Process Analysis Workshop 1995, Dallas, USA.

9.115 T.O. Passell, *J. Chromatogr. A* **671** (1994) 331.

9.116 G.E. Helmke, *Semiconductor Int.* **4** (1981) 119.

9.117 M.M. Plechaty, *LC-Magazin* **2** (1984) 684.

9.118 S. Heberling, "Advances in High Purity Water Analysis by Ion Chromatography." Presentation at the Symposium for Contamination Control in the Semiconductor and Electronics Industry, Sunnyvale, USA, 1996.

9.119 M. Wojtusik, "Advances in On-line Ion Chromatography for Semiconductor High Purity Water." Presentation at the Symposium for Contamination Control in the Semiconductor and Electronics Industry, Sunnyvale, USA, 1996.

9.120 E. Kaiser, J. Riviello, M. Rey and S. Heberling, *J. Chromatogr. A* **739** (1996) 71.

9.121 L.E. Vanatta, D.E. Coleman and R.W. Slingsby, *J. Chromatogr. A* **850** (1999) 107.

9.122 S. Heberling, B. Joyce and K. Haak, "Applications of Transition Metal Ion Chromatography to High Purity and Industrial Process Waters", Presentation at 5. *Semiconductor Pure Water Conference*, San Francisco, USA, 1986.

9.123 W. Locke, "On-Line Measurement of Parts-Per-Trillion Levels of Transition Metals in Semiconductor-Grade Water." Presentation at the Power/Semiconductor Process Analysis Workshop, Dallas, USA, 1995.

9.124 L. Vanatta, "Measuring Parts-Per-Trillion Anions and Transition Metals in Semiconductor Pure Water Using Grab Sampling and Laboratory Ion Chromatography." Presentation at the Symposium for Contamination Control in the Semiconductor and Electronics Industry, Sunnyvale, USA, 1996.

9.125 M.E. Potts, E.J. Gavin, L.O. Angers and E.L. Johnson, *LC-GC-Magazine* **4** (1986) 912.

9.126 J.D. Sinclair, *J. Electrochem. Soc.* **135** (1988) 89C.

9.127 W. Kern, *J. Electrochem. Soc.* **137** (1990) 1887.

9.128 M. Plat and J. DeLeo, *J. Chromatogr.* **546** (1991) 347.

9.129 D. Rathmann and L. Fabry, Proceedings of the Symposium on Cleaning Technology in Semiconductor Device Manufacturing 1992, p. 338 ff.

9.130 S. Tabrez, D. Yang, C. Lee, Y.-Y. Yang, E. Kaiser, S. Heberling and B. Newton, *Precision Cleaning* **6** (1998).

9.131 J. Thompson, T. Prommanuwat, A. Siriraks and S. Heberling, *Insight* May/June (1999) 24.

9.132 *Book of SEMI Standards 1993 Chemicals/Reagents Volume*, Semiconductor Equipment and Materials International, Mountain View, CA, USA 1993, p. 74.

9.133 E. Kaiser and M.J. Wojtusik, *J. Chromatogr. A* **671** (1994) 253.

9.134 E. Kaiser and J. Rohrer, *J. Chromatogr. A* **858** (1999) 55.

9.135 K.H. Viehweger, H. Schäfer, P. Walter and U. Waldburger, "Determination of Anions in High Purity Organic Solvents Used by the Semiconductor Industry." Poster at the International Ion Chromatography Symposium San Jose, CA, USA, 1999.

9.136 L.K. Jackson, R.J. Joyce, M. Laikhtman and P.E. Jackson, *J. Chromatogr. A* **829** (1998) 187.

9.137 M. Weidenauer, P. Hoffmann, and K.H. Lieser, *Fresenius, J. Anal. Chem.* **342** (1992) 333.

9.138 J.K. Sanders, *J. Chromatogr. A* **804** (1998) 193.

9.139 R. Iscoff, *Semicond. Int.* **16** (1993) 61.

9.140 K. Solder, *Material Safety Data Sheet for Water Soluble Soldering Flux*. MSDS Number 1452, Des Plaines, IL, USA 1996, p. 1.

9.141 C.W. Pearce and B.C. Chung, in: G.S. Higashi, M. Hirose, S. Raghavan and S. Verhaverberke (eds.), Proceedings of the Science and Technology of Semiconductor Surface Preparation Symposium, San Francisco, CA, USA 1997, Mater. Res. Soc., Pittsburgh, PA, USA 1997, p. 459 ff.

9.142 A. Siriraks, C.A. Pohl and M. Toofan, *J. Chromatogr.* **602** (1992) 89.

9.143 K. Watanabe and T. Ishizaki, Poster No. 66, International Ion Chromatography Symposium, Dallas, TX, USA, 1995.

9.144 J. Chen and M. Wu, *Micro* **15** (1997) 31.

9.145 M. Wu and J. Chen, *Micro* **15** (1997) 65.

9.146 E. Kaiser, J.S. Rohrer and K. Watanabe, *J. Chromatogr. A* **850** (1999) 167.

9.147 E. Kiser, R. Kaiser, J. Riviello and A. Siriraks, *LC-GC* **13** (1995) 888.

9.148 C.D. Chriswell, D.R. Mroch and R. Markuszewski, *Anal. Chem.* **58** (1986) 319.

9.149 J. Kerth and D. Jensen, *J. Chromatogr. A* **706** (1995) 191.

9.150 R. Rocklin, *Semiconductor International* Mai 1986.

9.151 T.R. Dulski, *Anal. Chem.* **51** (1979) 1439.

9.152 J.P. McKaveney, *Anal. Chem.* **40** (1968) 1276.

9.153 L.E. Vanatta and D.E. Coleman, *J. Chromatogr. A* **804** (1998) 161.

9.154 C.A. Jacobson, *Encyclopedia of Chemical Reactions.* Vol. 5, Chapman and Hall, New York 1953, p. 340.

9.155 R.M. Merrill, *LC-GC* **6** (1988) 416.

9.156 F.S. Becker, D. Pawlik, H. Schäfer and G. Staudigl, *J. Vac. Sci. Technol.* **B4** (1986) 732.

9.157 H. Schäfer and K.J. Budde, *Fresenius J. Anal. Chem.* **343** (1992) 782.

9.158 M.E. Cassinelli, *Am. Ind. Hyg. Assoc. J.* **47** (1986) 219.

9.159 S.J. Lue, T. Wu, H. Hsu and C. Huang, *J. Chromatogr. A* **804** (1998) 273.

9.160 R. Roehl and M. Doyle, "Development and Performance of an Automated Air Analyzer." Presentation at Pittsburgh Conference, New Orleans, USA 1998.

9.161 K. Haak, *Plat. Surf. Finish* **70** (1983) 34.

9.162 J. Weiß, *Galvanotechnik* **77** (1986) 2675.

9.163 J.W. Dini. In: F.A. Lowenheim (ed.). *Copper Pyrophosphate Plating in Modern Electroplating.* John Wiley & Sons, New York 1984; p. 204-223.

9.164 IC Application Note No. S-24, Metrohm AG, Herisau, Switzerland.

9.165 S. Heberling, D. Campbell and S. Carson, *Plating and Surface Finishing* **11** (1990) 58.

9.166 S. Sopok, *J. Chromatogr. A* **671** (1994) 265.

9.167 IC Application Note No. S-51, Metrohm AG, Herisau, Switzerland.

9.168 IC Application Note No. S-92, Metrohm AG, Herisau, Switzerland.

9.169 Y. Okinaka, *West. Electr. Eng.* **22** (1978) 72.

9.170 P. Wilkinson, *Gold. Bull.* **19** (1986) 75.

9.171 M. Powers, F. Hoppe and J. Gannotti, *Plating and Surface Finishing* **11** (1991) 76.

9.172 A. Maquieira, R. Puchades, F. de la Iglesia and J.A. Aparicio, *Metal Finishing* **10** (1992) 33.

9.173 IC Application Note No. O-12, Metrohm AG, Herisau, Switzerland.

9.174 D. Tench and J. White, *J. Electrochem. Soc.* **132** (1985) 831.

9.175 J. Reid, *Printed Circuit Fabrication* **11** (1987) 65.

9.176 J. Böcker and Th. Bolch, "Nickel Electroforming – Some Aspects for Process Control". Frauenhofer Institute for Production Control and Automation Stuttgart, Presentation at *Int. Symposium on Electroforming/Deposition*, Los Angelos, CA, USA 1983.

9.177 IC Application Note No. O-11, Metrohm AG, Herisau, Switzerland.

9.178 S. Shohat, E. Grushka and S. Glikberg, *J. Chromatogr.* **452** (1988) 503.

9.179 S. Heberling, D. Canmpbell and S. Carson, *Printed Circuit Fabrication* **8** (1989).

9.180 L. Longwell and W.D. Maniece, *Analyst (London)* **80** (1955) 167.

9.181 G.F. Longman, *The Analysis of Detergents and Detergent Products*. John Wiley & Sons, London 1975; p. 446, 455, and 480.

9.182 A. Hofer, E. Brosche and R. Heidinger, *Fresenius Z. Anal. Chem.* **253** (1971) 117.

9.183 p. 403 in [181].

9.184 K. Kiemstedt and W. Pfab, *Fresenius Z. Anal. Chem.* **213** (1965) 100.

9.185 J. Weiß, *Tenside Detergents* **23** (1986) 237.

9.186 p. 234 in [181].

9.187 J.M. Rosen and H.A. Goldsmith, *Systematic Analysis of Surface Active Agents*. 2nd Edition John Wiley & Sons, London 1972.

9.188 H. Waldhoff and P. Sladek, *Fresenius Z. Anal. Chem.* **320** (1985) 163.

9.189 J. Weiß, *Fresenius Z. Anal. Chem.* **320** (1985) 679.

9.190 J. Weiß, *Fresenius Z. Anal. Chem.* **328** (1987) 46.

9.191 E. Vaeth, P. Sladek and K. Kenar, *Fresenius Z. Anal. Chem.* **329** (1987) 584.

9.192 *Ullmanns Encyklopädie der technischen Chemie*. VCH Verlagsgesellschaft, Weinheim 1983, 4th edition, Vol. 24; p. 97.

9.193 H. Waldhoff, personal communication.

9.194 G. Janssen, Diploma thesis Fachhochschule Niederrhein, Krefeld WS 1987/88.

9.195 P. Edwards, *Food Technology* June (1983) 53-56.

9.196 R.D. Rocklin, *LC-Magazine* **1** (1983) 504.

9.197 E.J. Knudson and K.J. Siebert, *J. Am. Soc. Brew. Chem.* **42** (1984) 65.

9.198 G. Saccani, S. Gherardi, A. Trifirò, C. Soresi Bordini, M. Calza and C. Freddi, *J. Chromatogr. A* **706** (1995) 395.

9.199 S. Koswig, H.-J. Hofsommer, J. Weiß and D. Jensen, "Ionenchromatographische Untersuchung von Fruchtsäften." In: A. Kettrup, J. Weiß und D. Jensen (eds.), *Spurenanalytische Bestimmung von Ionen*. ecomed Landsberg 1997.

9.200 A. Trotzer, personal communication.

9.201 A. Weimar and H.W. Stuurman, *GIT Fachz. Labor* **8** (1993) 652.

9.202 A. Trifirò, G. Saccani, S. Gherardi, E. Vicini, E. Spotti, M.P. Previdi, M. Ndagijimana, S. Cavalli and C. Reschiotto, *J. Chromatogr. A* **770** (1997) 243.

9.203 Code of Practise for Evaluation of Fruit and Vegetable Juices, Reference Guideline for Apple Juice, Association of the Industry of Juices and Nectars from Fruit and Vegetables of the European Economic Community, Brussels, December 1993.

9.204 C. Tricard, M.H. Salagoity and P. Sudraud, *Ann. Fals. Exp. Chim.* **79** (1986) 303.

9.205 C. Junge and C. Spadinger, *Flüssiges Obst* **2** (1982) 57.

9.206 A. Trifirò, G. Saccani, A. Zanotti, S. Gherardi, S. Cavalli and C. Reschiotto, *J. Chromatogr. A* **739** (1996) 175.

9.207 B. Baumgärtner, Diploma thesis Fachhochschule Fresenius, Wiesbaden 1989.

9.208 A. Trotzer, H.-J. Hofsommer and K. Rubach, *Flüssiges Obst* **12** (1994) 581.

9.209 N.H. Low and K.W. Swallow, *Flüssiges Obst* **1** (1991) 13.

9.210 A. Wiesenberger, E. Kolb, M. Haug and D. Gürster, *Flüssiges Obst* **9** (1992) 547.

9.211 P.F. Cancalon, *J. of AOAC International* **76** (1993) 584.

9.212 H. Klein and R. Leubolt, *J. Chromatogr.* **640** (1993) 259.

9.213 R. Pecina, G. Bonn, E. Burtscher and O. Bobleter, *J. Chromatogr.* **287** (1984) 245.

9.214 S.H. Ashoor and J. Welty, *J. Chromatogr.* **287** (1984) 452.

9.215 S. Deseveaux, V. Daems, F. Delveaux and G. Derdelincx, "Analysis of Fermentable Sugars and Dextrins in Beer by HPAEC-PAD". In: W.J. Hurst (ed.) *Seminars in Food Analysis*. Chapman & Hall London 1997, Vol. 2, No. 1/2.

9.216 VO (EWG) No. 2676/90 der Kommission vom 17. September 1990 zur Festlegung gemeinsamer Analysenmethoden für den Weinsektor; AB1.(EG) No. L 272 vom 3.10.90, 1-192.

9.217 F.-J. Hüther and G. Gangluff, *GIT Fachz. Labor* **9** (1997) 882.

9.218 H. Otteneder and R. Marx, *Vitic. Enol. Sci.* **50** (1995) 67.

9.219 P. Ohs, Dissertation University Saarbrücken 1986.

9.220 IC Application Note No. S-12, Metrohm AG, Herisau, Switzerland.

9.221 IC Application Note No. C-10, Metrohm AG, Herisau, Switzerland.

9.222 H. Wünsche, P. Fodor, E. Szitha and R. Csenki, "Neue Erkenntnisse in der Weinklärung", Internationales Symposium Innovationen in der Kellerwirtschaft, Stuttgart 1986, p. 143.

9.223 P.L. Buldini, S. Cavallo and J. Lal Sharma, *J. Agric. Food Chem.* **47** (1999) 1993.

9.224 IC Application Note No. S-16, Metrohm AG, Herisau, Switzerland.

9.225 F. Gaucheron, Y. Le Graet, M. Piot and E. Boyaval, *Lait* **76** (1996) 433.

9.226 F. Gaucheron and Y. Le Graet, *J. Chromatogr. A* **893** (2000) 133.

9.227 IC Application Note No. S-44, Metrohm AG, Herisau, Switzerland.

9.228 G. Schwedt, *GIT Supplement* **3** (1987) 76.

9.229 AOAC Official Method 992.22, Offical Methods of Analysis of AOAC International, 16th edition; P. Cunniff (ed.); AOAC International Arlington, VA, USA, 1995; Vol. 2, Chapter 33, p. 29 ff.

9.230 F. Gaucheron and Y. Le Graet, *J. Chromatogr. A* **893** (2000) 133.

9.231 P. Walstra, R. Jenness, *Dairy Chemistry and Physics*. Wiley, New York 1984.

9.232 C. Karahadian and R.C. Lindsay, *J. Dairy Sci.* **70** (1987) 909.

9.233 C.V. Santos, P.M. Tomasula and M.J. Kurantz, *J. Food Sci.* **64** (1999) 400.

9.234 M.A.J.S. van Boekel, *Int. Dairy J.* **9** (1999) 237.

9.235 M. Courroye, J.L. Berdagué and O. Leray, *Lait* **69** (1989) 233.

9.236 C.O. Dasenbrock and W.R. LaCourse, *Anal. Chem.* **70** (1998) 2415.

9.237 A.I. Macintosh, *JAOAC Int.* **73** (1990) 880.

9.238 E. Kirchmann and L.E. Welch, *J. Chromatogr.* **633** (1993) 111.

9.239 D.J. Fletouris, J.E. Psomas and A.J. Mantis, *J. Agric. Food Chem.* **40** (1992) 617.

9.240 J. Boison and L. Keng, *Semin. Food Anal.* **1** (1996) 33.

9.241 K.E. Bach Knudsen and I. Heesov, *Br. J. Nutrit.* **74** (1995) 101.

9.242 J.-M. Durgnat and C Martinez, *Sem. Food Anal.* **2** (1997) 85.

9.243 J. Van Loo, P. Coussement, L. De Leenheer, H. Hoebregs and G. Smits, *Crit. Rev. Food Sci. Nutrit.* **35** (1995) 525.

9.244 R.G. Crittenden and M.J. Playne, *Trends Food Sci. Technol.* **7** (1996) 353.

9.245 E. Baluyot and C.G. Hartford, *J. Chromatogr.* **739** (1996) 217.

9.246 R. Grau and A. Mirna, *Fresenius Z. Anal. Chem.* **158** (1957) 182.

9.247 B. Schmidt and G. Schwedt, *Dtsch. Lebensm. Rdsch.* **80** (1984) 137.

9.248 B. Luckas, *Fresenius Z. Anal. Chem.* **318** (1984) 428.

9.249 D. Siu and A. Henshall, *J. Chromatogr. A* **804** (1998) 157.

9.250 E. Graf, *J. Am. Oil Chem. Soc.* **60** (1983) 1861.

9.251 P.T. Hawkins, D.R. Poyner, T.R. Jackson, A.J. Letcher, D.A. Lander and R.F. Irvine, *Biochem. J.* **294** (1993) 929.

9.252 K. Lee and J.A. Abendroth, *J. Food Sci.* **48** (1983) 1344.

9.253 B. Matthäus, *GIT Labor-Fachzeitschrift* **5** (1997) 448.

9.254 A.W. Fitchett and A. Woodruff, *LC-Magazin* **1** (1983) 48.

9.255 B.Q. Phillippy and M.R. Johnston, *J. Food Sci.* **50** (1985) 541.

9.256 A.R. De Boland, G.B. Garner and B.L. O'Dell, *J. Agric. Food Chem.* **23** (1975) 1186.

9.257 P. Talamond, S. Doulbeau, I. Rochette, J.-P. Guyot and S. Treche, *J. Chromatogr. A* **871** (2000) 7.

9.258 W. Frenzel, *GIT Labor-Fachzeitschrift* **7** (1997) 734.

9.259 IC Application Note No. C-28, Metrohm AG, Herisau, Switzerland.

9.260 L.A. Kaine and K.A. Wolnik, *J. Chromatogr. A* **804** (1998) 279.

9.261 R. Draisci, S. Cavalli, L. Lucentini and A. Stacchini, *Chromatographia* **35** (1993) 584.

9.262 J. Prodolliet, M. Bruelhart, F. Lador, C. Martinez, L. Obert, M.B. Blanc and J.-M. Parchet, *J. Assoc. Off. Anal. Chem. Int.* **78** (1995) 749.

9.263 J. Prodolliet, M.B. Blanc, V. Leloup, G. Cherix, C.M. Donelly and R. Viani, *J. Assoc. Off. Anal. Chem. Int.* **78** (1995) 761.

9.264 J. Prodolliet, E. Bugner and M. Feinberg, *J. Assoc. Off. Anal. Chem. Int.* **78** (1995) 768.

9.265 M.N. Clifford, Chemical and Physical Aspects. In: M.N. Clifford und K.C. Wilson (eds.) *Coffee: Botany, Biochemistry and Production of Beans and Beverages.* Croom Helm London 1985, p. 328 ff.

9.266 L.C. Trugo and R. Macrae, *Food Chem.* **15** (1984) 219.

9.267 C.J. Humphrey and R. Macrae, "Determination of Chlorogenic Acid in Coffee", presentation at the 12th International Colloquium on Coffee, Montreux 1987.

9.268 L.C. Trugo and R. Macrae, *Analyst* **109** (1984) 263.

9.269 H. Terada and Y. Sakabe, *J. Chromatogr.* **346** (1985) 333.

9.270 U. Zacke and H. Gründing, *Z. Lebensm. Unters. Forsch.* **184** (1987) 503.

9.271 A. Herrmann, E. Damawandi and M. Wagmann, *J. Chromatogr.* **280** (1983) 85.

9.272 T.A. Biemer, *J. Chromatogr.* **463** (1989) 463.

9.273 I. Goodall, M.J. Dennis, I. Parker and M. Sharman, *J. Chromatogr. A* **706** (1995) 353.

9.274 K.W. Swallow and N.H. Low, *J. Agric. Food Chem.* **38** (1988) 1828.

9.275 J. Senior, *Eur. Chromatogr. News* **2** (1988) 11.

9.276 A.C. Rychtman, *LC-GC Magazine* **7** (1989) 508.

9.277 E. Moro, M. De Angelis and B. Fugazza, *J. Chromatogr. A* **706** (1995) 451.

9.278 J.P. Waterworth and L.R. Skinner, *J. Chromatogr. A* **804** (1998) 211.

9.279 W.R. LaCourse, W.A. Jackson and D.C. Johnson, *Anal. Chem.* **61** (1989) 2466.

9.280 N.K. Jagota, A.J. Chetram, and J.B. Nair, *J. Chromatogr. A* **739** (1996) 343.

9.281 H.-J. Kim, G.Y. Park and Y.-K. Kim, *Food Technology* **1** (1987) 85.

9.282 R. Leubolt and H. Klein, *J. Chromatogr.* **640** (1993) 271.

9.283 N. O'Rourke, E. McCloskey, F. Houghton, H. Huss and J.A. Kanis, *J. Clin. Oncol.* **13** (1995) 929.

9.284 J.A. Kanis, E.V. McCloskey, P. Sirtori, D. Fern, K. Eyres, J. Aaron and M.N. Beneton, *Osteoporosis Int.* **3** (1993) S23.

9.285 J.D. Curry, J.B. Prentice, O.T. Quimby, D.A. Nicholson and J.B. Roy, *J. Organomet. Chem.* **13** (1968) 199.

9.286 S. Auriola, R. Kostiainen, M. Ylinen, J. Mökkönen and P. Ylitalo, *J. Pharm. Biomed. Anal.* **7** (1989) 1623.

9.287 V. Virtanen and L.H.J. Lajunen, *J. Chromatogr.* **617** (1993) 291.

9.288 G.E. Taylor, *J. Chromatogr. A* **770** (1997) 261.

9.289 J. Quitasol and L. Krastins, *J. Chromatogr. A* **671** (1994) 273.

9.290 P.E. Plotter, J.L. Meek and N.H. Neff, *J. Neurochem.* **41** (1983) 188.

9.291 H. Stadler and Th. Nesselhutt, *Neurochem. Int.* **9** (1986) 127.

9.292 P. Van Zonen, C. Gooijer, N.H. Velthurst, R.W. Frei, J.H. Wolf, J. Gerrits and F. Flentge, *J. Pharm. Biomed. Anal.* **5** (1987) 485.

9.293 D.N. Buchanan, F.R. Fucek and E.F. Domono, *J. Chromatogr.* **181** (1980) 328.

9.294 S. Chen, V. Soneji and J. Webster, *J. Chromatogr. A* **739** (1996) 351.

9.295 J.P. Waterworth, *J. Chromatogr. A* **770** (1997) 99.

9.296 J.S. Fritz, S.S. Yamamura and M.J. Richard, *Anal. Chem.* **29** (1957) 158.

9.297 F. Smith, A. McMurtrie and H. Galbraith, *Microchem. J.* **22** (1977) 45.

9.298 J. Weiss, J. Statler and A. Heckenberg, *S.T.P. Pharma Pratiques* **4** (1994) 372.

9.299 S.H. Ashoor, W.C. Monte and J. Welty, *J. Assoc. Off. Anal. Chem.* **67** (1984) 78.

9.300 J. Bianchi and R.C. Rose, *J. Micronutrient Anal.* **1** (1985) 3.

9.301 L.L. Lloyd, J.A. McConville, F.P. Warner, J.F. Kennedy and C.A. White, *LC-GC* **5** (1987) 338.

9.302 R.S.R. Robinett and W.K. Herber, *J. Chromatogr. A* **671** (1994) 315.

9.303 V.P. Hanko and J.S. Rohrer, *Anal. Biochem.* **283** (2000) 192.

9.304 L. Joergensen, A. Weimann and H.F. Botte, *J. Chromatogr.* **602** (1992) 179.

9.305 V.P. Hanko and J.S. Rohrer, *Gen. Eng. News* **19** (1999) 51.

9.306 R.G. Bell, *J. Chromatogr.* **546** (1991) 251.

9.307 R.S.R. Robinett, H.A. George and W.K. Herber, *J. Chromatogr. A* **718** (1995) 319.

9.308 R.P. Jones, *Process Biochem.* **21** (1986) 183.

9.309 E.D. Weinberg, in: A.H. Rose and J.F. Wilkinson (eds.), *Advances in Microbial Physiology*. Vol. 4, Academic Press New York 1970, p. 26-28.

9.310 J. Riviere, *Industrial Applications of Microbiology*. Masson Paris 1977, p. 81.

9.311 K. Ohsawa, Y. Yoshimura, S. Watanabe, H. Tanaka, A. Yokota, K. Tamura and K. Imaeda, *Anal. Sci.* **2** (1986) 165.

9.312 R. Salas-Auvert, J. Colmenarez, H. de Ledo, M. Colina, E. Gutierrez, A. Bravo, L. Soto and S. Azuero, *J. Chromatogr. A* **706** (1995) 183.

9.313 H.E. Boenigk, J.H. Lorenz and U. Juergens, *Nervenarzt* **56** (1985) 579.

9.314 U. Juergens, *LaborPraxis* **3** (1990) 126.

9.315 B. Rambeck and T. May, *Ther. Drug Monit.* **6** (1984) 164.

9.316 W.G. Robertson, M. Peacock, P.J. Heyburn, R.W. Marshall, A. Rutherford, R.E. Williams and P.B. Clark, "The significance of mild hyperoxaluria in calcium stone-formation". In: G.A. Rose, W.G. Robertson, and R.W.E. Watts (eds.). *Oxalate in Human Biochemistry and Clinical Pathology*. The Wellcome Foundation Ltd., London 1979; p. 173.

9.317 H.A. Moye, M.H. Malagodi, D.H. Clarke and C.J. Miles, *Clin. Chim. Acta* **144** (1981) 173.

9.318 B.G. Wolthers and M. Hayer, *Clin. Chim. Acta* **120** (1982) 87.

9.319 H. Hughes, L. Hagen and R.A.L. Sutton, *Anal. Biochem.* **119** (1982) 1.

9.320 C.J. Mahle and M. Menon, *J. Urol.* **127** (1982) 159.

9.321 W.G. Robertson, D.S. Scurr, A. Smith and R.L. Orwell, *Clin. Chim. Acta* **126** (1982) 91.

9.322 R.L. Baranowski and Ch. Westenfelder, *Kidney International* **30** (1986) 113.

9.323 Applications Report No. 11/82/2, Dionex GmbH Idstein, Germany.

9.324 C. Neuberg, E. Strauss and L.E. Lipkin, *Arch. Biochem.* **4** (1944) 101.

9.325 Ch. Reiter, S. Müller and Th. Müller, *J. Chromatogr.* **413** (1987) 251.

9.326 P. de Jong and M. Burggraaf, *Clin. Chim. Acta* **132** (1983) 63.

9.327 Z. Khalkhali and B. Parsa, "Measurement by non-destructive neutran activation analysis of bromine concentrations in the secretion of nursing mothers". In: *Nuclear activation techniques in the life sciences*. Vienna: Int. Atomic Energy Agency 1972, p. 461-66.

9.328 L.-O. Plantin, p. 466 in [327].

9.329 P. Duvaldestin, *Anesthesiology* **46** (1977) 375.

9.330 R. Rautu, V. Lupea and Gh. Negut, *Nahrung* **18** (1974) 13.

9.331 S. Moncada, M.A. Marletta, J.B. Hibbs, jr. and E.A. Higgs (eds.), *The Biology of Nitric Oxide.* Vol. 2, Portland Press London 1992.

9.332 R.M.J. Palmer, A.G. Ferrige and S. Moncada, *Nature* **327** (1987) 524.

9.333 K. Kikuchi, T. Nagano, H. Hayakawa, Y. Hirata and M. Hirobe, *J. Biol. Chem.* **268** (1993) 23106.

9.334 P.C. Ford, D.A. Wink and D.M. Stanbury, *FEBS Lett.* **326** (1993) 1.

9.335 J.B. Hibbs, C. Westenfelder, R. Taintor, Z. Vavrin, C. Kablitz, R.L. Baranowski, J.H. Ward, R.L. Menlove, M.P. McMurray, J.P. Kushner and W.E. Samlowski, *J. Clin. Invest.* **89** (1992) 867.

9.336 L.C. Green, D.A. Wagner, J. Glogowski, P.L. Skipper, J.S. Wishnok and S.R. Tannenbaum, *Anal. Biochem.* **126** (1982) 131.

9.337 S.A. Everett, M.F. Dennis, G.M. Tozer, V.E. Prise, P. Wardman and M.R.L. Stratford, *J. Chromatogr. A* **706** (1995) 437.

9.338 M.R.L. Stratford, M.F. Dennis, R. Cochrane, C.S. Parkins and S.A. Everett, *J. Chromatogr. A* **770** (1997) 151.

9.339 J.M. Monaghan, K. Cook, D. Gara and D. Crowther, *J. Chromatogr. A* **779** (1997) 143.

9.340 P. Pastore, I. Lavagnini, A. Boaretto and F. Magno, *J. Chromatogr.* **475** (1989) 331.

9.341 P.R. Haddad, Marheni and A.R. McTaggart, *J. Chromatogr.* **546** (1991) 221.

9.342 H. Mackie, S.J. Speciale, L.J. Throop and T. Yang, *J. Chromatogr.* **242** (1982) 177.

9.343 P.L. Annable and L.A. Sly, *J. Chromatogr.* **546** (1991) 325.

9.344 L.M. Thienpont, J.E. Van Nuwenborg and D. Stöckl, *Anal. Chem.* **66** (1994) 2404.

9.345 L.M. Thienpont, J.E. Van Nuwenborg and D. Stöckl, *J. Chromatogr. A* **706** (1995) 443.

9.346 L.M. Thienpont, J.E. Van Nuwenborg H. Reinauer and D. Stöckl, *Clin. Biochem.* **29** (1996) 501.

9.347 J.E. Van Nuwenborg, D. Stöckl and L.M. Thienpont, *J. Chromatogr. A* **770** (1997) 137.

9.348 S.S. Brown (ed.), *Clinical Chemistry and Clinical Toxicology of Metals.* Elsevier Amsterdam 1979.

9.349 R.B. Passey, K.C. Maluf and R. Fuller, *Anal. Biochem.* **151** (1985) 462.

9.350 S. Maret and R.I. Henkin, *Clin. Chem.* **17** (1971) 369.

9.351 C.N. Ong, H.Y. Ong and L.H. Chua, *Anal. Biochem.* **173** (1988) 64.

9.352 G.K. Grimble and A.M. Adam, *Chromatogr. Anal.* February (1985) 5.

9.353 S.C. Fleming, J.A. Kynaston, M.F. Laker, A.D.J. Pearson, M.S. Kapembwa and G.E. Griffin, *J. Chromatogr.* **640** (1993) 293.

9.354 R.E. Smith, S. Howell, D. Yourtree, N. Premkumar, T. Pond, G.Y. Sun and R.A. MacQuarrie, *J. Chromatogr.* **439** (1988) 83.

9.355 R.E. Smith and R.A. MacQuarrie, *Anal. Biochem.* **170** (1988) 308.

9.356 A.M. Vogt, C. Ackermann, T. Noe, D. Jensen and W. Kübler, *Biochem. Biophys. Res. Comm.* **248** (1998) 527.

9.357 P. Jandik, J. Cheng, J. Evrovski and N. Avdalovic, *J. Chromatogr. A* **759** (2001) 145.

9.358 K.S. McCully, *Nat. Med.* **2** (1996) 386.

9.359 A. Pastore, K. Massoud, C. Motti, A.L. Russo, G. Fucci, C. Cortese and G. Federici, *Clin. Chem.* **44** (1998) 825.

9.360 C.M. Pfeifer, D.L. Huff and E.W. Gunter, *Clin. Chem.* **45** (1999) 290.

9.361 J. Evrovski, M. Callaghan and D.E.C. Cole, *Clin Chem.* **41** (1995) 757.

9.362 T.L. Kirley, *Anal. Biochem.* **180** (1989) 231.

9.363 B.-M. Eriksson, S. Gustafsson and B.-A. Persson, *J. Chromatogr.* **278** (1983) 255.

9.364 C. Sarzanini, E. Mentasti and M. Nerva, *J. Chromatogr. A* **671** (1994) 259.

9.365 R.M. Riggin and P.T. Kissinger, *Anal. Chem.* **49** (1977) 2109.

9.366 A. Kobata, *Glycobiology* **1** (1990) 5.

9.367 T.W. Rademacher, R.B. Parekh and D.A. Dwek, *Ann. Rev. Biochem.* **57** (1988) 785.

9.368 N. Tsuchiya, T. Endo, K. Matsuka, S. Yoshinoya, T. Aikawa, E. Kosuge, F. Takeuchi, T. Miyamoto and A. Kobata, *J. Rheumatology* **16** (1989) 285.

9.369 M. Weitzhandler, M. Hardy, M.S. Co and N. Avdalovic, *J. Pharm. Sci.* **83** (1994) 1670.

9.370 Technical Note 30, Dionex Corporation, 1994.

9.371 T. Mizuochi, J. Hamako and K. Titani, *Arch. Biochem. Biophys.* **257** (1987) 387.

9.372 M. Tandai, T. Endo, S. Sasaki, Y. Masuho, N. Kochibe and A. Kobata, *Arch. Biochem. Biophys.* **261** (1991) 339.

9.373 Y.-T. Li and S.C. Li, In: *The Glycoconjugates*. M.I. Horowitz and W. Pigman (eds.), Academic Press New York 1977, p. 53.

9.374 R. Kadnar and J. Rieder, *J. Chromatogr. A* **706** (1995) 301.

9.375 R. Kadnar, *J. Chromatogr. A* **804** (1998) 217.

9.376 R. Kadnar and J. Rieder, *J. Chromatogr. A* **706** (1995) 339.

9.377 R. Kadnar, *J. Chromatogr. A* **850** (1999) 289.

9.378 S. Rydholm, *Pulping Processes*. Wiley New York 1965.

9.379 S. Utzman and D. Campbell, *LC-GC* **9** (1991) 300.

9.380 S. Utzman, *J. Chromatogr.* **640** (1993) 287.

9.381 Official Test Method from the Technical Association of the Pulp and Paper Industry – TAPPI, "Analysis of Pulping Liquors by Suppressed Ion Chromatography", T699 om-87, 1987.

9.382 J. Weiss, S. Reinhard, C. Pohl, C. Saini and L. Narayaran, *J. Chromatogr. A* **706** (1995) 81.

9.383 J. Sullivan and M. Douek, *J. Chromatogr. A* **804** (1998) 113.

9.384 L. Van Nifterik, J. Xu, J.L. Laurent and J. Mathieu, *J. Chromatogr.* **640** (1993) 335.

9.385 J. Sullivan and M. Douek, *J. Chromatogr. A* **671** (1994) 339.

9.386 R.C. Sheridan, *J. Chromatogr.* **371** (1986) 383.

9.387 K. Bussau, personal communication.

9.388 J. Bruins, B. Hillebrecht, H. Monien and W. Maurer, *Fresenius Z. Anal. Chem.* **331** (1988) 611.

9.389 F. Wirsching, "Die Anwendung neuerer Analysenverfahren zur Charakterisierung von Gipsen", Presentation at *Eurogypsum*, Salzburg 1987.

9.390 D. Hildebrand, Vulcan Chemical, Wichita, Kansas, USA, personal communication.

9.391 B. Knipping and F. Türck, *J. Chromatogr.* **640** (1993) 279.

9.392 H.M. Kingston, *Anal. Chem.* **50** (1978) 2064.

9.393 J. Riviello, "Determination of Transition Metals by Ion Chromatography Using Chelation Concentration". Preliminary Report, Dionex Corporation Sunnyvale 1988.

9.394 L.R. Adams, "PPB level transition metals (Fe, Cu, Ni, Mn) in 50% Caustic by Ion Chromatography Using Chelation Concentration". Presentation at the Symposium *"Advances in Ion Exchange Chromatography and Electrochemical Detection*. Newport Beach, CA, USA 1989.

9.395 A. Siriraks, H.M. Kingston and J.M. Riviello, *Anal. Chem.* **62** (1990) 1185.

9.396 E.M. Heithmar, T.A. Hinners, J.T. Rowan and J.M. Riviello, *Anal. Chem.* **62** (1990) 857.

9.397 J.W. McLaren, D. Beauchemin and S.S. Berman, *Anal. Chem.* **59** (1987) 610.

9.398 D. Beauchemin, J.W. McLaren and S.S. Berman, *Spectrochim. Acta*, Part B, **42B** (1987) 467.

9.399 P.R. Haddad, *J. Chromatogr.* **482** (1989) 267.

9.400 R. Bagchi and P.R. Haddad, *J. Chromatogr.* **351** (1986) 541, cited in [99]

9.401 P.F. Kehr, B.A. Leone, D.E. Harrington and W.R. Bramstedt, *LC-GC Magazine* **4** (1986) 1118.

9.402 P.E. Jackson, P.R. Haddad and S. Dilli, *J. Chromatogr.* **295** (1984) 471.

9.403 R.A. Hill, *J. High Resolut. Chromatogr. Chromatogr. Commun.* **6** (1983) 275.

9.404 D.D. Siemer, *Anal. Chem.* **59** (1987) 2439.

9.405 P.L. Buldini, S. Cavalli and A. Mevoli, *J. Chromatogr. A* **739** (1996) 167.

9.406 F.A.J. Armstrong, P.M. Williams and J.D.J. Strickland, *Nature* **211** (1966) 481.

9.407 M. Kolb, P. Rach, J. Schäfer and A. Wild, *Fresenius J. Anal. Chem.* **342** (1992) 341.

9.408 H. Albrich, M. Müller and H. Pinhack, *GIT* **8** (1994) 837.

9.409 H. Albrich and M. Müller, *LaborPraxis* **9** (1995) 62.

9.410 K. Michels, R. Siegfried and D. Jensen, *GIT Sonderheft Chromatographie* **9** (1998) 78.

9.411 K. Ebert and I. Melzer, *Agribiol. Res.* **50** (1997) 146.

9.412 L.M. Busman, R.P. Dick and M.A. Tabatabai, *Soil Sci. Soc. Am. J.* **47** (1983) 1167.

9.413 J.P. Senior, *Anal. Proc.* **27** (1990) 116.

9.414 J.C. Umali, G.M. Moran and P.R. Haddad, *J. Chromatogr. A* **706** (1995) 199.

9.415 J.K. Gard, D.R. Gard and C.F. Callis, *Am. Chem. Soc. Symp. Series* **486** (1992) 41.

9.416 P.A. Blackwell, M.R. Cave, A.E. Davis and S.A. Malik, *J. Chromatogr. A* **770** (1997) 93.

9.417 P.R. Haddad and S. Laksana, *J. Chromatogr. A* **671** (1994) 131.

9.418 J.M. Pettersen, H.G. Johnsen and W. Lund, *Talanta* **35** (1988) 245.

9.419 M. Nonomura, *Anal. Chem.* **59** (1987) 2073.

9.420 C.J. Hill and R.P. Lash, *Anal. Chem.* **52** (1980) 24.

9.421 A. Siriraks and J. Stillian, *J. Chromatogr.* **640** (1993) 151.

9.422 R.M. Montgomery, R. Saari-Nordhaus, L.M. Nair and J.M. Anderson, jr., *J. Chromatogr. A* **804** (1998) 55.

9.423 D. Jensen, *GIT, Fachz. Labor* **4** (1995) 332.

9.424 R.W. Frei and K. Zech, *Selective sample handling and detection in high-performance liquid chromatography.* Elsevier, Amsterdam 1988.

9.425 D.L. Strong and P.K. Dasgupta, *J. Membr. Sci.* **57** (1991) 321.

9.426 D.L. Strong, P.K. Dasgupta, K. Friedman and J.R. Stillian, *Anal. Chem.* **63** (1991) 480.

9.427 H. Small and J. Riviello, *Anal. Chem.* **70** (1998) 2205.

9.428 H. Small, Y. Liu and N. Avdalovic, *Anal. Chem.* **70** (1998) 3629.

Index

Numbers in front of the page numbers refer to Volume 1 or 2, respectively.

a

Accuracy 2/551
Acebutolol 1/449
Acesulfam K 2/752
Acetaldehyde 2/622
Acetate 1/44, 1/168 f, 1/199
Acetic acid 1/74, 1/362, 2/627, 2/650, 2/701, 2/714, 2/716, 2/786
N-Acetylchitobiose 1/241
Acetylcholine 1/302, 1/420, 1/451, 2/742
N-Acetylgalactosamine 1/241
N-Acetylglucosamine 1/238
N-Acetylmannosamine 1/251
N-Acetylneuraminic acid 1/238, 1/242, 2/799
Acid blue 40 1/459
Acid blue 113 1/459
Acid red 114 1/459
cis,trans-Aconitate 1/73
Acridine orange 1/459
Acrylic acid 1/363, 1/373
Activity coefficient 1/28, 1/466, 1/467
Additives
 – inorganic 1/406
 – organic 1/100, 2/690 ff, 2/731
Adenosine 1/178
 – diphosphate 1/179, 2/794
 – monophosphate 1/179, 2/794
 – triphosphate 1/179, 2/794
Adipic acid 1/372
Adsorption process, non-ionic 1/3, 1/359
Aerosol 2/622
Alanine 1/258, 1/387
Alcohol 1/364, 1/379, 1/485, 2/774
Aldehyde 1/379, 1/485
Alizarin red S 1/459
Alkali metal 1/280, 1/286, 1/299, 1/318, 1/518, 2/628, 2/718, 2/780, 2/787

Alkaline-earth metal 1/281, 1/286, 1/299, 1/303, 1/322, 1/501, 1/518, 2/628, 2/718, 2/742, 2/780, 2/787, 2/819
Alkaloid 1/402, 1/418, 1/455
Alkanesulfonate 1/404, 1/424
Alkanolamine 1/418, 1/491
Alkenesulfonate 1/424
Alkylamine 1/302, 1/320, 1/493
Alkylbenzene sulfonate 1/424, 1/430, 2/709
Alkylbenzyl-dimethyl-ammonium 2/706
Alkyl-dimethyl-benzylammonium chloride 1/435
Alkylether sulfate 2/709
Alkyl sulfate 1/425
Alkyl sulfonate 1/425, 2/706
Alkyltrimethylammonium 1/433
Allobarbital 1/442, 1/453
Allylsulfonic acid 2/689
Aluminum 1/341, 1/501, 2/649, 2/789
Amidosulfonic acid 1/170, 2/647
Amine 1/418, 1/487, 1/491, 1/514
 – aliphatic 1/288, 1/324, 2/759
 – aromatic 1/356, 1/418
 – biogenic 1/295, 1/302, 1/346, 1/451, 2/742, 2/747
 – catechol- 1/421, 1/451, 1/455, 2/796
 – di- 1/295, 1/345
 – hydrophobic 1/295
 – poly- 1/344 ff, 1/514
 – polyvalent 1/295
Amino acid 1/257 ff, 1/382 ff, 1/487, 1/493, 1/514
 – carbohydrate interference 1/259
 – detection 1/389 ff
 – dissociation behavior 1/99, 1/384
 – elution order 1/258
 – O-phosphorylated 1/264

– physiological   1/388
– sample preparation   1/391 f
p-Aminobenzoic acid   2/770
2-[(Aminocarbonyl)oxy]-N,N,N-trimethyl-1-propanaminium chloride (Bethanechol)   2/766
7-Aminocephalosporic acid   1/458
5-Amino-2,3-dihydro-1,4-phthalazindione (Luminol)   2/632
2-(2-Aminoethyl)pyridine   1/357
p-Aminohippuric acid   2/782
3-Amino-1-hydroxypropylidene-1,1-bisphosphonate (Pamidronate)   2/765
Aminomethanephosphonic acid   1/517
Aminomethanesulfonic acid   1/516
Aminopolycarboxylic acid   2/63, 1/181, 1/443
Aminopolyphosphonic acid   1/63, 1/181, 1/183, 1/184
– di-N-oxide   1/185
– mono-N-oxide   1/184
2-Aminopyridine   1/357
Amino sugar   1/206, 1/214 f, 1/243
– N-acetylated   1/214
– de-N-acetylated   1/243
AminoTrap column   1/215, 1/243
Aminotris-(methylenephosphonic acid) (ATMP)   1/191
Ammonium   1/283, 1/293, 1/315, 1/318, 1/514, 2/598, 2/609, 2/615, 2/621, 2/634, 2/690, 2/734, 2/815
Ammonium compound, quaternary   1/355, 1/418, 1/433
Amobarbital   1/442, 1/453
Amperometry
– integrated   1/391, 1/487 f
– pulsed   1/480 ff
Ampicillin   1/487, 2/735
Amylamine   2/759
Amylopectin   1/235
Amylose   1/234
Analysis
– air   2/675
– – clean room   2/675
– of chemicals   1/378 ff, 2/648 ff,
– function   2/560
– online   2/650, 2/652
– quantitative   2/549 ff
– time   1/7
Aniline   1/356

Anion
– halide   1/121 f, 1/133
– inorganic
– – survey   1/121 f
– – UV measuring wavelengths   1/498
– non-metal   1/125
– non-polarizable   1/94
– organic   1/168 ff
– oxyhalide   1/121
– peroxohydroxide   2/701
– polarizable   1/61, 1/66, 1/72, 1/79, 1/94, 1/122, 1/159 ff, 1/413, 2/601
– polyvalent   1/63, 1/163, 1/181 ff, 1/509
– standard   1/39
– – limit of determination   2/628
– surface-active   1/72, 1/424 ff
– surface-inactive   1/409 ff
Anion exchanger
– high capacity   1/35
– – cellulose   1/35
– – polymer-based   1/35
– – silica-based   1/35
– low capacity
– – latexed   1/54 ff
– – polymethacrylate   1/47 ff
– – polyvinyl   1/54
– – PS/DVB   1/35 ff
– – silica-based   1/81 ff
Anion-trap column (ATC)   1/196
Anomer   1/224
Anserine   1/388
9-Anthryldiazomethane (ADAM)   1/520
Antibody, monoclonal   1/240, 1/269, 1/271
AOX/AOS   2/624
Application
– building material industry   1/302
– chemical industry   2/814 f
– clinical chemistry   2/781 ff
– electroplating industry   2/678 ff
– environmental analysis   2/588 ff
– food and beverage industry   2/711 ff
– household product industry   2/697 ff
– mineralogy   2/815
– petrochemical industry   1/302, 2/803 ff
– pharmaceutical industry   2/756 ff
– power plant chemistry   1/342, 2/626 ff
– pulp & paper industry   2/810 ff
– semiconductor industry   1/336, 1/370, 2/651 ff
Approximation test, acc. to Mandel   2/565

Aprobarbital 1/442
Arabinose 1/213, 2/726, 2/750, 2/774
Arabitol 1/211, 2/774
Arginine 1/258, 1/262, 1/387
Arogenic acid 1/267
Arsenate 1/125 f, 2/622, 2/649
Arsenazo I
 see [o-(1,8-dihydroxy-3,6-disulfo-2-naphthylazo)benzenearsonic acid]
Arsenazo III
 see bis-(2-arsono-benzeneazo)-2,7-chromotropic acid
Arsenite 1/124, 1/370, 1/475
Arsenobetain 1/528
Arsenocholine 1/528
Arylalkylamine 1/418
Aryl sulfonate 1/427
Ascorbic acid 2/714, 2/722, 2/741, 2/769
Asparagine 1/258, 1/387
Aspargic acid 1/258
Aspartame 2/731
Aspirin 1/454
Asymmetry factor 1/14
Atropine 1/402, 1/421
Audit Trail 2/556
AutoNeutralization 2/639, 2/648, 2/667, 2/833
Azide 1/125, 1/152, 1/512, 1/787

b

Band broadening
 see peak broadening
Barbital 1/423, 1/442, 1/453
Barbiturate 1/418, 1/423, 1/441, 1/450
Barbituric acid 1/423, 1/453
Barium 1/283, 1/287, 1/305
Bath, galvanic
 – chromic acid 2/682, 2/690
 – copper, electroless 2/687
 – copper pyrophosphate 2/679, 2/689
 – copper sulfate, electrolytic 2/679
 – gold 2/686
 – nickel, electroless 2/680
 – nickelborohydride 2/693
 – nickel/iron 2/686, 2/692
 – nickel sulfamate 2/682
 – nickel/zinc 2/682
 – tin/lead 2/673
 – zinc 2/688
Benzalkonium chloride 2/761
Benzene 1/452
1,2-Benzenedicarboxylate 1/450
1,3-Benzenedicarboxylate 1/450
1,4-Benzenedicarboxylate 1/450
m-Benzene disulfonate 1/451
o-Benzene disulfonate 1/451
Benzenehexacarboxylate
 see phytic acid
Benzenepentacarboxylate 1/450
Benzene sulfonate 1/427
 1,2,4,5-Benzenetetracarboxylate
  (Pyromellitate) 1/448, 1/450
1,2,3-Benzenetricarboxylate 1/450
1,2,4-Benzenetricarboxylate 1/450
1,3,5-Benzenetricarboxylate 1/448, 1/450
1,3,5-Benzenetrisulfonate 1/451
Benzoate 1/410, 1/448, 2/731, 2/746, 2/762
Benzylalcohol 1/452
4-Benzylpyridine 1/357
Beryllium 2/615
Bethanechol chloride
 see 2-[(aminocarbonyl)oxy]-N,N,N-trimethyl-1-propanaminium chloride 2/766
Beverage 2/713 ff
2,2'-Bipyridine 1/357
Bis-(2-arsono-benzeneazo)-2,7-chromotropic acid (Arsenazo III) 1/502
Blank value 2/562, 2/627
Bleaching activator 2/701
Bleaching agent 2/697
Boltzmann's law of energy distribution 1/467
Borate 1/153, 1/366, 1/369, 2/633, 2/648, 2/686
BorateTrap column 1/218
Borophosphorosilicate glass film 1/131, 2/672
Brightener 2/694
Brine 2/817
Bromate 1/40, 2/65, 2/67, 1/136, 1/503, 1/544, 2/591 ff
Bromide 1/130, 1/411, 1/498, 2/588, 2/624, 2/781
p-Bromobenzoate 1/449
Bromocresol purple 1/459
Builder 2/697 f
Butabarbital 1/442, 1/450

4-Butene-18-crown-6 1/88
2,3-Butanediol 2/773
1,4-Butanedisulfonic acid 2/757
Butanesulfonic acid 1/428
n-Butanol 1/381
iso-Butylamine 1/353
n-Butylamine 1/353
sec.-Butylamine 1/353
tert.-Butylamine 1/353, 2/759
tert.-Butyl-ethane-1,2,2-trisphosphonic acid 1/189
n-Butyric acid 1/74, 1/362, 1/443

C

Cadaverine 1/344, 2/747
Cadmium 1/331, 1/335, 2/613
Caffeic acid 1/10
Caffeine 1/455, 1/457, 2/731
Caffeoylquinic acid 2/752
Calcium 1/283, 2/598, 2/758
Calibration 2/560 ff
– area normalization 2/569
– basic 2/560, 2/565
– external standard 2/571
– function 2/560 ff
– internal standard 2/570
– standard addition 2/572 f
Capacity factor 1/17, 1/29, 1/30, 1/319
Capacity ratio 1/21, 1/155
Capillary electrophoresis 1/3
Capric acid 1/443
Caproic acid 1/362, 1/443
Caprylic acid 1/443
Carbohydrate 1/62, 1/205 ff, 1/364, 1/485, 2/736, 2/790
– acidity 1/246
– phosphoric acid ester 1/224
Carbonate 1/365, 1/369, 1/511, 2/591, 2/633, 2/686, 2/698, 2/701
Carbonylphosphonate 2/764
Carboxylic acid
– aliphatic 1/168, 1/448, 2/615
– aromatic 1/448
– di- 2/615
– hydroxy- 1/168, 2/614
– keto- 2/615
Carboxypeptidase B 1/270
Carnithine 1/388
Carprofen 1/454
Carrez precipitation 2/740

Casein 2/738
Cation
– exchange process 1/279, 1/301
– exchanger 1/280 ff
– – latexed 1/298 ff
– – silica-based 1/303 ff
– – solvent influence 1/284
– – surface-sulfonated 1/280
– – weak acid 1/282, 1/324
– surface-active 1/433 ff
– surface-inactive 1/418 ff
– simultaneous analysis 1/282 ff, 1/304, 1/323, 2/598
Cavity effect 1/122, 1/143
Cefadroxil 1/458
Cefazolin 1/458
Cefotaxim 1/458
Cell
– constant 1/469
– pulsed amperometric 1/496
Cellobiose 1/221, 1/227, 2/774
Cellotriose 2/812
Cellulose 1/235
Cement analysis 2/815
Cephalexin 1/458
Cephaloridine 1/458
Cephalosporin 1/453, 1/458
Cephalosporin C 1/458
Cephalotin 1/458
Cephapirine 1/487, 2/736
Cerium 1/344
Cerium(III)/Cerium(IV) 1/518
Cesium 1/287
Cetylpyridinium
see hexadecylpyridinium
Chlorate 1/40, 2/65, 1/409, 1/546, 2/602 f, 2/811
Chloride 1/59 ff, 2/588, 2/624, 2/811, 2/816
Chlorine dioxide 2/602
Chlorite 1/65, 1/73, 1/504, 1/546, 2/596, 2/602 f, 2/811
p-Chlorobenzenesulfonate 1/449, 1/533
Chlorocholine 1/420
Chlorogenic acid 1/10, 2/752
Chloromethyl methylether 1/37
Choline 1/302, 1/419, 1/451, 2/742, 2/767
Chromate 1/159, 1/336, 1/508
Chromatogram 1/13
Chromatography
– affinity 1/267

- gas 1/1
- high performance liquid 1/1
- ion 1/1
- – multi-dimensional 1/5, 1/444
- – online 2/650, 2/652
- – two-dimensional 1/369
- ion-exchange 1/2, 1/3
- – anion exchange 1/27 ff
- – cation exchange 1/279 ff
- ion-exclusion 1/4, 1/359 ff
- ion-pair 1/4, 1/393 ff
- – retention model 1/394 ff
- liquid 1/1
- paper 1/1
- reversed-phase 1/4, 1/403
- thin layer 1/1

Chromazurol S 1/459, 2/618
Chromium
- chromium(III) 1/336
- chromium(III)/(VI) ratio 2/683
- chromium(IV) 1/508, 2/645
- simultaneous analysis 1/336 f, 1/524

α-Chymotrypsinogen 1/269
Cinchonin 1/457
Citrazinic acid
 see 2,6-dihydroxyisonicotinic acid
Citric acid 1/76, 1/174, 1/363, 1/413, 1/448, 2/704, 2/714, 2/731, 2/738, 2/762, 2/786
Citrus juice 2/714
Clodronate 2/763
Cluster, molybdenum-sulfur 1/418
Cobalt 1/306, 1/331, 1/334, 2/613, 2/631
Cocaine 1/421
Codeine 1/421
Colchicine 1/457
Collagen hydrolysate 1/266
Column
- body, material 1/5
- coefficient 1/21
- dead time 1/155
- efficiency 1/17
- length 1/17
- maintenance 2/835
- poison 2/614, 2/835
- temperature 1/208, 1/291, 1/400, 1/408

Complex
- aluminum halide 1/98
- cryptand-cation 1/93
- formation coefficient 1/327
- Gd-DTPA 1/414
- indicator 1/330
- iron cyanide 1/407, 1/415
- lanthanide-PDCA 1/343
- metal-chloro 1/163, 1/498
- metal-cyano 1/128, 1/498
- metal-DTPA 1/414
- metal-EDTA 1/163, 1/414, 2/687
- metal-oxalate 1/333
- – stability constant 1/335
- metal-PDCA 1/333
- – stability constant 1/335
- molydenum disulfido 1/419
- multimetal 1/416
- stability
- – kinetic 1/330
- – thermodynamic 1/329
- Zn-EDTA 1/500

Complexing agent 1/327, 1/510, 2/641
Complexon 1/509
Computer, personal 2/555
Concentrator column 2/628
- IonPac AG5 2/653
- MetPac CC-1 2/594, 2/819, 2/821
- TAC-1 2/653
- TAC-LP1 2/678
- TBC-1 1/368, 2/633
- TCC-LP1 2/677

Conditioning agent 2/634
Conductivity
- background 1/101, 1/103, 1/110, 1/469
- coefficient 1/469
- electrical 1/462
- equivalent 1/462, 1/464
- – table 1/466
- of electrolyte solution 1/462 ff

Confidence interval 2/559, 2/564
Control card 2/578 ff
- blank value 2/580
- $\bar{x}$-R-Combination 2/583
- Cusum 2/583
- differences 2/583
- mean value 2/580
- quality 2/578
- recovery 2/580
- Shewhart 2/579
- Span 2/582

Copper 1/332, 1/334, 2/613, 2/789
Correlation coefficient 2/562

Cotrell equation 1/478
Coulomb explosion 1/542
Coupling
   – HPICE/HPIC 1/359, 1/375 ff
   – IC/AAS 1/522
   – IC/DCP-AES 1/526
   – IC/ICP 1/523 ff
   – IC/ICP-MS 1/525
   – IC/ICP-OES 1/125, 1/526
   – IC/MS 1/504, 1/530 ff, 2/607
o-Cresol 1/441
Cross-linking 1/35
12-Crown-4 1/87
15-Crown-5 1/87
18-Crown-6
   (1,4,7,10,13,16-Hexaoxa-
   cyclooctadecane) 1/86
Crown ether 1/87
Cryptand 1/92, 1/179
   – binding constant 1/93
Cumene sulfonate 1/424, 1/427, 2/703
Current, electrical 1/463
   – charging 1/478
   – diffusion 1/476
   – Faraday 1/478
Cutter agents 2/741
Cyanate 1/124, 2/833
Cyanic acid 2/622
Cyanide 1/63, 1/126, 1/370, 1/475, 1/479, 2/810
   – easily releasable 1/129
   – free 1/128 f
   – reaction to cyanate 2/833
   – total 1/130
Cyanocobalamine 2/770
p-Cyanophenol 1/29, 1/62
Cyclamate 2/752
Cyclic voltammetric stripping (CVS) 2/692
Cyclitol 1/210
Cyclodextrin 1/232
Cyclohexylamine 1/350
Cystathionine 2/795
Cysteic acid 1/391
Cysteine 1/387, 1/392, 2/795
Cysteinylglycine 2/795
Cystine 1/258, 1/392
Cytidine 1/178
   – 5'-monophosphate 1/181
Cytochrome c 1/269, 1/272

d

Dansyl chloride 1/383
Dead time 1/13
Debye-Hückel theory 1/466 f
1,10-Decanediamine 1/354
Decanesulfonic acid 1/428
Decyl-2.2.2 1/93
Decyl-dimethyl-benzylammonium 1/435
Decyl sulfate 1/427
Decyltrimethylammonium 1/355, 1/434
Degree
   – of cross-linking 1/361
   – – effective 1/144, 1/448
   – of freedom 2/559
Deoxyadenosine-5'-monophosphate 1/181
Deoxycytidine-5'-monophosphate 1/181
2-Deoxyglucose 1/211, 1/217, 1/224, 1/242
2'-Deoxyguanosine 1/178
   – 5'-monophosphate 1/181
2'-Deoxy-D-ribose 1/212
Depolarisator 1/475
Deproteination 1/391, 2/736, 2/758, 2/783, 2/790 f
Derivatization
   – post-column 1/182, 1/499 ff
   – – two-step 1/510, 1/515
   – pre-column 1/383
Desalter, membrane 1/252
Detection 1/461 ff
   – amperometric 1/461, 1/473 ff
   – – integrated 1/487 f
   – – pulsed 1/480 ff, 2/696
   – – with fixed working potential 1/478
   – chemiluminescence 2/632
   – choice 1/9
   – conductivity 1/461 ff
   – – application form 1/469 ff
   – – direct 1/469
   – – indirect 1/102, 1/471
   – criterium 2/574
   – element-specific 1/125, 1/524
   – fluorescence 1/461, 1/514 ff
   – – indirect 1/518
   – phosphorus-specific 1/165, 1/190, 1/510, 2/643, 2/699
   – photometric 1/411, 1/461, 1/498 ff
   – – after derivatization 1/499 ff
   – – direct 1/498 f
   – – indirect 1/510 ff
   – refractive index 1/461, 1/521

– radioactivity 1/521
Detector
 – amperometric 1/6
 – – pulsed amperometric 1/496
 – – cell 1/496
 – choice 1/9
 – conductivity 1/6
 – fluorescence 1/6
 – performance criteria 1/6
 – szintillation 1/522
 – UV/Vis 1/6
Dextran 1/232
 – hydrolysate 1/232
Di-*N*-acetylchitobiose 1/241
Dialkyldimethylammonium 1/433
Dialkyl sulfosuccinate 1/431
Dialysis 2/733, 2/832
 – Donnan 2/832
 – electro 2/832
 – passive 2/832
1,2-Diaminopropane 1/347
o-Dianisidine (ODA) 1/505, 2/597
Dibenzo-18-crown-6 1/87, 1/88
Dibromoacetic acid 1/171
Dibromomonochloroacetic acid 1/170, 1/171
Dibutylphosphate 1/176
Dicaffeoylquinic acid 2/752
Dicarboxylic acid 2/615
Dichloroacetic acid 1/171, 1/413
Dichlorobenzyl-alkyl-dimethylammonium 1/436
Dichloromethylenebisphosphonic acid (Clodronate) 2/763
Dichloromonobromoacetic acid 1/171
2,6-Dichloro-4-nitroaniline 1/356
2,4-Dichloro-3-nitrophenol 1/441
2,4-Dichlorophenoxyacetic acid (2,4-D) 1/176
Diethanolamine 1/296, 1/302
Diethylamine 1/287, 1/321, 1/325
2-Diethylaminoethanol 1/290, 1/350
N,*N*′-Diethylaniline 1/356
Diethylene glycol 1/383
Diethylenetriamine 1/348
Diethylenetriaminepentaacetic acid (DTPA) 1/183
Diethylenetriaminepentamethylenephosphonic acid (DTPP, DEQUEST 2060) 1/183
N,*N*-Diethylethylenediamine 1/347

Diethyltolueneamide 1/452
Diffusion
 – current 1/476
 – Eddy 1/19
 – lateral 1/19, 1/20
 – longitudinal 1/19, 1/20
Diflunisal 1/454
Digalacturonic acid
 – mass spectrum 1/539
Digestion
 – bomb 2/828
 – combustion 2/827
 – dry ashing 2/827
 – fusion 2/831
 – oxygen 2/830
 – Schöninger 2/768
 – wet 2/827
 – Wickbold 2/828
Dihydrolipoic acid 2/770
4,5-Dihydroxy-1,3-benzenedisulfonic acid-disodium salt (Tiron) 1/501
3,4-Dihydroxybenzylamine 1/422, 2/797
o-(1,8-Dihydroxy-3,6-disulfo-2-naphthylazo)benzenearsonic acid (Arsenazo I) 1/501
2,6-Dihydroxyisonicotinic acid (Citrazinic acid) 1/175
D-Dihydroxyphenylglycine 1/458
Diisobutyl-[2-(2-phenoxyethoxy)ethyl]-dimethylbenzyl-ammonium chloride (Hyamine 1622) 1/436, 2/698
Dimethylamine 1/289, 1/321
3-Dimethylaminopropylamine 1/349
2-Dimethylaminopyridine 1/357
N,*N*′-Dimethylaniline 1/356
Dimethylarsinic acid 1/528
N,*N*-Dimethyl-2-mercaptoethylamine-hydrochloride (Thiofluor) 1/515
2,4-Dimethylphenol 1/441
Dimethylphosphate 1/546
1,2-Dimethylpropylamine 1/351
Dinaphthal-18-crown-6 1/87
3,4-Dinitrobenzoate 1/449
Diphenhydramine 1/402
1,5-Diphenylcarbazide (DPC) 1/508
Diphenylhydantoin 1/453
N,*N*′-Diphenylthiourea 2/693
Diphosphonic acid
 – geminal 1/185
 – vicinal 1/185

1,1-Diphosphonopropane-2,3-dicarboxylic
  acid (DPD) 1/191
Di-*n*-propylamine 1/351
Disaccharide 1/226
  – non-reducing 1/226
  – reducing 1/226
Distribution
  – coefficient, Nernst 1/17, 1/30, 1/32,
    1/38, 1/326
  – equilibrium 1/17
Disulfate 1/414
Disulfite 1/414
5,5′-Dithiobis(2-nitrobenzoic acid)
  (DTNB) 1/390
Dithiomolybdate 1/418
Dithionate 1/414, 2/646
Diuretic 1/455
Divinylbenzene (DVB) 1/35
1,12-Dodecanediamine 1/354
Dodecyl-dimethyl-benzylammonium 1/435
Dodecyl ether sulfate 1/429
Dodecylpyridinium chloride 1/437
Dodecyl sulfate 1/425, 1/427, 2/692,
  2/698, 2/710
Dodecyl sulfonate 1/425
Dodecyltrimethylammonium 1/355, 1/434
Donnan effect 1/104
Donor atom 1/86
Dopamine 1/422, 2/797
Double layer, electrical 1/398
Dulcitol 1/209
Dye
  – azo 1/451, 1/542
  – reactive 2/814
Dysprosium 1/344

e

Electrode
  – counter 1/474
  – recession 1/484
  – reference 1/474
  – types 1/475
  – working 1/474
Electron affinity 1/537
Electronic signature 2/556
Electrophoretic effect 1/468
Electro-transfer 2/798
Eluant
  – amino acid 1/99
  – aminoalkylsulfonic acid 1/100

  – ammonium sulfate/sulfuric acid 1/341,
    1/501
  – benzoic acid 1/145, 1/149, 1/366,
    1/470
  – borate/gluconate 1/47, 1/51, 1/145
  – boric acid 1/49
  – carbonate hydroxide
    see sodium carbonate/sodium
    hydroxide
  – cerium(III) nitrate 1/309
  – choice 1/124, 1/145
  – citrate buffer 1/386
  – citric acid 1/147, 1/151, 1/326
  – citric acid/tartaric acid 1/332
  – concentration and pH value 1/137 ff,
    1/156 ff
  – copper sulfate 1/513
  – *p*-cyanophenol 1/100
  – ethylenediamine/oxalic acid 1/309
  – ethylenediamine/tartaric acid 1/281,
    1/309, 1/330
  – flow rate 1/123
  – for anion exchange chromatography
    1/98 ff
  – for cation exchange chromatography
    1/308 ff
  – for ion-exclusion chromatography
    1/365
  – formic acid 1/304
  – for non-suppressed systems 1/100
  – fumaric acid 1/151
  – heptafluoropropanoic acid
    (perfluorobutyric acid) 1/366
  – Hi-Phi buffer 1/386
  – hydrochloric acid 1/280, 1/308, 1/365
  – hydrochloric acid/2,3-diaminopropionic
    acid 1/282, 1/300, 1/309
  – hydroxide/benzoate 1/51
  – *p*-hydroxybenzoic acid 1/49, 1/51,
    1/146, 1/521
  – α-hydroxyisobutyric acid (HIBA)
    1/331, 1/343
  – iodide 1/511
  – isonicotinic acid 1/309
  – lithium hydroxide 1/95
  – lithium sulfate/sulfuric acid 1/491
  – methanesulfonic acid 1/283, 1/308,
    1/310
  – methanesulfonic acid/mannitol 1/368
  – 2-methyllactic acid 1/502

- nicotinic acid 1/151, 2/787
- nitric acid 1/182, 1/280, 1/308, 1/370
- nitric acid/dipicolinic acid 2/618
- octanesulfonic acid 1/366
- octanesulfonic acid/mannitol 1/366
- oxalic acid 1/49, 1/326, 1/333
- oxalic acid/citric acid 1/332
- oxalic acid/diglycolic acid 1/343
- oxalic acid/ethylenediamine 1/309
- perchloric acid 1/381
- perfluorobutyric acid
    see heptafluoropropanoic acid
- perfluoroheptanoic acid
    see tridecafluoroheptanoic acid
- m-phenylenediamine-dihydrochloride 1/282, 1/309
- phthalic acid 1/49, 1/51, 1/83, 1/102, 1/470, 1/511, 1/521, 2/787
- picolinic acid 1/298
- potassium chloride/EDTA 1/165, 1/348
- potassium hydrogenphthalate 1/145, 1/162
- potassium hydroxide 1/47, 1/94, 1/102, 1/196 f, 1/470
- pyridine-2,6-dicarboxylic acid 1/326, 1/333
- pyridine-2,6-dicarboxylic acid/oxalic acid 1/310
- salicylic acid 1/151, 1/521
- silver nitrate 1/309
- sodium benzoate 1/102
- sodium carbonate 1/66 f, 1/126, 1/162
- sodium carbonate/sodium bicarbonate 1/39, 1/51, 1/57, 1/99
- sodium carbonate/sodium dihydrogenborate/ethylenediamine 1/126
- sodium carbonate/sodium hydroxide 1/58
- sodium hydroxide 1/69, 1/77, 1/94, 1/100, 1/195
- – addition of barium acetate 1/221
- – addition of zinc acetate 1/228
- – preparation 1/196, 1/220
- sodium hydroxide/sodium acetate 1/232
- sodium p-hydroxybenzoate 1/39
- sodium nitrate 1/176
- sodium perchlorate 1/98
- sodium phenolate 1/99
- sodium phthalate 1/511
- sodium sulfate/sulfuric acid 1/493
- sodium tetraborate 1/100, 1/135, 1/199
- succinic acid 1/151
- o-sulfobenzoate 1/101, 1/511, 1/521
- sulfuric acid 1/288, 1/308, 1/365
- tartaric acid 1/49, 1/84, 1/296, 1/305, 1/326
- tartaric acid/oxalic acid 1/306
- tartaric acid/pyridine-2,6-dicarboxylic acid 1/304, 1/310, 1/323
- tetrabutylammonium salicylate 1/411
- tridecafluoroheptanoic acid (perfluoroheptanoic acid) 1/366
- trifluoroacetic acid 1/290
- trimesic acid 1/102, 1/151, 1/511
- tyrosine 1/100, 1/133
- vanillic acid/N-methyldiethanolamine 1/51
- water 1/90, 1/307, 1/365, 1/381

Eluant generator 1/196 f, 1/206, 1/223, 1/295, 2/591, 2/627, 2/750, 2/814
Endcapping 1/89
EndoF2 2/801
EndoH 2/801
Endoplasmatic reticulum 1/239
Endothall
    see 7-oxabicyclo[2,2,1]heptane-2,3-dicarboxylic acid
Enthalpy
- hydration 1/122
- sorption 1/33
- – free 1/31
Entropy
- configuration 1/34
- mixing 1/34
- sorption 1/32 ff
Ephedrine 1/421, 1/453
Epimerisation 1/251
Epinephrine 1/421, 2/797
Erbium 1/344
Error
- α- 2/574
- β- 2/574
- random 2/557
- statistical 2/558
- systematic 2/557
Erythritol 1/211, 2/774

Erythropoietin (rEPO)   1/240, 1/255
Etching solution   2/670
Ethane-1,2-bis(P-methyl-phosphinic acid)   1/186
1,2-Ethanediphosphonic acid   1/186
1,2-Ethanedisulfonic acid   2/684
Ethane-1,1,2,2-tetrakis(P-methyl-phospinic acid)   1/186
Ethane-1,2,2-tris(P-methyl-phosphinic acid)   1/186
Ethanol   1/365, 1/381, 2/722, 2/773
Ethanolamine   1/306, 1/419, 1/491, 2/634, 2/807
Ethylamine   1/320, 1/324, 1/419
2-Ethylaminoethanol   1/290
Ethylenediamine   1/345
Ethylenediaminediacetic acid (EDDA)   1/183, 1/443
Ethylenediaminetetraacetic acid (EDTA)   1/183
Ethylenediaminetetramethylenephosphonic acid (EDTP, DEQUEST 2041)   1/183
Ethylenediaminetriacetic acid (EDTriA)   1/183, 1/443
Ethylene glycol
   see monoethylene glycol
Europium   1/344
Exclusion
   – Donnan   1/4, 1/359
   – steric   1/4, 1/359
   – volume   1/360
Extraction cartridge   2/599, 2/751, 2/826

f

Faraday constant   1/477
Fatty acid   1/520
   – long-chain   1/365, 1/441
   – short-chain   1/76, 1/80, 1/362
Fatty alcohol ether sulfate   1/424, 1/428, 2/709
Fatty alcohol polyglycolethersulfosuccinate   1/432
Fatty alcohol sulfate   1/427
Fenbufen   1/454
Fermentation   2/772 ff
Ferulic acid   1/10
Feruloylquinic acid   2/752
Fetuin, bovine   1/243
Fibrinogen   1/243
Filler   2/697, 2/702

Flue gas
   – desulfurization   2/644
   – denitrification   2/647
   – scrubber solution   2/644
Fluorenylmethyloxycarbonyl chloride (FMOC)   1/383
Fluorescein   1/459
Fluoride   1/39, 1/44, 1/73, 1/76, 1/78, 1/133 ff, 1/369, 2/589
Formaldehyde   1/381, 2/622, 2/680
Formamidinium   1/320
Formate   1/44
Formic acid   1/74, 1/168, 1/362, 2/628, 2/650, 2/688, 2/786
Fronting effect
   see leading effect
Fructose   1/213, 2/718
D-Fructose-1,6-diphosphate   1/226
β-D-Fructose-2,6-diphosphate   1/226
D-Fructose-1-phosphate   1/226
D-Fructose-6-phosphate   1/226
Fruit juice analysis   2/714 ff
F-test   2/566
Fucose   1/212, 1/238, 2/750
Fucosidase   2/801
Fumaric acid   1/72, 1/172, 1/363, 2/716, 2/786

g

Gadolinium   1/344
β-1,4-Galactan   1/539
Galactinose   1/230
Galactitol   2/774
D-Galactosamine   1/214
α-D-Galactosamine-1-phosphate   1/226
Galactose   1/213, 1/239, 2/750
α-D-Galactose-1-phosphate   1/226
α-D-Galactose-6-phosphate   1/226
β-Galactosidase   2/801
Galacturonic acid   1/76, 1/374, 2/714
Gallium   1/336
Gentiobiose   1/227
Gentisate   1/449
α-1,6-Glucan   1/232
Gluconic acid   2/688, 2/704, 2/775, 2/812
α-D-Glucopyranosido-1,6-mannitol   1/228, 2/754
α-D-Glucopyranosido-1,6-sorbitol   1/228, 2/754
D-Glucosamine   1/214

α-D-Glucosamine-1-phosphate   1/226
α-D-Glucosamine-6-phosphate   1/226
Glucose   1/213, 1/219, 1/224, 2/718
 – oxidation mechanism   1/481
α-D-Glucose-1,6-diphosphate   1/226
α-D-Glucose-1-phosphate   1/226
β-D-Glucose-1-phosphate   1/226
D-Glucose-6-phosphate   1/226
Glucuronic acid   1/373, 2/812
α-D-Glucuronic acid-1-phosphate   1/226
Glutamic acid   1/258, 1/387
Glutamine   1/258, 1/387
Glutaric acid   1/74, 1/363
Glutaric dialdehyde   1/381
Glutathione   2/795
Glycerol   1/210, 1/381 f, 2/707, 2/722, 2/762
α-/β-Glycerophosphate   1/177
Glycine   1/258, 1/387
Glycolic acid   1/44, 1/76, 1/168, 1/199, 1/363, 2/627, 2/650
Glycolipid   1/237
N-Glycolylneuraminic acid   1/242, 2/799
N-Glycopeptidase F   1/246, 1/256, 2/801
Glycoprotein   1/236
 – antifreeze   1/240
 – hydrolysate   1/242
Glycosylation   1/251
Glycosyltransferase   1/239
Glyoxal   1/381
Glyphosate
  see [N-(methylphosphono)glycine]
Gold(I)/(III)   1/416, 2/686
Golgi apparatus   1/239
Gradient
 – aminoalkylsulfonic acid   1/201
 – capacity   1/92, 1/194
 – carbonate   1/106
 – composition   1/106, 1/112, 1/194
 – concentration   1/112, 1/194, 1/283, 1/312
 – – optimization   1/201 f
 – p-cyanophenol   1/201
 – 3-(N-cyclohexylamino)-1-propanesulfonic acid (CAPS)   1/201
 – inverse   1/213
 – pH   1/194
 – profile   1/233
 – step   1/191, 1/283, 1/325
 – taurine   1/201

 – tetraborate   1/44, 1/199
Gradient elution   1/67, 1/191
 – choice of eluants   1/194 f
 – isoconductive technique   1/202 ff
 – of amines   1/292, 1/349
 – of anilines   1/356
 – of carbohydrates   1/214
 – of ethanolamines   1/352
 – of ethylamines   1/352
 – of inorganic anions   2/627
 – of inorganic and organic anions   1/69, 1/72, 1/192, 2/627
 – of inorganic and organic cations   1/349 ff
 – of inositol phosphates   2/792
 – of lanthanides   1/343
 – of methylamines   1/351
 – of organic acids   2/715
 – of polyphosphates   1/195
 – of pyridines   1/357
 – of water-soluble vitamins   2/769
 – theoretical aspects   1/192 ff
Grafting   1/42 f
Griess method   2/785
Growth hormone, human   1/271
Guanidinium   1/320
Guanosin   1/178
 – 5'-monophosphate   1/181
Guanyl urea   1/321
Gypsum analysis   2/817

h

Hafnium   1/340
Haloacetic acid   1/170, 2/606
Heat of sorption   1/33
1,7-Heptanediamine   1/354
Heptanesulfonic acid   1/428
Heptyltriethylammonium   1/355
Hexacyanoferrate(II)/(III)   1/407, 1/415
Hexadecylpyridinium   1/436
Hexadecyl sulfate   1/427
Hexadecyltrimethylammonium   1/355, 1/434
Hexafluorosilicate   2/682
Hexametaphosphate   2/739
Hexamethylenediamine-tetramethylene-phosphonic acid (DEQUEST 2051)   1/184
1,6-Hexanediamine   1/354
Hexanesulfonic acid   1/428

1,4,7,10,13,16-Hexaoxacyclooctadecane (18-crown-6) 1/86
Hexasaccharide 1/226
Hexitol 1/209
Hexobarbital 1/423
Hexosaminidase 2/801
Hippuric acid 1/171, 2/786
Hirudine 1/271
Histamine 1/347, 2/748
Histidine 1/258, 1/387, 2/748
Holmium 1/344
Homocysteine 2/794
Homocystine 1/388, 1/391
Homovanillic acid 1/422
Hull probe 2/692
Humic acid 2/599
Hyamine 1622
    see diisobutyl-[2-(2-phenoxy-ethoxy)ethyl]-dimethylbenzyl-ammonium chloride
Hydration
    – primary 1/143
    – secondary 1/143
Hydrazine 1/475, 2/680
Hydrazinium compounds 1/439
Hydrazinolysis 1/245
Hydrogen
    – bonding 1/363, 1/404
    – chloride 2/624
    – cyanide 2/622
Hydrophobicity 1/398
Hydrotrope 1/424
Hydroxyalkane sulfonate 1/426
p-Hydroxybenzene sulfonate 1/449, 1/451
p-Hydroxybenzoate 1/449
Hydroxycarboxylic acid, aliphatic 1/363, 2/614
1-Hydroxyethane-1,1-diphosphonic acid (HEDP, DEQUEST 2010) 1/186
Hydroxyethyl-ethylenediaminetriacetic acid (HEDTA) 1/183
5-Hydroxy-3-indolylacetic acid 1/422, 2/797
Hydroxyisobutyric acid 1/363
α-Hydroxyisocaproic acid 1/170
Hydroxylamine 1/319, 2/705
Hydroxylaminedisulfonic acid (HADS) 2/645, 2/647
Hydroxylaminetrisulfonic acid 2/647
Hydroxylysine 1/386, 1/388
Hydroxymethanesulfonic acid 1/130
4-Hydroxymethylbenzo-18-crown-6 1/88

Hydroxymethylene cation 1/38
D-Hydroxyphenylglycine 1/458
Hydroxyproline 1/266, 1/386
2-Hydroxypropyltrimethyl ammonium (2-HPTA) 2/766
Hypobromite 2/591
Hypophosphite 1/126, 2/680

i

Ibuprofen 1/454
2-Imidazolidinthion 2/693
Imino acid 1/390
Iminodiacetic acid 1/443
Iminodisulfonic acid 2/647
Immunoglobulin 1/240, 1/243, 1/269, 2/797
Indigocarmine 1/459
Indomethacin 1/454
Injection volume 1/5
Injector, loop 1/5
Inosin 1/178
Inositol 1/210, 2/720
    – 1,4-diphosphate 2/793
    – 4,5-diphosphate 2/793
    – monophosphate 1/210, 2/791
    – 1,2,5,6-tetraphosphate 2/793
    – 1,3,4,5-tetraphosphate 2/793
    – 1,4,5-triphosphate 2/791, 2/793
    – 1,5,6-triphosphate 2/793
Integrator, digital 1/6, 2/554
Interaction
    – adsorptive 1/398
    – Coulomb 1/398, 1/464
    – electrostatic 1/56, 1/298, 1/398, 481
    – interionic 1/466 ff
    – ion-dipole 1/122, 1/143
    – ion-water molecule 1/122
    – ion-molecule 1/467
    – non-ionic 1/29, 1/213
    – π-π 1/29, 1/176, 1/187, 1/364, 1/373, 1/410, 1/439
    – sorption 1/29
    – van-der-Waals 1/56, 1/276, 1/298, 1/398
Interface
    – electrospray 1/504, 1/542 ff, 2/607
    – particle beam 1/531 ff
    – thermospray 1/251, 1/534 ff
    – user 2/555
Interleukin-1 1/271

Inulin   1/234, 2/736
Inverted sugar   2/720, 2/754
Iodate   1/136, 1/504, 1/506, 1/546, 2/596
Iodide   1/61, 1/74, 1/133, 1/159, 1/413,
   1/479, 1/498, 1/506, 1/516, 2/608, 2/621,
   2/624, 2/733, 2/745, 2/818
Ion
   − cloud   1/466
   − exclusion process   1/359 f
   − interaction model   1/397
   − lipophilic   1/393, 1/400
   − mobility   1/462
   − pair   1/397, 1/457, 1/464
   − polarizable   1/3
   − radius   1/122
   − reflux   2/836
   − strength   1/89, 1/97, 1/466
   − suppression mode   1/441
   − surface-active   1/424 ff
   − surface-inactive   1/401, 1/409 ff
   − velocity   1/462
Ion chromatograph
   − computer controlled   2/555
   − process   2/651
   − schematics   1/5
Ion chromatography
   − advantage   1/7
   − method   1/3
Ion-exchange
   − capacity   1/34, 1/38, 1/57
   − function   1/3
   − model   1/394
   − process   1/3, 1/27, 1/279
Ionization
   − atmospheric pressure (API)   1/542
   − chemical (CI)   1/531
   − electron impact (EI)   1/531
   − electrospray   1/542
   − fast atom bombardment (FAB)   1/532
   − ion spray   1/543
   − thermospray   1/535
Ion-pair
   − formation   1/394
   − reagent   1/401
Iron   1/332, 1/334, 2/789
Isoascorbic acid   2/769
Isobutyric acid   1/374
Isocitric acid   1/76, 1/174, 1/448, 2/714
Isoelectric focussing (IEF)   1/255
Isoelectric point   1/99

Isoleucine   1/258, 1/387
Isomaltose   1/227, 2/754
Isomaltulose   1/228, 2/754
Isopropylethylphosphonic acid   1/73
Isopropylmethylphosphonic acid (IMPA)
   2/621
Itaconic acid   1/373

j

Jet-Separator   1/531

k

1-Kestose   1/230, 2/738
6-Kestose   1/230
Ketocarboxylic acid   2/615
α-Ketoglutaric acid   1/364
α-Ketoisocaproic acid   1/170
α-Ketoisovaleric acid   1/376
Ketomalonate   1/73
Knox plot   1/23, 1/397
Kohlrausch square root law   1/464
Kraft process   2/810

l

Lactic acid   1/74, 1/168, 1/363, 1/376, 2/615,
   2/682, 2/714, 2/716, 2/738, 2/786
Lactose   1/221, 1/227, 2/736, 2/745
Lactulose   1/227, 2/790
Lanthanide   1/325, 1/336, 1/342 ff, 1/502
Lanthanum   1/343
Latex
   − anion exchanger   1/54 ff
   − − overview   1/57 ff
   − cation exchanger   1/298 ff
   − particle   1/55, 1/299
Lauric acid   1/443
Lauryl sulfate
   see dodecyl sulfate
Lead   1/331, 1/333
Leading effect   1/15, 1/44
Leucine   1/258, 1/387
Leucrose   1/228
Ligand   1/336
   − choice   1/329
   − concentration
   − − complexing   1/329
   − − effective   1/328
   − − free   1/329
   − − total   1/329

– exchange 1/336
Limit of coverage 2/574
Limit of detection 2/574 f
Limit of determination 2/574, 2/577
  – for standard anions 2/588, 2/628
  – for standard cations 2/628
  – for transition metals 2/653
Lincomycin 1/489, 2/735
Lipoic acid 1/487, 2/770
Lithium 1/287, 2/619, 2/651
Loading capacity 1/60, 2/64
Lobry-de-Bruyn-von-Ekenstein rearrangement 1/208, 1/250
Low method 2/720
Luminol
  see 5-amino-2,3-dihydro-1,4-phthalazindione
Lutetium 1/343
Lysine 1/258, 1/387
Lysozyme 1/269

m

Magnesium 1/283, 2/598, 2/645
Malachite green 1/459
Maleic acid 1/73, 1/172, 1/372, 2/786
Malic acid 1/72, 1/174, 1/363, 2/714
Malonic acid 1/72, 1/73, 1/172, 1/174, 1/372, 2/786
Maltitol 2/762
Maltodecaose 2/725
Maltodextrin 2/745
Maltoheptaose 1/230, 2/725
Maltohexaose 1/230, 2/725
Maltononaose 2/725
Maltooctaose 2/725
Maltopentaose 1/230, 2/725
Maltose 1/206, 1/227, 2/722
  – oligomers 1/230
Maltotetraose 1/230, 2/722
Maltotriose 1/230, 2/722
Mandelic acid 1/171, 1/372, 830
Manganese 1/284, 1/324, 1/331, 2/619, 2/631, 2/778
Mannitol 1/209, 2/720, 2/754, 2/762, 2/790
Mannose 1/210, 1/218, 1/239, 2/750
Mass transfer
  – effect 1/20
  – resistance to 1/19 f, 1/22
Matrix elimination, inline 1/378, 2/639, 2/641, 2/659, 2/670

Mean, arithmetic 2/558
Measuring cell, amperometric 1/474, 1/496
Mecoprop
  see 2-(2-methyl-4-chlorphenoxy)-propionic acid
Median 2/551
Melibiose 1/227
Melizitose 2/755
Mellitate 1/448
Membrane
  – desalting 1/252, 1/537
  – Donnan 1/360
  – polyvinylenedifluoride (PVDF) 1/257, 2/798
  – reactor 1/325, 1/501
Mephobarbital 1/442, 1/453
Metanephrine 2/797
(4-Methacryloylamino)-benzo-15-crown-5 1/88
Methanedisulfonic acid 2/684
Methanesulfonic acid 1/170, 1/428, 2/682
Methanol 1/365, 1/381, 2/773
Metharbital 1/453
Methionine 1/258, 1/387, 2/794
Methohexital 1/453
Methylamine 1/320, 1/350
N-Methylaniline 1/356
Methylarsonic acid 1/528
3-Methylbutanol-1 1/381
2-(2-Methyl-4-chlorophenoxy)propionic acid (Mecoprop) 1/176
N-Methyldiethanolamine 1/302, 2/809
Methylenebisphosphonic acid 2/764
Methylene blue 1/459
4,4'-Methylenedianiline 1/356
(1S,2R)-(+)-N-Methylephedrine 1/453
3-O-Methylglucose 2/790
Methyl green 1/459
o-/p-Methylhippuric acid 1/171
N-Methyloctylammonium-p-toluene sulfonate 1/513
Methyl orange 1/451
Methylose 1/212
Methylphosphonic acid (MPA) 2/621
N-(Methylphosphono)glycine (Glyphosate) 1/515
2-Methylpropanol-1 1/381
Methyl red 1/451
Methyl sulfate 2/756
Methylsulfonate 1/73

Micro-extraction technique   2/657
Modifyer
   – inorganic   1/406
   – organic   1/400, 1/403 f
Molybdate   1/159, 1/524
Molybdenum sulfur cluster   1/418
Moment
   – central   2/551
   – first   2/551
   – second   2/551
   – zero   2/551
Monobromoacetic acid   1/171
Monobromomonochloroacetic acid   1/171
Monobutyl phosphate   1/176
Monochloroacetic acid   1/171
Monochloromethylenebisphosphonic acid   2/764
Monoethanolamine   1/289, 1/291, 1/296, 1/302
Monoethylamine   1/321, 1/325
Monoethylene glycol   1/381, 1/383
Monofluoro phosphate   2/671, 2/672, 2/710
Monoisopropyl sulfate   2/612
Monomethylamine   1/289, 1/321
Monomethyl phosphate   1/546
Monomethyl sulfate   1/546
Monosaccharides   1/206, 1/212 ff, 2/799
   – Alditol   1/244
Morphine   1/421, 1/457
Morpholine   1/288, 1/349, 2/634
Mucin   1/240
Myoglobin   1/269
Myristic acid   1/443

n

Naphthaline-2,3-dialdehyde (NDA)   1/515
Naphthaline-2-sulfonic acid   1/403
α-Naphthol   1/441
1,3,6-Naphtholenetrisulfonate   1/451
β-Naphthol orange   1/451
Naproxen   1/454
Negative ion mode   1/532, 1/537, 1/539, 1/543
Neodymium   1/344
Neo-kestose   1/230
Nernst equation   1/476
Neuraminic acid   1/274
Neuraminidase   1/246, 2/801
Neurotensin   1/273
Neutralization   2/825

Nickel   1/306, 1/332, 1/334, 2/613
Nicotine   1/457
Nicotinic acid   2/770
Nicotinic acid amide   2/770
Nile blue   1/459
Ninhydrin   1/389
Nitrate   1/65, 1/409, 1/411, 1/498, 2/588, 2/609, 2/621, 2/679, 2/733, 2/739
Nitrilotriacetic acid (NTA)   1/183, 2/701
Nitrilotri(methylenephosphonic acid) (NTP, DEQUEST 2000)   1/183
Nitrilotrisulfonic acid   2/647
Nitrite   1/475, 1/479, 1/498, 1/516, 2/588, 2/609, 2/611, 2/621, 2/733, 2/739, 2/784
2-Nitroaniline   1/356
4-Nitroaniline   1/356
Nitrogen oxide   2/622
1,9-Nonanediamine   1/354
Norepinephrine   1/422, 2/797
Norleucine   1/258
Normetanephrine   2/797
Nucleic acid   1/275 ff
Nucleoside   1/177
Nucleotide   1/177
   – phosphates   1/178
Nystose   1/230, 2/738

o

Obstruction factor   1/20
1,8-Octanediamine   1/354
Octanesulfonic acid   1/428
Octyl sulfate   1/427
Ohm's law   1/462
Olefin sulfonate   1/426
Oligonucleotide   1/275
   – secondary structures   1/276
   – use   1/275
Oligosaccharide   1/226 ff, 1/237
   – derived from glycoproteins   1/236 ff
   – – analysis   2/797 ff
   – – complex type   1/238
   – – data bank   1/254
   – – hybride type   1/238
   – – mannose type   1/238
   – – retention behavior   1/247
   – – separation and detection   1/246 ff
   – – sialylated   1/249
   – – structural analysis   1/240 ff
   – – structural isomer   1/246
   – – structure   1/238

– mass spectra 1/538
Orange I 1/451, 1/459
Orange II 1/451, 1/459
Orange G 1/459
Ornithine 1/258, 1/388
Orthophosphate 1/126, 1/195, 1/371, 2/588, 2/591, 707, 2/830
Orthophosphite 1/126, 1/195, 2/673
Orthosilicate 1/153, 1/371, 1/510, 2/591, 2/633, 2/653, 2/698
Osmate 2/604
Ostwald dilution law 1/150
Outlier 2/567
Outlier test 2/567 ff
  – acc. to Grubbs 2/568
  – acc. to Nalimov 2/567
Ovalbumine 1/240, 1/274
  – phosphorylation pattern 1/275
Overvoltage 1/497
  – passage 1/497
  – concentration 1/497
7-Oxabicyclo[2,2,1]-heptane-2,3-dicarboxylic acid (Endothall) 1/176
Oxalacetic acid 1/364
Oxalic acid 1/173, 2/674, 2/782, 2/810
Oxamic acid 1/170
Oxyhalide 1/78 f, 1/121
Ozonation, of drinking water 2/591

p

Palatinitol 1/228, 2/754
Palatinose
  see isomaltulose
Palmitic acid 1/443
Pamidronate
  see 3-amino-1-hydroxypropylidene-1,1-bisphosphonate
Papain 1/269
Papaverin 1/402, 1/424, 1/457
PAR
  see 4-(2-pyridylazo)resorcinol
Parameter
  – experimental retention-determining 1/400 ff
  – – of non-suppressed systems 1/145 ff
  – – of suppressed systems 1/123 ff
  – information 2/550
Peak 1/13
  – area 2/551, 2/552
  – asymmetry 1/14
  – broadening 1/17
  – Gauß curve 1/13, 1/15, 1/18
  – height 2/551, 2/552
  – form 1/14
  – system 1/51, 1/84, 1/147, 1/153 ff, 1/513
  – variance 1/18
  – width 1/19
Pectin hydrolysate 1/214
D-Penicillamine 1/391
D-Penicillaminesulfonic acid 1/391
Penicillin V 2/771
n-Pentanol 1/381
Pentasaccharide 1/226
Pentitol 1/210
Perborate 1/184, 2/701
Perchlorate 1/74, 1/79, 1/94, 1/413, 2/599
Peroxohydroxide anion 2/669, 2/701
Peroxide bleach 2/701
Peroxoborate 2/701
Peroxodisulfate 1/149, 1/414
Peroxomonosulfate 1/149
Phase
  – aluminum oxide 1/95 ff
  – aminopropyl 1/411, 1/498
  – cyanopropyl 1/411, 1/498
  – crown ether 1/86 ff
  – – polyamide 1/88
  – – synthesis 1/88 f
  – cryptand 1/92 ff
  – multimode 1/445
  – octadecyl 1/4, 1/10
  – volume ratio 1/17, 1/31
Phenobarbital 1/423, 1/442, 1/450
Phenol 1/441, 1/475
Phenoxyacetic acid 1/542, 2/771
Phenoxycarboxylic acid 1/175, 1/542
Phenylalanine 1/258, 1/387
1-Phenyl-ethane-1,2-diphosphonic acid 1/186
1-Phenyl-ethane-1,2,2-triphosphonic acid 1/186
1-Phenyl-ethane-1,2,2-tris(P-methyl-phosphinic acid) 1/186
1-Phenyl-ethene-1-phosphonic acid 1/186
$trans$-1-Phenyl-ethene-2-phosphonic acid 1/186
Phenylglyoxylic acid 2/783
Phenylisothiocyanate (PITC) 1/383
Phenylphosphonate 1/449
Phenylthiohydantoine (PTH) 1/383

Phosphate
  see orthophosphate
Phosphite
  see orthophosphite
Phosphonium compound   1/438
2-Phosphonobutane-1,2,4-tricarboxylic
  acid (PBTC) 1/185
Phosphonopropanetricarboxylic acid   1/189
Phosphorothioate   1/275
Photolysis   2/729, 2/827
Phthalate   1/73, 1/448
o-Phthaldialdehyde (OPA)   1/383, 1/390,
  1/514
Phytic acid   1/448, 1/450, 2/743
4-Picoline   1/356
Piperazine   2/809
pK value
  – acetic acid   1/372
  – acrylic acid   1/372
  – adipic acid   1/372
  – aliphatic dicarboxylic acids   1/173
  – ammonium   1/317
  – aromatic monocarboxylic acids   1/172
  – arsenite   1/125
  – azide   2/787
  – barbital   1/423
  – barbituric acid   1/423
  – benzoic acid   1/150, 1/172
  – boric acid   1/368
  – iso-butyric acid   1/372
  – n-butyric acid   1/372
  – caffeine   2/731
  – carbohydrates   1/62, 1/205
  – carbonic acid   1/138
  – citric acid   1/372
  – cyanate   2/833
  – cyanide   2/833
  – 2,3-diaminopropionic acid (DAP)
      1/282, 1/309
  – diethylamine   1/317
  – formic acid   1/372
  – fumaric acid   1/372
  – glycolic acid   1/372
  – hippuric acid   1/172
  – hydroxide substituted dicarboxylic
      acids   1/174
  – p-hydroxybenzoic acid   1/158
  – lactic acid   1/372
  – maleic acid   1/372
  – malic acid   1/174

  – malonic acid   1/173, 1/372
  – mandelic acid   1/172, 1/372
  – monoethylamine   1/317
  – monomethylamine   1/317
  – morpholine   1/317
  – nicotinic acid   1/151
  – orthophosphoric acid   1/138
  – oxalic acid   1/173, 1/372
  – phthalic acid   1/158
  – poly(butadiene-maleic acid)   1/304
  – propionic acid   1/372
  – pyruvic acid   1/372
  – saccharin   772
  – salicylic acid   1/151
  – succinic acid   1/151, 1/173, 1/372
  – tartaric acid   1/174, 1/372
  – triethanolamine   1/317
  – trimethylamine   1/317
Plasma etching   2/654
Plate
  – hight   1/17, 1/19, 1/24
  – – effective   1/18
  – – reduced   1/23
  – – theoretical   1/17, 1/23, 1/24
  – number   1/17, 1/19
  – – effective   1/18
Polyalcohol, cyclic
  see cyclitol
Poly(butadiene-maleic acid) (PBDMA)
  1/282, 1/304, 1/323
Polydeoxyadenosine   1/276
Polydeoxyguanosin   1/277
Polyether
  – bicyclic   1/179
  – cyclic   1/86
Polyfructan   1/234
Polyphosphate   1/80, 1/164, 1/509, 1/529,
  2/738
Polyphosphinic acid   1/186
  – retention behavior   1/187 f
  – stereoisomer   1/188
  – structural isomer   1/187
Polyphosphonic acid   1/186, 1/509, 2/641,
  2/698
  – diastereomer   1/189
  – retention behavior   1/187 f
  – rotational isomer   1/188
  – stereoisomer   1/188
  – structural isomer   1/188
Polysaccharide   1/230 ff

Poly(thiometalate) 1/416
Polythionate 1/414, 2/646
Positive ion mode 1/532, 1/536, 1/543
Post-column
 – addition, of NaOH 1/210, 1/213
 – derivatization 1/182
Potassium 1/283, 2/598, 2/645
Potential
 – Galvani 1/497
 – – equilibrium 1/497
 – limit 1/496
 – standard 1/476
 – working 1/476
Praseodymium 1/344
Precision 2/551, 2/566
Pre-concentration
 – technique 2/628
 – via chelation 2/821
Preservative 2/746
Proline 1/258, 1/387
1,2-Propanediamine 1/353
Propanesulfonic acid 1/428
1,2,3-Propanetricarboxylate 1/448
2-Propanol 1/381
Propionaldehyde 1/486
Propionic acid 1/74, 1/362
n-Propylamine 1/353
Propylene glycol 1/382, 2/762
Protein 1/267 ff
 – deamidation 1/271
 – hydrolysis 1/392
 – microheterogeneous 1/273
 – recombinant 1/271
 – therapeutic 1/270
Proteoglycan 1/237
(1R,2R)-(-)-Pseudoephedrine 1/453
(1S,2S)-(+)-Pseudoephedrine 1/453
Pullulan 1/236
Pullulanase 1/236
Pulse sequence
 – for amino acids 1/494
 – with three potentials 1/483
 – with four potentials 1/484
 – multicyclic 1/487 f, 775
Pump, analytical 1/5
Purine base 1/178
Putrescine 1/344, 2/747
Pyridine 1/357
Pyridoxine 2/770
4-(2-Pyridylazo)resorcinol (PAR) 1/499

Pyrimidine base 1/178
Pyrogallic acid 1/441
Pyroglutamate aminopeptidase 1/272
Pyromellitate
 see 1,2,4,5-benzenetetracarboxylate
Pyrophosphate 1/163, 1/195, 2/769, 2/830
Pyrophosphoric acid 1/284
Pyruvic acid 1/73, 1/364, 1/376, 2/738, 2/786

q

Qualification
 – installation (IQ) 2/556
 – operation (OQ) 2/556
 – performance (PQ) 2/556
 – vendor (VQ) 2/556
Quantity
 – chromatographic 1/13
 – statistical 2/557
 – thermodynamic 1/30
Quinic acid 1/73, 1/74, 2/715
Quinine 1/457

r

Radiostrontium analysis 1/521
Raffinose 1/206, 1/228
Reagent
 – 9-anthryldiazomethane (ADAM) 1/520
 – Arsenazo I 1/501
 – Arsenazo III 1/340, 1/502
 – cerium(IV) 1/516
 – delivery 1/500
 – o-dianisidine (ODA) 1/505
 – 1,5-diphenylcarbazide (DPC) 1/508
 – ion-pair 1/4
 – iron(III) nitrate 1/510
 – Luminol 2/632
 – naphthaline-2,3-dialdehyde (NDA) 1/515
 – PAR 1/339, 1/499
 – PAR/ZnEDTA 1/332
 – o-phthaldialdehyde (OPA) 1/383, 1/514
 – sodium molybdate 1/510
 – Tiron 1/341, 1/501
Recovery 2/572
Reddening agent 2/741
Regenerant
 – delivery 1/113 f
Regeneration

- continuous 1/108, 1/110, 1/113, 1/115, 1/311, 1/313, 1/367, 1/399 f
- periodic 1/104, 1/311, 1/399

Regression
- coefficient 2/561
- linear 2/561
- weighed 2/567

Relaxation
- constant 1/469
- effect 1/468
- time 1/468

Residual 2/566
- analysis 2/566
- standard deviation 2/563, 2/565

Resin
- divinylbenzene 1/393
- Dowex 1x10 1/310
- ethylvinylbenzene/divinylbenzene 1/35, 1/61, 1/282 f
- functionalization 1/36, 1/37 f
- – amination 1/37
- – chloromethylation 1/37
- gel-type 1/35
- hydroxyethylmethacrylate 1/49
- iminodiacetic acid 2/821
- ionic form 1/36
- laurylmethacrylate 1/83
- macroreticular 1/35, 1/37
- mesoporous 1/37
- microporous 1/35
- polyamide crown ether 1/88
- poly(benzo-15-crown-5) 1/307
- polymethacrylate 1/35, 1/47 f, 1/296
- polyvinyl 1/35, 1/47 f, 1/296
- polyvinylpyrrolidone 2/599
- porosity 1/35
- shrinking process 1/36
- swelling process 1/36
- styrene/divinylbenzene 1/35, 1/280
- XAD-1 1/38

Resistivity 1/462
Resolution 1/15
Resorcinol 1/441
Result 2/559
Retention
- temperatur dependance 1/32
- time
- – gross 1/13
- – solute 1/13

Rhamnose 1/213, 2/726, 2/750, 2/774

Rhodamine B 1/459
Riboflavin 2/770
Ribonuclease B 1/240, 1/269
- crystal structure 1/272

D-Ribose 1/221, 2/750
α-D-Ribose-1-phosphate 1/226
Rubidium 1/287
Ruhemann's purple 1/389

S

Saccharin 2/692, 2/731, 2/752, 2/762
Sarcosine 1/388
Salicylic acid 2/688
Saliva analysis 2/781
Samarium 1/344
Sample
- loading capacity 1/65
- preparation 2/550, 2/823 ff
- – for amino acid analysis 1/391 f
- – on-column 1/449
- storage 2/550

Sampling 2/549
Scatter 2/559
Schiff base 1/205
Schöninger flask 2/769
SDS Polyacrylamide gel electrophoresis (SDS-PAGE) 1/255
Secobarbital 1/442, 1/453
Selectivity 1/8, 1/16
- coefficient 1/28, 1/38, 1/326, 1/472
- of latexed anion exchangers, overview 1/57 ff
- solvent influence 1/143

Selenate 1/126
Selenite 1/73, 1/126
Selenometalate 1/416
Sensitivity 1/473, 2/560
- of the detection system 2/560
- of the method 2/561, 2/588

Separator column
- Aminex 50W-X4 1/365
- Aminex HPX-85H 1/379
- Aminex HPX 87H 2/723
- AminoPac NA-1 1/386
- AminoPac PA-1 1/385
- AminoPac PA-10 1/257, 1/385
- AminoPac PC-1 1/385
- AN1 1/41
- AN2 1/41
- AN300 1/41

Index

- BTC 2710 1/385
- CarboPac MA1 1/210, 1/245, 2/762
- CarboPac PA1 2/62, 1/129, 1/136, 1/169
- CarboPac PA10 1/212, 1/242
- CarboPac PA20 1/212, 1/219
- CarboPac PA-100 1/233, 1/539
- DNAPac PA-100 1/273, 1/276
- efficiency 1/17 ff
- ExcelPak ICS A23 1/40
- Fast-Sep Anion 1/64
- Fast-Sep Cation 1/299
- Hypersil 5 MOS 1/429, 2/710
- IC Pak A 1/47, 1/146
- Inertsil ODS II 1/438
- ION-100/110 1/52
- ION-200/210 1/296, 1/519
- ION-300 1/364
- IonPac AS1 1/57
- IonPac AS2 1/58
- IonPac AS3 1/58, 1/126
- IonPac AS4 1/60
- IonPac AS4A 1/60
- IonPac AS4A-SC 1/61
- IonPac AS5 1/61, 1/160
- IonPac AS5A 1/69
- IonPac AS6
  see CarboPac PA1
- IonPac AS7 1/63, 1/163, 1/182, 1/527
- IonPac AS9 1/65
- IonPac AS9-SC 1/66, 1/162, 2/593
- IonPac AS9-HC 1/66, 1/504, 2/609
- IonPac AS10 1/77, 1/125, 2/641
- IonPac AS11 1/70, 1/164, 1/196, 2/715
- IonPac AS11-HC 1/73
- IonPac AS12A 1/78, 1/126, 1/169, 2/812
- IonPac AS14 1/42
- IonPac AS14A 1/45
- IonPac AS15 1/45, 1/199
- IonPac AS15A 1/46
- IonPac AS16 1/79, 2/601
- IonPac AS17 1/80, 2/590
- IonPac Cryptand A1 1/93
- IonPac CS1 1/280
- IonPac CS2 1/281, 1/341
- IonPac CS3 1/298, 1/341, 1/345
- IonPac CS5 1/62, 1/332
- IonPac CS5A 1/338, 2/620
- IonPac CS10 1/283, 1/300, 1/324
- IonPac CS11 1/302
- IonPac CS12 1/282, 2/599
- IonPac CS12A 1/285
- IonPac CS14 1/288, 1/346, 1/493
- IonPac CS15 1/291, 1/351, 2/634,
- IonPac CS16 1/292, 2/806
- IonPac CS17 1/295, 1/346
- IonPac ICE-AS1 1/361
- IonPac ICE-AS6 1/361, 2/615
- IonPac ICE-Borate 1/369
- IonPac NS1 1/446
- LCA A01 1/39
- LCA A03 1/48
- LCA A04 1/39
- LCA K01 1/281
- LCA K02 1/296, 1/332
- length 1/123
- LiChrosil IC CA 1/306
- LiChrosorb RP18 1/411
- maintenance 2/835 f
- MCI Gel SCA04 1/51
- MCI Gel SCK01 1/281
- Metrosep Anion Dual 1 1/49, 1/163
- Metrosep Anion Dual 2 1/49
- Metrosep Anion SUPP 1 2/663
- Metrosep Anion SUPP 4 1/54
- Metrosep Anion SUPP 5 1/54
- Metrosep Cation 1-2 1/304, 2/599
- Metrosep Organic Acids 2/689
- Mikropac MCH10 1/443
- Novosep A-1 1/107
- Nucleosil 5 SA 1/303
- Nucleosil 10 SA 1/503
- Nucleosil 10 Anion 1/86
- Nucleosil 10 C_8 1/431
- OmniPac PAX-100 1/445
- OmniPac PAX-500 1/445, 2/554, 2/750
- OmniPac PCX-100 1/303, 1/445, 2/768
- OmniPac PCX-500 1/445, 2/794
- ORH 801 1/364
- Polygosil-60-D-10 CN 1/412
- Polyspher CHCA 2/726
- Polyspher IC AN-1 1/49
- Polyspher OA-HY 2/615
- ProPac PA1 1/271
- ProPac SAX-10 1/274, 2/738
- ProPac SCX-10 1/269
- ProPac WCX-10 1/269
- PRP-1 1/401
- PRP-X100 1/39

- PRP-X200 1/281
- PRP-X300 1/364
- Rezex RPM Monosaccharide 1/211
- SAR-40-0.6 1/157
- Shimpack IC-A1 1/48, 1/149
- Shimpack IC-C1 1/281, 1/322
- Shimpack IE 2/716
- Spherisorb A5Y 1/98
- Spherisorb ODS 2 2/769
- stability 1/9
- Star Ion A300 IC Anion 1/41
- Star Ion A300 HC 1/42
- Supersep Anion 1/480
- TSK Gel 620 SA 1/471
- TSK Gel IC Cation 1/281
- TSK Gel IC Cation SW 1/303
- TSK Gel IC-PW 1/47
- TSK Gel IC-SW 1/84
- Universal Anion 1/50
- Universal Cation 1/307
- Vydac 300 IC 405 1/83, 1/163
- Vydac 302 IC 4.6 1/83
- Vydac 400 IC 405 1/303, 1/320
- Vydac C8 (208TP5451) 1/490
- Waters C18 Radial-Pak 1/412
- Wescan 269-001 1/84
- Wescan 269-029 1/147
- Wescan 269-031 1/152
- Zorbax-NH$_2$ 1/413

Serine 1/258, 1/387
Serotonin 1/422, 2/797
Serum analysis 2/783
Shikimic acid 1/373
Sialic acid
　see N-acetylneuraminic acid
Silicate
　see orthosilicate
Silvex
　see 2-(2,4,5-trichlorophenoxy)-propionic acid
Sodium 1/283, 2/598, 2/635, 2/645
Soil analysis 2/619
Sorbic acid 2/76, 1/442, 2/746
Sorbitol 1/209, 1/211, 2/718, 2/726, 2/754, 2/762
Speciation 1/523, 2/630
- arsenic 1/526
- chromium 1/524, 1/526
Spermidine 1/344, 2/747
Spermine 1/344

Stachyose 1/206, 1/228
Standard
- addition 2/572 f
- external 2/571
- internal 2/570
Standard deviation 2/558
- method 2/563, 2/565
- - relative 2/563, 2/565
- residual 2/563, 2/565
- theoretical 2/558
Stannate 1/524
Stearic acid 1/442
Strontium 1/287, 1/502, 2/619
Strychnine 1/457
Student factor 2/559
Succinic acid 1/74, 1/172, 1/363, 2/615, 2/688
Succinylcholine 2/767
Sucrose 1/226, 2/718, 2/745
- mass spectrum 1/538
Sugar
- alcohol 1/209 ff
- substitute 2/753
Sulfadiazin 1/457
Sulfadimethoxin 1/457
Sulfamate 2/682
Sulfamerazin 1/457
Sulfamethazin 1/457
Sulfanilamide 1/457
Sulfanilic acid 1/455
Sulfate 1/65, 2/588, 2/644, 2/702, 2/771
Sulfathiazole 1/457
Sulfide 1/63, 1/126, 1/475, 1/479, 2/810
- free 1/127
Sulfisoxazole 1/457
Sulfite 1/65, 1/130, 1/370, 2/722, 2/728, 2/761
- purity determination 1/131
- stabilization 1/130
α-Sulfofatty acid methyl ester 1/424
Sulfonamide 1/455, 1/542
Sulfonium compound 1/437
5-Sulfosalicylate 1/451
Sulfosuccinic acid ester 1/431
Sulfur dioxide 2/622
Sulfur-nitrogen compound 2/645, 2/647
Supporting electrolyte 1/102, 1/475
Suppressor
- capacity 1/112

- column   1/2, 1/103 ff, 1/310, 1/366, 1/399
- – regeneration   1/104, 1/311, 1/399
- DS-Plus™   1/105
- ERIS   1/105
- hollow fiber membrane
- – for anion exchangechromatography   1/108 ff
- – for ion-exclusionchromatography   1/367
- – for ion-pair chromatography   1/399
- – for cation exchange chromatography   1/311
- – regeneration   1/110, 1/311, 1/367, 1/399
- – schematics   1/109
- micromembrane
- – DCR™ mode   1/115
- – for anion exchange chromatography   1/111 ff
- – for ion-exclusion chromatography   1/367
- – for ion-pair chromatography   1/400
- – for cation exchange chromatography   1/312
- – regeneration   1/113, 1/115, 1/313, 1/367, 1/400
- – schematics   1/111
- monolithic
- – for anion exchange chromatography   1/120
- – for cation exchange chromatography   1/318
- – schematic   1/120
- MSM   1/104
- reaction   1/103, 1/366
- self-regenerating
- – converter mode   1/315 f
- – for anion exchange chromatography   1/115 ff
- – for cation exchange chromatography   1/313 f
- – operation modes   1/118 f, 1/314
- system   1/103
- void volume   1/104, 1/106, 1/109, 1/111

Surfactant   2/698 ff
- anionic   1/542, 2/698, 2/709
- cationic   2/698

Sweetener   1/228, 2/752

t

Tailing effect   1/14, 1/44
Tartaric acid   1/72, 1/172, 1/363, 2/688, 2/714
Tartronic acid   1/174
Taurine   1/258, 1/388
Terbium   1/344
Terephthalate   1/449
Tetraacetyl-ethylenediamine (TAED)   1/184, 2/701
Tetrabutylammonium   1/355, 1/421, 1/434
Tetradecylpyridinium   1/436
Tetradecyl sulfate   1/427
Tetraethylammonium   1/355
Tetraethylenepentamine   1/348
Tetrafluoroborate   1/131, 2/672, 2/833
Tetraheptylammonium   1/355
Tetrahexylammonium   1/355
Tetrametaphosphate   1/166, 2/740
Tetramethylammonium   1/355, 1/420
Tetrapentylammonium   1/355, 1/434
Tetrapolyphosphate   1/166 f, 1/195, 2/740
Tetrapropylammonium   1/355, 1/420, 1/434
Tetrasaccharide   1/226
Tetrathionate   1/414
Theobromine   1/455
Theophyllin   1/455
Thiamine   2/770
Thiamylal   1/453
Thiocyanate   1/61, 1/74, 1/159
Thiofluor
   see N,N-dimethyl-2-mercaptoethylamine-hydrochloride
Thioglycolic acid   1/408
Thiometalate   1/416
Thiomolybdate   1/417
Thiosulfate   1/61, 1/74, 1/159, 1/475, 1/479, 1/516
Thorium   1/339, 1/503
Threonine   1/258, 1/387
Thulium   1/344
Thymidine   1/178
- 5′-monophosphate   1/181
Thymol   1/441
Thymol blue   1/459
Tiron
   see 4,5-dihydroxy-1,3-benzene-disulfonic acid-disodium salt
Tissue plasminogen activator (tPA)   1/240
Tolmetin   1/454

Toluene sulfonate 1/424, 1/427, 1/451, 2/703
p-Toluenesulfonic acid 1/403
3-Toluidine 1/356
Transferrin, human serum 1/239, 1/254, 1/273
Transglucosidase 1/232
Transition metal 1/325 ff, 2/630, 2/690, 2/778, 2/821
 – limit of detection 2/653
 – separation 1/326 ff
Transport number 1/465
Trehalose 1/226, 1/228, 2/726, 2/774
Triangulation 2/551, 2/553
Tribromide 1/504, 2/596
Tribromoacetic acid 1/170 f
Tributylmethylammonium 1/355, 1/434
Tricarballylate 1/73
Trichloroacetic acid 1/171
2,4,5-Trichlorophenoxyacetic acid (2,4,5-T) 1/176
2-(2,4,5-Trichlorphenoxy)propionic acid (Silvex) 1/176
Triethanolamine 1/296, 1/302, 2/704
Triethylamine 1/321, 1/325
Triethylene glycol 1/383
Triethylenetetramine 1/348
Trifluoroacetate 1/76, 2/756
Triiodide 1/507, 2/596
Trimesate 1/449
Trimetaphosphate 2/740, 2/830
Trimethoprim 1/457
Trimethylamine 1/289, 1/321
Triphenylarsonium compound 1/439
Triphenyl-mono(β-jonylidene-ethylene)-phosphonium chloride 1/438
Triphenylphosphonium compound 1/439
Tripolyphosphate 1/163, 1/195, 2/698, 2/708, 2/740, 2/769
Trisaccharide 1/226
Tropaeolin O 1/459
Tryptophane 1/265, 1/388, 1/392
t-Test 2/569
Tungstate 1/73, 1/159
Turanose 1/227
Two-phase titration 1/436, 2/698
Tyrosine 1/258, 1/387

u

Ultrafiltration 1/392
Ultracentrifugation 1/392

Uranium 1/336, 1/339, 1/503
Uranyl cation 1/340, 1/514
Uridine 1/178
 – 5′-monophosphate 1/181
Urine analysis 2/782

v

Valency 1/122, 371
Valeric acid 1/362
Validation 2/556
Valine 1/258, 1/260, 1/387
Vanadate 1/524
Vanadium 1/336
van Deemter
 – curves 1/24 f
 – equation 1/19, 1/22
 – theory 1/19 ff, 1/24
van't Hoff plot 1/32
Variance
 – homogeneity 2/561, 2/566
 – inhomogeneity 2/561, 2/566
4-Vinylbenzo-18-crown-6 1/88
Vinylsulfonic acid 2/689
Vitamin, water-soluble 1/455, 2/769
V-Mask 2/585
Voltammetry 1/475
 – basics 1/475 ff
 – cyclic 1/481
 – hydrodynamic 1/475
 – pulsed 1/475
Voltammogram 1/476
 – cyclo
 – – of formaldehyde 1/486
 – – of glucose 1/481
 – – of propionaldehyde 1/486
Volume
 – breakthrough 2/634
 – dead 1/17
 – exclusion 1/360
 – totally permeated 1/360

w

Water analysis
 – conditioned water 2/634
 – cooling water 1/294, 2/640
 – drinking water 2/589
 – feed water 2/627
 – formation water 2/803

- ground water 2/598
- ice 2/598
- landfill leachate 1/532, 2/614
- rain water 2/598
- sea water 2/615, 2/786
- seepage water 2/614
- snow 2/598
- surface water 2/608
- swimming pool water 2/598
- ultra-pure water 2/626, 2/631, 2/651
- wastewater 2/609

Weight distribution coefficient 1/29
Western transfer 1/257
Working range 2/561

x

Xanthine 1/455
β-1,4-Xylan 1/539
Xylene sulfonate 1/424, 1/427, 1/533
Xylitol 1/211, 2/720
Xylose 1/212, 1/218, 1/239, 2/750

y

Ytterbium 1/343

z

Zeolith A 2/698
Zinc 1/306, 1/331, 1/334, 2/613, 2/789
Zirkonium 1/340
ZnEDTA 1/332, 1/500